THE EDITOR

Thomas Bainbrigge Fletcher (1878-1950) was an English entomologist. He worked in the Royal Navy; he became an expert on the "microlepidoptera" and was appointed as the second Imperial Entomologist in India, succeeding Harold Maxwell-Lefroy at the Imperial Agricultural Research Institute at Pusa. He was a meticulous naturalist and very careful on matters of systematics and taxonomic nomenclature.

He produced a list of publications on India Entomology and Catalogue of Indian Insects. He also wrote general works on entomology including Some South Indian Insects, and a Veterinary Entomology for India and hints on collecting and preserving insects. He was Fellow of the Linnean Society, the Royal Entomological Society, Zoological Society of London and a President of the Cotteswold naturalists field club.

Catalogue
Indian Insect

Catalogue of Indian Insects

Volume 5

–Editor–

Thomas Bainbrigge Fletcher

2016

Daya Publishing House®

A Division of

Astral International Pvt. Ltd.

New Delhi – 110 002

Catalogue of Indian Insects (5 Volumes Set)

Publisher's note:

The book "***Catalogue of Indian Insects***" by Thomas Bainbrigge Fletcher was published in 1921. But presently complete set not available anywhere. We have tried to reproduced the complete valuable set from different sources in a form as close to the original possible. This historical reproduction fulfill this long felt needs of students, teachers and researchers engaged in the field of entomology, zoology and other related discipline.

Cataloging in Publication Data—DK
Courtesy: D.K. Agencies (P) Ltd. <docinfo@dkagencies.com>

Catalogue of Indian insects / editor, Thomas Bainbrigge Fletcher.

volumes cm
First published in 1921.
ISBN: 978-93-86071-10-1 (Vol. 5)
ISBN: 978-93-5130-922-2 (Set) (International Edition)

1. Insects—Catalogs and collections—India. I. Fletcher, T. B. (Thomas Bainbrigge), 1878-1950, editor.

QL468.2.C38 2016 DDC 595.70954 23

Published by : **Daya Publishing House®**
A Division of
Astral International Pvt. Ltd.
– ISO 9001:2008 Certified Company –
4760-61/23, Ansari Road, Darya Ganj
New Delhi-110 002
Ph. 011-43549197, 23278134
E-mail: info@astralint.com
Website: www.astralint.com

Printed at : **Replika Press Pvt. Ltd.**

डा. एस. अय्यप्पन
सचिव एवं महानिदेशक
Dr. S. AYYAPPAN
SECRETARY & DIRECTOR GENERAL

भारत सरकार
कृषि अनुसंधान और शिक्षा विभाग एवं
भारतीय कृषि अनुसंधान परिषद
कृषि एवं किसान कल्याण मंत्रालय, कृषि भवन, नई दिल्ली 110 001

GOVERNMENT OF INDIA
DEPARTMENT OF AGRICULTURAL RESEARCH & EDUCATION
AND
INDIAN COUNCIL OF AGRICULTURAL RESEARCH
MINISTRY OF AGRICULTURE AND FARMERS WELFARE
KRISHI BHAVAN, NEW DELHI 110 001
Tel.: 23382629; 23386711 Fax: 91-11-23384773
E-mail: dg.icar@nic.in

Foreword

With rapidly increasing human population that hardly matched by the rate of increase in agriculture production, the per capita availability of food and other products is bound to fall short. Therefore, increasing production per unit area is the most plausible alternative to sustain self sufficiency in food, fodder and fibre. Insects have got both positive and negative impacts in agriculture. Insects become pests when they interfere with our welfare, aesthetics or profits. A reduction or loss in yield and quality of crops is partly due to result of insect feeding activities and as vector/host to pathogen transmissions. On the other hand these small creatures serve us a lot as pollinating agents of many important agricultural crops, as biological controlling agents of pests and weeds as well as by producing honey, silk and many other materials essential to us.

This book 'Catalogue of Indian Insects' is divided into 5 volumes and covers 29 parts. This historical compilation is a dedicated effort to present the different aspects of the insects in a very concise and relevant manner. I believe that the students and scientist working in the area of entomology, agriculture, fisheries, zoology, botany and veterinary science will find this book as a source of useful and authentic information.

I appreciate the efforts of Daya Publishing House, A Division of Astral International (P) Ltd. New Delhi.

(Dr. S. Ayyappan)
Secretary (DARE) & Director General (ICAR)
Krishi Bhawan, New Delhi – 110001

Preface

There is a immense number of insect species on the planet. The class Insecta is a large and diverse as well as important ecologically. When compared to other groups, the number of insect species is much larger and represents more than three quarters of all known animal species. Most insects are either beneficial or harmless. Yet there are large numbers which are economically important as pests, pollinators, natural enemies and other beneficial insects like honey bees, silkworms, etc. Insects are fascinating to investigate, their classification is complex. It important to learn the main differences among insects so you can distinguish one group from another. Knowledge of insect classification is of primary importance in carrying out proper control procedures and quality pest management. Proper insect classification and identification is extremely important. If a beneficial or harmless insect is improperly identified, the resultant unnecessary and undesirable pesticide application usually disrupts natural control agents.

Insects deliver a host of ecological services fundamental to the survival of humankind. Insects play an important role in plant reproduction. Thousands of pollinator species have been identified and almost all of these are insects. Over 90 percent of the flowering plant species depend on pollinators. Insects play an equally vital role in waste biodegradation. They clean up dead plant matter, breaking down organic matter until it is fit to be consumed by fungi and bacteria. Insects also play a significant role in decomposing animal carcasses and manure. Insects have a large array of natural enemies, predators and parasitoids, keeping them under economic threshold levels. Due to large number of insects species it is necessary that these catalogued so as to enable locating the species into correct classification. One shall also know what are the species knows from a specific group. Especially in case of Indian Insects where large number of species are known. These are important Catalogue of Indian Insects as a compilation serves this purpose.

"Catalogue of Indian Insects" is divided into 5 volumes and covers 29 parts. Volume 1 deals with Part I to Part IX. It covers Acrydidae, Culicidae, Bombyliidae, Trypetidae, Nitidulidae, Stapylinidae, Lasiocampidae, Amatidae and Zygaenidae. Volume 2 gives a detailed account of Part X to Part XVII. This volume covers Stephanidae, Brenthidae, Tabanidae,

Cicindelidae, Palpicornia, Cecidomyidae, Cosmopterygidae and Yponomeutidae. Volume 3 is devoted to Part XVIII i.e. Carabidae. Volume 4 covers in detail of Part XIX to Part XXV. This volume deals with Gyrinoidea, Alucitidae, Lycidae, Phalonidae and Chlidanotidae, Chalcidoidea, Evanidae and Thysanoptera. Volume 5 of the book describes the Part XXVI to Part XXIX. This volume deals with Serphoidea, Isoptera, Anthribidae and Asilidae.

This historical reproduction fulfill the long felt needs of students, teachers, researchers engaged in the field of entomology, agriculture, ecology, fisheries, forestry, paleontology, zoology, botany, plant science, medical and veterinary science, towards knowing insects.

We are extremely grateful to Dr. S. Ayyapppan, Secretary and Director General, Indian Council of Agricultural Reserarch, New Delhi for providing a foreword. The cooperation and support of Dr. V.V. Ramamurthy, Principal Scientist (Retd.), Division of Entomology, IARI, New Delhi is duly acknowledged.

All suggestions for future improvement for this book will be highly solicited.

Publisher

Contents

VOLUME 3

VOLUME 4

VOLUME 5

PART XXVI - SERPHOIDEA

INTRODUCTION

The superfamily Serphoidea, though comparatively smaller than the Chacidoidea, is perhaps of greater economic importance in that, unlike the Chalcidoidea, all the species comparising it are parasitic. The group has, however, been very much neglected both in India and abroad and hardly 3,000 species are known from the whole world, of which only about one hundred occur in India. In this catalogue all the species recorded from British indian and also from the neighbouring regions like Ceylon, Burma, Malaya Siam (Thailand) and Java are included. Naturally no claim is made that the catalogue exhausts all the species occurring the Indian region, as numerous new forms are being constantly found with the increase of interest in the study of parasites of important insect pests.

The superfamily, as at present recognised comprises the following five families:—

1. Scelionidae (=Scelionidae s. str. + Platygasteridae).
2. Calliceratidae (=Ceraphronidae).
3. Diapriidae (= Diapriidae s. str. + Belytidae).
4. Serphidae (=Proctotrupidea = Serphidae s. str. + Heloridae).
5. Pelecinidae.

A key to the families, adopted from Ashmead and Kieffer, is given below. Detailed keys to subfamilies, Indian genera and species of the various families are also included in the catalogue, in accordance with the recommendations of the Committee on Catalogue of Indian Insects, appointed by the Third Entomological Meeting held at Pusa in 1919 *(vide Rep. Proc. Third Ent. Meet. Pusa*, III, p. 998).

The compilation of the catalogue was started while I was working in the Zoological Survey of India, Indian Museum, Calcutta and completed in the Imperial Agricultural Research Institute, New Delhi. My sincere thanks are due to Dr. Hem Singh Pruthi, Imperial Entomologist, New Delhi and Dr. Baini Prashad, Director, Zoological Survey, Calcutta, for facilities for preparing the catalogue.

M. S. MANI.
Department of Entomology,
Imperial Agricultural Research Institute,
New Delhi

September 11th, 1940.

SERPHOIDEA

Key to families

1. Winged forms	2
wingless forms	6
2. Trochanter uniarticulate, fore wings with a stigma more or less developed	Pelecinidae.
Trochanter usually biarticulate	3
3. Antennae inserted far below in the middle of face	4
Antennae inserted far below the middle of face, *i.e.* at the clypeus near the mouth	5
4. Fore wings without a stigma	Diapriidae.
Fore wings with a stigma.	Serphidae.
5. Abdomen margined and acute at the sides	Scelionidae.
Abdomen normal.	Calliceratidae.
6. Antennae inserted in the middle of face	7
Antennae inserted near the mouth	8
7. Mandibles dentate; tip of abdomen not stylate	Diapriidae.
Mandibles edentate; tip of abdomen stylate	Serphidae.
8. Abdomen margined and acute laterally	Scelionidae.
Abdomen normal	Calliceratidae.

Family I. **SCELIONIDAE**

Oxyuri (partim), Latreille, *Hist. Crust. Ins.*, XII, p. 224 (1805). Haliday, *Ent. Mag.*, I, p. 269 (1833).

Proctotrupii, Latreille, *Hist. Crust. Ins.*, XIII, p. 225 (1805).

Cynipsera (partim), Latreille, *Gen. Crust. ins.*, IV, p. 33. (1809).

Pteromalini (partim), Nees, *Hymen. Monogr.*, II, p. 285 (1834).

Platygasteridae (partim), Förster, *Hymen. Stud.*, II, pp. 27, 28 (1856).

Sparasionidae, Dahlbom, *Öfv. Akad. Förh.*, XIV. p. 290 (1858).

Proctotrypidae (Partim), Marshall, Cat. *Brit. Mus.*, pp. 1, 13, 17 (1873).

Scelioninae, Ashmead, *Bull. U. S. N. Mus.*, XLV, pp. 136, 247 (1893). Dalla Torre, *Cat. Hymen.*, V, p. 494 (1898).

Scelionidae, Ashmead in Smith's Ins. *of New Jersey*, p. 545 (1899); *J. N. Y. ent. Soc.*, XI, pp. 86-94 (1903), [Tables of subfamilies and genera]. Brues, *Gen. Ins.*, fas. 80, pp. 1-59 (1908), [Revised tables of genera]. Kieffer, *Gen. Ins.*, fas. 80 B, pp. 62-112 (1910), [Supplement to Brues's work]. Bures, *Bull Conn. St. Geol. Nat. Hist. Surv.*, No. 22, p. 541 (1916), [Connecticut forms]. Kiefer, *Broteria* (Zool), XV, pp. 50-62 (1917), [Philippinese forms]; *Philip J. Sci.*, XX, p. 102 (1922), [List of Philippines species]; *Das Tierreich.* Lief. XLVIII, pp. 1-556 (1926), [Monographic revision of the world forms, with tables of sub-families, genera and species]. Nixon, *Ann Mag. Nat. Hist.*, (10) XII, pp. 288-304, 305-325, 465-479, 549-563, pls. ii, fig. 11, (1933), [African forms].

Key to Subfamilies

1. Abdomen acute and margined laterally, with a groove ventrally in which the leteral margins of the tergites are curved under	2
Abdomen not acute lateraelly, tergites not bent underneath laterally; antennae usually with 12 segments in male and 11 or seldom 10-12 segments in the female	Telenominæ.
2. Fore wings with marginal and stigmal veins, except in some genera which have 12 antennal segments in both the sexes, rarely apterous or nearly so; fore tibial spur not trifurcate	3
Fore wings without marginal or stigmal veins, mostly with out any distinct venation ; fore tibial spur trifurcate; antennae usually with 10 segments, rarely 8, 9 or 11	Platygasterinæ.
3. Marginal vein very long, three to six times to stigmal vein, which is very short, at the most two to three times as long as broad; postmarginal vein not developed. Antennae with 12 segments	Tleasinae.

in male and female in the female with a distinct jointed club; third abdominal tergite the longest, except in *Gyron,* in which the second is the longest

Marginal vein thrice the length of the stigmal vein, which is also mostly long; in forms with vestigial or obsolete wings the third abdominal tergite is not the longest or the club of antennae is unjointed — 4

4. Antennae with 7-8 distinctly separated segments in the female, terminal segments forming a thick, unjointed club; with 12 of soldom 11 segments in the male. marginal vein very short, postmarginal vein mostly obsolete. Abdomen almost oval, first, second or third tergite the longest — Baeinae.

Antennae with 12 distinctly separated segments in both the sexex (exceptions are *Microbaeus* female, *Microbaeoides* female with 11; *Scelio* male, *Discelio* male, *Ceratobaeoides* female and *Romilius* female with 10 segments). Abdomen often long, linear or fusiform, with the basal three tergites mostly almost equal, often third or seldom second or the last tergite the longest — Scelioninae.

Subfamily I. **TELENOMINAE**

Telenomini, Thomson, *öfv. Vet.-Akad. Förh.,* XVII. p. 169 (1860). Ashmead, *Bull U. S. N. Mus.,* No. 45, p. 138 (1893).

Telenominae, Ashmead, *J. N. Y. ent Soc.,* XI, p. 87 (1903). Bruies, *Gen. Ins.,* fas. 80, p. 3 (1908). Kieffer *Gen. Ins.,* fas. 80 B, p. 105 (1910); *Das Tierreich,* Lief. XLVIII, p. 15 (1926), Nixon, *Trans. Roy ent. Soc.,* LXXXIII, pp. 73-103, pl. i, figs. 14 (1935), [Revision of African Forms]; *Ann. Mag. Nat. Hist.* (10) XX, pp. 113-128 (1937), [Asiatic species]; *ibid.,* pp. 444-475 (1937), [Key to Asiatic species).

Key to genera

1. Body uncommonly flattened — *Nardo.*

Body not uncommonly flattened — 2

2. *Eyes hairy — 3

---

* *Protelenomus* differs from all genera under this couplet in its obsolate post marginal vein.

	Eyes not hairy	6
3.	Parapsidal furrows develeoped a least posteriorly	4
	Parapsidal furrows absent	5
4.	Antennae with 11 segemnts in the female and 12 segments in the male	*Dissolcus.*
5.	Antennae with 10 segments in the female; vein usually short	*Aholcus.*
	Antennae with 11 segments in the female; vein long; second abdominal tergite longitudinally striate at least anteriorly	*Telenomus.*
6.	Antennae with 11 segments in the female; postmarginal vein mostly longer than the stigmal vein	7
7.	Second abdominal tergite without longitudinal striations Second abdominal tergite with longitudinal striations	*Liophanurus.* *Microphanurus.*

1. **Nardo** Nixon.

Nardo, Nixon, *Ann Mag. Nat. Hist.,* (11) I, p. 279 (1938), [Genotype : *N. Cumaeus* Nix.]

Key to species.

Fore wings rounded apically, marginal ciliation short	*N. cumaeus.*
Fore wings pointed apically, marginal ciliation long	*N. Phaeax.*

cumaeus Nixon.

Nardo cumaeus, Nixon, *Ann. Mag. Nat. Hist.,* (11) I, p. 279 ♀ ♂ (1938)[1]. Mani, *Indian J. Ent.,* I, p. 93 (1939)[2]. [Host : Eggs of *Macropes excavatus* Dist.].

[1]Lyallpur.
[2]Karnal.

phaeax Nixon.

Nardo phaeax, Nixon, *Ann. Mag. Nat. Hist.,* (11) I, p. 283 ♀ ♂ (1938)[1]. Mani, *Indian J. Ent.,* I, p. 93 (1939)[2]. [Host : Eggs of *Macropes excavatus* Dist.].

[1]Lyallpur.
[2]Karnal.

2. **Protelenomus** Kieffer.

Protelenomus Kieffer, *Bull. Soc. Hist. Nat. Metz,* XXV, p. 6 (1908), [Genotype : *P. flavicornis* Kieff. ♀].

flavicornis Kieffer.

Protelenomus flavicornis kieffer, *Bull. Soc. Hist. Nat. Metz,* XXV, p. 7♀ (1908)[1]; *Das Tierreich,* Lief. XLVIII, p. 22 ♀ (1926).

[1]Bhamo : Burma.

3. **Telenomus** Haliday.*

SYSTEMATICS.

Telenomus, (Partim), Haliday, *Ent., Mag.,* I, p. 271 (1833), [Genotype : *T. brachialis* Halid.]. Dalla Torre, *Cat. Hymen.,* V, p. 513 (1898). Ashmead, *J. N. Y. ent. Soc.,* XI, p. 87 (1903). Brues, *Gen. Ins.,* fas. 80, p. 6 (1908). Kieffer, *Gen. Ins.,* fas 80 B, p. 106 (1910). Brues, *Bull. Conn. St. Geol. Nat. Hist. Surv.,* No. 22, p. 544 (1916), [Connocticut fauna]. Kieffer, *Das Tietrreich.* Lief. XLVII, p. 24 (1926). Fouts, *Mem. Soc. ent. Ital.,* XIII, p. 104 (1935), [Keys, descriptions, etc., of African species.] Nixon, *Ann. mag. Nat. Hist.,* (10) XX, pp. 113-127 (1937), [Status, Scope, Synonymy; Ceylonese and Malayan species]; *ibid.,* (11) I, pp. 584-593 (1938), [New Asiatic spp.]; *Ibid.,* (11) VI, pp. 497-512 (1940), [Keys to Asiatic spp., new spp., etc.].

Phanurus, Thomson, *öfv. Svensk. Vet. -Akad. Förh.,* XVII, pp. 169, 172 (1860). Dalla Torre, *Cat. Hymen.,* V, p. 521 (1898). Ashmead, *J. N. Y. ent. Soc.,* XI, p. 87 (1903). Brues, *Gen. Ins.,* fas. 80, 3 (1908). Kieffer, *Das Tierreich,* Lief XLVIII, p. 49 (1926). Nixon, *Ann. Mag. Nat. Hist.,* (10) XX, pp. 113-127 (1937).

BIOLOGICAL.

Telenomus, Regnier, R., *Rev. Bot. appl. & Agric. Colon.,* III (19) pp. 169-185 (1923), [*Telenomus acrobates* as a parasite of *Chrysopa vulgaris* in France]. Gahan, *Proc. ent. Soc. Washington,* XXVII (9) p. 188 (1925), [*Telenomus nawai* Ashm. as an egg-parasite of *Prodenia litura* in Fiji]. Uichanco, L. B., *rep. Comm. Cane Varieites, Diseases, & Fertilizers, 6th Ann. Conv. Philip. Sugar Ass.,* reprint pp. 1-16 (1928). [*T. nawai,* originally introduced into Hawaii from Japan,

---

* The student should consult Nixon [*Ann. mg. Nat. Hist.,* (11) VI, pp. 497-501 (1940) for key to most of the species recorded in this catalogue.

introduced from there into Philippines against *Spodoptera mauritia]*. Paine, R. W. & Surridge, H. R., *Ann. Bull. Div. Reps. Dept. Agric. fiji*, 1931, pp. 1-8 (1932), [introduced into Fiji from Java against *Tirathaba trichogramma]*. Winburn T. F. & Painter, R. H., *J. Kansas ent. Soc.*, V (1) pp. 1-28 (1932). [*T. heliothidis* and its. biology as an egg-parasite of *Heliothis obsoleta* in Kansas]. Yanagihara, M., *J. Formosan Sugar Pl. Ass.*, XII (1-3) reprint pp. 1-53 (1934), [In Japanese. Bionomics as an egg-parasite of *Sesamia inferens* in Formosa].

Phanurus, Underhill, G. W. *Bull. Virginia agric. Expt. Sta.*, No. 251, pp. 1-21 (1926), [Probable egg-parasite of *Phtorimea operculella* in Virginia]. Box, H. E., *Rep. Agric. Dept. St. Kitts-Nevis*, 1931, pp. 9-12 (1932), *[T. (=Prophanurus) alecto* and its bionomics as an eggparasite of *Diatraea saccharalis* in St. Kitts]; *Bull. Imp. Inst.*, XXX, pp. 185-197 (1932), [Bionomics as an egg-parasite of *D. saccharalis* in Antigua]. Jaynes, H. A., *J. econ. Ent.*, XXV, pp. 64-68 (1932), [Bionomics as an eggparasite of *D. saccharalis* in S. America].

anwari Mani.

Telenomus anwari, Mani, *Indian J. Ent.*, I, p. 94 ♀ (1939).[1] [Host. : Eggs of anuknown bug on sugarcane leaf].

[1]Delhi.

Attachi Nixon.

Telenomus attaci, Nixon, *Ann. Mag. Nat. Hist.*, (10) XX, p. 456, fig. 3 g, ♀ ♂ (1937).[1] [Host : *Attacus atlas*].

[1]Kuala Lumpur: Malaya ; Bangkok Siam.

beneficiens (Zehntner) Nixon.

Ceraphron beneficiens, Zehntner, *Archif voor de Java-Suikerindustrie*, IV, p. 487 (1896);[1] in *Krüger's die Zuckerrohr und seine Kultur aus Java*, p. 350 (1899).

[1]Java

Phanurus beneficiens, DODD, *Canad. Ent.*, XLVI, 293 (1914); *Archiv Naturgesch.*, LXXX, A, p. 164 (1914). Kieffer, *Das Tierreich*, Lief. XLVIII, p. 57 (1926).

Telenomus beneficiens Nixon, *Ann. Mag. Nat. Hist.*, (10) XX, p. 465 ♀ ♂ (1937). Mani, *Indian J. Ent.*, I, p. 93 (1939).

BIOLOGICAL

Phanurus beneficiens, Dodd, *Canad. Ent.*, XLVI, p. 293 (1914). [Bionomics as an egg-parasite of *Diatraea venosata (=striatalis)*.[1] Goot, P. van Der, *Meded. Proefst. voor de Java-Suikerindustrie*, V (4) pp. 125-176 (1914). [Bionomics as an egg-parasite of *Daiatraea venosata* and *Chilo infuscatellus*].[2] Ishida, *Ibid.*, pp. 1-17 (1914), [Considered an effective egg-parasite of *D. venosata* owing to the preponderance of females over male].[3] Urich, F.W., *Bull. Dept. Agric. Trinidad & Tobago, Port-of-Spain*, XIV (5) pp. 156-161 (1915), [Possible egg-parasite of *D. saccharalis D. canella & D. lineolata*].[4] Annonymous, *Hawaii. Planter's Rec.* XXVIII (4) pp. 443-446 (1924), [Introduced into Formosa by Ishida against the satlk-borer of sugarcane]. Goot, P. Van der, *Meded. Inst. Plantenziekten*, 66, pp. 1-308 (1925). [Bionomics as an egg parasite of *Scirpchpaga innotata*].[5] Annonymous, *Abstracts Sugar Expt. Sta., Formosa*, pp. 272-294 (1928), [Bionomics as egg-parasite of *Chilo infuscatellus & D.venosata*].[6] Hazelhoff, E.H.,*Voordracht gehouden voor het Tiende Suiker Congres te Soerabaia op 17 April 1928*, pp. 1-15 (1928), [Biology as eggparasite of *D. venosata*].[6] Ishida M., *Proc. II Conf. Internat. Soc. Sugarcane Technol. Havana Cuba Sugar Club (1927)*, pp. 78-79 (1927), [Introduced into Formosa from Java against *Diatraea venosata* and *Chilo infuscatellus*]. Nakayama, s., *Dobutsugaku Zasshi (Zool. Mag.)*, XL, (482) p. 481 (1928), [Bionomies as egg-parasite of *Chilo simplex*].[7] Uichanco, L. B. *Rep. Comm. cone variaties, diseases & fertilizers, 6th Ann. Conv. Philip. Sugar Ass.*, reprint pp. 16 (1928), [Bionomics as egg-parasite of *D., venosa'a* & *Scirpophaga intacta*].[8] Van Zwaluwenburg, R. H., Rust, E. W. & Rosa, J. S., *Hawaii. For & Agric.*, XXV, (3) pp. 79-82 (1928). [introduced into Hawaii from Japan against *Chilo simplex*]. Kuwana, L., *Proc. IV Pacific Sci. Congr.*, Java, 1929, pp. 379-384 (1929). [Bionomics as egg-parasite of *Chilo simplex* & *Schonobius incertellus*].[9] Nakayama, S., *Ann. Agric. Expt. Sta. Chosen*, IV(4) pp. 248-250 (1929), [Bionomics as egg-parasite of *Chilo Simplex*].[10] Fullaway, *Hawaii. For & Agric.*, XXVII (1) pp. 45-50

[1,2,3]Java

[4]Trinidad.

[5]Java

[6]Dutch E. Indies.

[7]Japan : Korea.

[8]Philippines.

[9,10]Japan.

(1930), [Bionomics, liberations in field for the control of *Chilo simplex*][11]. Matsumura, S., *Proc. IV Pac. Sci. Congr. Java, 1929*, pp. 139-142 (1930), [*Chilo infuscatellus* recommended as a laboratory host in the absence of eggs of Diatraea venosata]. Pagden, H. T. *Bull. Dept. Agric. S.S. & F. M. S.*, (Sci. Ser.) (1), pp.1-30 (1930), [Bionomics as egg-parasite of *Schoenobius incertellus* and *Diatraea auricilia* in Malaya]. Lopez, A. W. *Ann. Rep. Res. Bur. Philip. Sugar Ass.*, 1930-31, pp. 227-273 (1931), [Bionomics as egg-parasite of *C. infucatellus & Sciropaphaga intacta* in philipines]. Hart, P. C., *Versl. Cultuarafd. Proefsta. Java-Suikerindustrie,* 1931, pp. 48-64 (1932), [Method of studying percentage of parasitism on *Scirpophaga intacta* in Java]. Pang Hiva Tsai, *Zeitschr. angew. Ent.*, XIX (4) pp. 608-614 (1932) [Bionomics as egg-parasite of *Schoenobius bipunctifer* and *Chilo simplex* in China.] Corbett, G. H., *Malayan Agric. J.*, XXI (8) p. 362-378 (1933), [Natural parasitism on *Schoenobius bipunctifer* in Malava]. Hart, P.C., *Arch. Suikerind. Ned.-Indië*, 1933, pp. 731-801; *Meded. Proefsta. Java Suikerind.*, 1933, No. 16 (1933), [Bionomics as parasite of *Scirpophaga intacta* in Java]. Van Dillewijn, C., *Meded. Onderafd. Cheribod Proefsta. Java-suikerind.*, 1932, pp. 1-9 (1933), [Egg-parasite of *Scipophaga intacta* in Java]; *Versl. Onderafd. Cheribon Profesta, Java-Suikerind.*, 1932, pp. 1-21 (1933), [Bionomics in Java]. Okada, J. & Maki, T., *Nojikairyoshiryo*, No. 69 (2) 2, pp. 1-78 (1934), [Biological studies as parasite of *Chilo simplex* in Japan. In Japanese with an English summary]. Pagden, H. T., *Sci. Ser. Dept. Agric. S. S. & F. M. S.* No. 15, pp. 1-13 (1934), [Bionomics as egg- parasite of *Schoenobius bipunctifer* in Malaya]. Subramanyam, T.V., *Rep. Mysore Agric. Dept.*, 1932-33, pp. 57-62 (1934),[12] [Bionomics as egg-parasite of *Argyira sticticraspis* and *Scirpophaga nivella]*. Takano, S., *J. Formosan Sug. Pl. Ass.* XI (10) pp. 454-466 (1934). [Bibliography and host-list in Formosa]. Ramkrishna & Margabhandhu, *Agric Live-stk. India*, V (5) pp. 503-521 (1935),[13] [Bionomics as egg-parasite of *Argyria sticticraspis].* Vanvuuren, L., *Ent. Meded. Ned-Indië*, I (2) pp. 29-33 (1935), [Bionomics as egg-parasite of *Schoenobius bipunctifer* in Java]. Hart., P. C., *Hand. 7 Ned.-Ind. natuurw. Cong.*, pp.

[11]Hawaii.

[12]Mysore: S. India.

[13]Madras.

415-423 (1936), [Difficulty of utilising against *Scirpophaga intacta* in the Dutch East Indies]. Pruthi, H. S., *Rep. Imp. Ent., Sci. Rep. Agric. Res. Inst., Pusa* 1935-36, pp. 123-137 (1937), [Egg-parasite of *Scirpophaga* in India]. Ramakrishna & Ananta-narayanan, *Agric. Live-stk. India,* VII (2) pp. 171-179 (1937), [Bionomics as egg-parasite of *Schoenobius bipunctifer* in South India.]

Prophanurus beneficiens, De Charmoy, D. D'F, Ann. *Rep. Dept. Agric. Colony of Mauritius* for 1914 (1915), [Egg-parasite of sesamia vuteria (=*nonagroides)* in mauritius]. Shiraki, *Agric. Expt. Sta. Govt. Formosa. Thaikhoku,* 1917, pp. 1-256 (1917); [Egg-parasite of *Chilo simplex, Diatraea venosata & Schoenobius incertellus* in Formosa].

beneficiens var. **elongatus** (Ishida).

Phanurus beneficiens var. *elongatus* Ishida, *Proc. II Conf. Internat. Soc. Sugarcane Technol.,* pp. 78-79 Havana, Cuba Sugar Club (1927), [Introduced from Java Into Formosa against *Scirpophaga].* Hazelhoff, E. H. *Jaarsversl. Proefsta. Java-Suikerind.,* 1928, pp. 96-111 (1929), [Bionomics as egg-parasite of *Scirpophaga intacta* in java]. Moritsugu, A., *Rep. Dept. Agric. Govt. Res. Inst. Formosa,* No. 50, pp. 1-56 (1931), [Bionomics as egg-parasite of *Scipophaga nivella* in Formosa]. Hazelhoff, E. H., Arch. *Suikerind. Ned.-Ind.,* III, p. 361-463 (19310, [Bionomics as egg-parasite of *Scirpophaga intacta* in Java]. Takano, S., *J. formosan Sug. Pl. Ass.,* XI (10) pp. 454-466 (1934), [Bibliography, hostlist from Formosa].

colemani Crawford.

Telenomus colemani, Crawf., *Proc. U. S. N Mus.,* XLII, p. 2 (1920).[1] Ramakrishna, *Bull. ent. Res.,* XVIII, p. 76 (1927), [Host : Eggs of *Dolycoris indicus].* — [1]Mysore.

cyrus Nixon.

Telenomus cyrus, Nixon, *Ann. Mag. Nat. Hist.,* (10) XX, p. 448 ♀ (1937).[1] [Host : Eggs of *Nezara viridula].* — [1]Lembang : Java.

dignoides Nixon.

Telenomus dignoides, Nixon, *Ann. Mag. Nat. Hist.,* (1) XX, p. 463 ♀ ♂ (1937).[1] [Host : Eggs of *Scipophaga auriflua* var. *intacta*]. — [1]Cheribon : Java

dignus Nixon.

Telenomus dignus, Nixon, *Ann. Mag. Nat. Hist.,* (10) XX, p. 462♀ ♂ (1937).[1] [Hosts : *Schoenobius incertellus* and *Scirpophaga innotata].* — [1]Luzon: Philippines, Canton: China, Tandjeong: Java.

javensis Dodd.

Telenomus javensis, Dodd, *Arch. Naturgesch.,* LXXX, A, (5) p. 163♀ (1914).[1] Kieffer, *Das Tierreich,* Lief. XLVIII, p. 40♀ (1926)[1]. Ahmad T., *Indian J. agric., Sci..* IX, p 622., pl. XXVIII, fig. 7 (1939), [Hosts : Tineid in Java. Eggs of *Lixus truncatulus* in India]. — [1]Java. [1]Pusa: India.

latisulcus Crawford.

Telenomus latisulcus, Crawford, Ramakrishna, *Bull. ent. Res.,* XVIII, p. 76. (1927),[2] [Host : *Coptosoma cribraria.* Also bred from brood cells of *Xylocopa].* — [1]Coimbatore, Mysore.

lelus Nixon.

Telenomus lelus, Nixon, *Ann. Mag. Nat. Hist.,* (10) XX, p. 450♀ (1937).[1] — [1]Kuala Lumpur: Malaya

lucullus Nixon.

telenomus lucullus, Nixon, *Ann. Mag. Nat. Hist.,* (10) XX, p. 45♀ (1937),[1] [Host : Eggs of a noctuid moth). — [1]Java.

manolus Nixon.

Telenomus manolus, Nixon, *Ann. Mag. Nat. Hist.,* (10) XX, p. 454 ♀ ♂ (1937),[1] [Host : Lepidopterous egg on coffe]; *ibid.,* (11) VI, p. 512 (1940),[2] [egg of some pest of Uncaria gambir]. — [1]Kuala Lumpur. Malaya. [2]Sumatra.

mesillae (Cockerell).

handronotus mesillae, Cockerell, *Canad. Ent.,* XXIX, p. 25 (1897). Kieffer, *Das Tierreich,* Lief. XLVIII, p. 464 (1926).

Telenomus mesillae, Gahan, *Ann. ent. Soc. Amer.,* XXV, p. 757 (1932),[1] [Hosts : *Pentatoma ligata, P. sayi & Euchistus servus].* — [1]New Mexico, Calexico. California.

Telenomus ashmeadi, Morrill, *Amer Nat.* XLI. p. 419 ♀ ♂ (1907).[2] Brues, *Gen. Ins.,* fas. 80, p. 6 (1908). Morrill, — [2]Texas.

Bull. Dept. Agric., Ent., No. 86, p. 62 (1910). Kieffer, *Das Tierreich*, Lief. XLVIII, p. 48 (1926). Mani, *Rec. Indian Mus.*, XXXVIII, p. 335 (1936),[3] [Identification only provisional]. — [3]Pareshnath : Bengal.

ochus Nixon.

Telenomus ochus, Nixon, *Ann. Mag. Nat. Hist.*, (11) I, p. 590 (1937),[1] [Host : *oreta cornea*]. — [1]Sumatra.

olynthus Nixon.

Telenomus olynthus, Nixon, *Ann. Mag. Nat.* Hist., (11) I, p. 586 (1937),[1] [Host : Eggs probably of a pyralid on *Uncaria gambir].* — [1]Sumatra.

otones Nixon.

Telenomus otones, Nixon, *Ann. Mag. Nat.* Hist., (11) VI, pp. 510-512 ♀ ♂ (1940),[1] [Hosts : bred from eggs of eggs of *Taru cus theophrastus* Linn. on lac in company with *Telenomus pegasus* Nix.]. — [1]Namkum : Ranchi.

pegasus Nixon.

Telenomus pegasus, Nixon, *Ann. Mag. Nat.* Hist., (11) VI, lpp. 509-510 ♀ ♂ (1940),[1] [Hosts : Bred from eggs of *Tarucus theophrastus* Linn. on lac in company with *Telenomus transversiceps* Nix. and *T. tones* Nix.]. — [1]Namkum : Ranchi.

periparetus Nixon.

Telenomus periparetus, Nixon, *Ann. Meg. Nat. Hist.*, (11) 1. p, 588 (1937),[1] [Host : Eggs of a geometrid on *Uncoria gambir*]. — [1]Sumatra.

Perplexus Nixon.

Telenomus perplexus, Nixon, *Ann. Mag. Nat.* Hist., (10) XX, p. 470 ♀ ♂ (1937),[1] [Host : Lepidopterous eggs]. — [1]Ulu Sali Road : Malaya.

proditor Nixon.

Telenomus proditor, Nixon, *Ann. Mag. Nat.* Hist., (10) XX, p. 456 ♀ (1937),[1] [Host : Eggs of a month on (?) *Gmelia arborea].* — [1]Dehra Dun : United Prov.

remus Nixon.

Telenomus remus, Nixon, *Ann. Mag. Nat.* Hist., (10) XX, p. 471♀ ♂ (1937),[1] [Host : *Spodoptera mauritia* & — [1]Malaya.

egg-masses Probably of *Prodenia* sp. on *Pithecolobium saman.]*.

rowani (Gahan).

Phanurus rowani, Gahan, *Philip, J. Sci.*, p. 106 (1925),[1] [Hosts: *Schoenobius incertellus]*. — [1]Luzon: Phillippines.

Telenomus rowani, Nixon, *Ann. Mag. Nat. Hist.*, (10) XX, p. 460 ♀ ♂ (1937),[2] [Host : *Schoenobius incertellus & Scirpophaga innotata* bred along with *Telenomus dignus]*. — [2]Java.

saccharicola Mani.

Telenomus saccharicola, Mani, *Indian J. Ent.*, III, i. 26♀ (1941),[1] [Host : Pentatomid on sugarcane]. — [1]Delhi.

sagei Mani.

Telenomus sagei, Mani, *Rec. Indian Mus.*, XXXVIII, p. 335 (1936),[1] [Host : Pentatomid]. — [1]Dharamsala : W. Himalayas.

sorus Nixon.

Telenomus sorus, Nixon, *Ann. Mag. Nat., Hist.*, (10) XX, p 458♀ (1937),[1] [Host: Eggs of *Philo* sp.]. — [1]Malaya.

spodopterae Dodd.

Telenomus spodopterae, Dodd, Archiv Naturgesch., LXXX, A, (5), p. 164♀ ♂ (1914).[1] Kieffer, *Das tierreich*, Lief. XLVIII, p. 40 (1926). [Host : *Spodoptera* sp. (?) on sugarcane]. — [1]Java.

tirathabae Fierrière.

Telenomus tirathabae, Ferrière, *Stylops*, II, p. 106 (1933),[1] Paine, R. W., *Bull. Dept. Agric. Fiji*, N. 18, pp. 1-30 (1935),[2] [Used in the control of *Tirathaba trichogramma]*. — [1]Buitenzorg; Java. [2]Fiji.

transversiceps Nixon.

Telenomus transversiceps, Nixon, *Ann. Mag. Nat. Hist.*, (11) VI, pp. 506-508, figs. 13, 15, ♀ ♂ (1940), [1] [Host: eggs of *Tarucus theophrastus* Linn. on lac.] — [1]Namkum: Ranchi.

triptus Nixon.*

Teleenomus triptus, Nixon, *Ann. Mag. Nat. Hist.,* (10) XX, p. 452♀ ♂ (1937),[1] [Host : eggs of *Scotinophara coarctata].* — [1]Kajang: Malaya.

usipetes Nixon.

Telenomus usipetes, Nixon, *Ann. Mag. Nat. Hist.,* (11), I, p. 584 (1938),[1] [Host : Eggs of *Hapalia machaeralis].* — [1]Pyinmana: Toungoo; Myohla: Burma.

vandergooti Dodd.

Telenomus vandergooti, Dodd. *Archiv Naturgesch.,* LXXX, A, (5) p. 164♀ (1914).[1] Kieffer, *Das. Tierreich,* Lief. XLVIII, p. 40 (1026). — [1]Java.

4. Aholcus Kieffer.*

Aholcus, Keiffer, in Alluad & jeanel *voy. Afr. Or. Hymen.,* I. p. 4 (1933). [Genotype *A. monticola* Kieff. ♀]; *Das Tierreich,* Lief. XLVIII, p. 117 (1926). Nixon ; *Ann. Mag. Nat. His.,* (10) XX, p. 116 (1937). [Keys to Ceylonese and Malayan species].

Telenomus (Aholcus). Nixon, *Ann. Mag. Nat. Hist.,* (10) XX, pp. 113-127 (1937), [Status and scope of genus discussed; considered to be of subgeneric rank only].

adenyus Nixon.

Telenomus (Aholcus) adenyus, Nixon, *Ann. Mag. Nat. Hist.,* (10) XX, p 122♀ ♂ (1937),[1] [Host : Eggs of a lepidopterous insect on lima beans]. — [1]Peradynia : Ceylon

euproctiscidis Mani.

Aholcus euproctiscidis, Mani, *Indian J. Ent.,* I, p. 95♀ ♂ (1939),[1] [Host : Eggs of *Euproctis lunata*]. — [1]New Delhi.

incommodus Nixon.

Telenomus (Aholcus) incommodus, Nixon, *Ann. Mag. Nat. Hist.,* (10) XX, p. 123♀ ♂ (1937),[1] [Host : Eggs — [1]Batu Gajan : Malaya.

---

* Since the second proof of this catalogue left my hands, I hve received specimens of this species from the Director. Zoological Survey of India, Calcutta, bred from eggs of *Scotinophara lurida* (Burma) in Ceylon (Note added in final proof).

* The student should consult Nixon [*Ann. Mag. Nat. Hist.,* (10) XX, pp 116-117 (1937)] for key to the species recorded in this catalogue.

of a lepidopteron, the larva of which feeds on banana leaves].

molochrus Nixon.

Telenomus (Aholcus) molochrus, Nixon, *Ann. Mag. Nat. Hist.,* (10) XX, p. 117♀ ♂ (1937),[1] [Host : *Lenodora vittata*]. — [1]Pussellawa: Ceylon.

stigis Nixon.

Telenomus (Aholcus) stigis, Nixon, *Ann. Mag. Nat. Hist.,* (10) XX, p. 125♀ ♂ (1937),[1] [Host : Eggs of *Acherontia Styx*]. — [1]Kuala Lumpur, Malay.

talaus Nixon.

Telenomus (Aholcus) talaus, Nixon, *Ann. Mag. Nat. Hist.,* (10) XX. p. 127♀ ♂ (1937),[1] [Host : Eggs of *Papilio agamemnon*]. — [1]Kuala Lumpur: Malaya.

turbatae Nixon.

Telenomus (Aholcus) turbatae, Nixon, *Ann. Mag. Nat. Hist.,* (10) XX, p. 121♀ ♂ (1937),[1] [Host: *Orgyiaturbata*]. — [1]Kuala Lumpur : Malaya.

ulusalus Nixon.

Telenomus (Aholcus) ulusalus, Nixon, *Ann. Mag. Nat. Hist.,* (10) XX, p. 119 ♀ ♂ (1937),[1] [Host : Eggs of a Lepidopteron]. — [1]Ulu Sali Road: Malaya.

urios Nixon.

Telenomus (Aholcus) urios, Nixon, *Ann. May. Nat. Hist.,* (11) I, p. 591 (1938),[1] [Host : *Hidari irava*]. — [1]Malaya.

5. **Liophanurus** Kieffer.

Telenomus (partim), Walker, *Ent. Mag.,* III, p. 348 (1836), Ashmead, *Bull. U. S. N. Mus.,* No. 45, p. 151 (1893).

Liophanurus, Kieffer, in Andrè, *Spec. Hymen, Eur.,* XI, p. 61 (1912), [Genotype: *L. spilosomatis* (Ashm.) ♀ ♂]; *Das Tierreich, Lief.* XLVIII, p. 64 (1926).

saccharalis (Dodd).

Telenomus saccharalis, Dodd, *Canad. Ent.,* XLVI, p. 293♀ ♂ (1914).[1]

Liophanurus saccharalis, Kieffer, *Das Tierreich,* Lief. XLVIII, p. 71♀ ♂ (1926),[2] [Host : Eggs of Pentatomid on sugarcane leaf].	[1, 2]Java.

6. **Microphanurus** Kieffer.*

Teleas (partim), Nees, *Hymen. Ichneum, affin. Monogr.,* II, p. 290 (1834). Ratzeburg, *Ichneum.,* II, p. 142 (1848).

Telenomus (partim), Thomson, *Öfv. Vet.-Akad. Förh.,* XVI, p. 170 (1860). Mayr, *Verh. Zool-bot. Ges. Wein,* XXIX, p. 704 (1860) (?).

Aphanurus, (nec Loos 1907) Kieffer, in Andrè, *Spec. Hymen. Eur.,* XI, p. 69 (1912).

Microphanurus, Kieffer, *Das* Tierreich, Lief. XLVIII, p. 91 (1926), [Genotype : *M. semistriatus (Nees)*]. Chiormonte, A., *Atti. 1° Cong. Studl. colon., Firenze,* 1931, reprint pp. 1-51 (1931). [Sepcies from Italian Somaliland; Biology as eggprasites of Rhynchota]. Priesner, H., *Bull. Soc. R. ent. Egypte,* 1931, fas. 3-4, pp. 137-139 (1931), [Bionomics of *M. megacephalus* Ashm. as egg-parasite of *Nezaraviridula* in Egypt]. Fouts, *Mem. Soc. ent. ital., Genoa,* XIII, p. 106 (1934), [keys to Ethiopean species]. Nixon, Trans. Roy. ent. soc., LXXXIII, p. 96 (1935), [Keys to spp., Synonymies, etc.]; *Ann. Mag. Nat. Hist.,* (11), (8) pp 122-139 (1938), [Monograph of Asiatic species, Keys, description, etc.].

artabazus Nixon.

Microphanurus artabazus, Nixon, *Ann. Mag. Nat. Hist.,* (11), II, p. 131♀ ♂ (1938),[1] [Host : Eggs of *Scotinophara* sp. and a pentatomid].	[1]Malaya.

barrowi (Dodd).

Telenomus barrowi, Dodd, Trans. Roy. Ent. Soc., 1919, p. 356 ♀ (1920),[1] [Host : Eggs of a sphingid].	[1]Dalhousie : India.

carinifrons (Cameron).

Immsia carinifrons, Cameron, *Indian For. Rec.,* IV, p. 105 (1913).[1]	[1, 2, 3]Dehra Dun: India.

* The student should consult Nixon [*Ann. Mag. Nat. Hist.,* pp 123-125 (1937)] for key to the species recorded in this catalogue.

Telenomus carinifrons, Dodd, *Trans. Roy. ent. Soc.* 1919, p. 355 (1920).[2]

Microphanurus carinifrons, Nixon, *Ann., Mag. Nat. Hist.,* (11) II, p. 138 (1938).[3]

Priapus Nixon.

Microphanurus priapus, Nixon, *Ann. Mag. Nat. Hist.,* (11) II, p. 133♀ (1938),[1] [Host : Eggs of *Chrysocoris atricapillus;* also bred in company with *Microphanurus vindicius* Nix. from eggs of *Dasynus manilae*].

[1]Buitenzorg : Java.

seychellensis (Kieffer).*

Telenomus seychellensis, Kieffer, *Bull. Soc. ent. Fr.,* P. 294 ♀ ♂ (1910); 1 *Trans. Linn. Soc. London,* (2) XV, p. 51 (1912).

Microphanurus seychellensis, Kieffer, *Das Tierreich,* Lief. XLVIII, p. 105 ♀ ♂ (1926). Nixon, *Trans. ent. Soc. London.* LXXXIII, p. 97 (1935); *Ann. Mag. Nat. Hist.,* (11) p. 125 (1938),[2] [Hosts: *Cantheconidea robusta* Dist., *Antestia orbitalis* Westw. (= *variegata* Thunb.), Antestia orbitalis var. lineaticollis stal, *Agonoscelis pubescens* Thunb. (= *versicolor* Fabr.)].

[1]Seychelles Islands.
[2]Passara: Ceylon; Abyssinia to Cape in Africa.

stoicus Nixon.

Microphanurus stoicus, Nixon, *Ann. Mag. Nat. Hist.,* (11) II, p. 135 ♀ ♂ (1938)[1].

[1]Malaya.

striaticeps Dodd.

Microphanurus striaticeps, (Dodd) Nixon, *Trans. ent. Soc. London,* p. 100 (1935); *Ann. May. Nat. Hist.,* (11) II, p. 125 (1938)[1], [Hosts : Egss of a pentatomid on brinjal leaf[1]; *Acanthomia brevirostris* Stal[2]].

[1]Lyallpur: Punjab.
[2]Nyasaland Br. Badan Nigeria, Cape; Africa.

sulmo Nixon.

Microphanurus sulmo, Nixon, *Ann. mag. Nat. Hist.,* (11) II, p. 126♀ (1938)[1], [Host : Eggs of *Cantheconidea robusta* Dist.].

[1]Talawakelle: Ceylon.

---

* This is probably the species parasitic in the eggs of *Cantheconidea furcellate* (Wolff) in the Madras Presidency.

trophonius Nixon.

Microphanurus trophonius, Nixon, *Ann. mag. Nat. Hist.,* (11) II, p. 127♀ (1938),[1] [Host : Eggs of a reduvid on *Uncaria gambir*]. — [1]Asahan : Sumatra.

vindicius Nixon.

microphanurus vindicius, Nixon, *Ann. Mag. Nat. Hist.,* (11) II, p. 128♀ (1938)[1]. — [1]Buitenzong: Java.

7. **Dissolcus** Ashmead.

Dissolcus, Ashmead, *Bull. U. S. N. Mus.,* XLV, p. 164, pl. iv, fig. 8 (1893), [Genotype: *D. nigricornis* Ashm.♀]. Dalla Torre, *Cat. Hymen.,* V, p. 512 (1898). Ashmead, *J.N.Y. Ent. Soc.,* XI, p. 87 (1903). Brues, *Gen. Ins.,* fas. 80, P. 11 (1908). Kieffer, Gen. Ins., fas. 80 B, p. 105 (1910); *Das Tierreich,* Lief. XLVIII, p. 124 (1926).

indicus Mani.

Dissolcus indicus, Mani, *Rec. Indian Mus.,* XXXVIII, p. 335 (1936).[1] — [1]Calcutta.

Subfamily II. BAEINAE.

Scelioninae (partim), Dalla Torre, *Cat. Hymen.,* V, p. 494 (1898)

Baeini, Ashmead, *Bull. U.S.N. Mus.,* No. 45, P. 167 (1893).

Baeinae, Ashmead, *J. N. Y. ent. Mus.,* No. 45, p. 167 (1893).

Baeinae, Ashmead, *J. N. Y. ent. Soc.,* XI, p. 88 (1903). Brues, *Gen. Ins.,* fas. 80, p. 14 (1908). Kieffer, *Gen. Ins.,* fas. 80 B. p. 100 (1910); *Das Tierreich,* Lief. XLVIII, p. 132 (1926).

Key to genera

Second abdominal tergite longest ; marginal and stigmal veins obsolete — *Aneurobaeus*

Second abdominal tergite not the longest but the third is so; marginal and stigmal veins developed — *Dissacolus.*

8. **Aneurobaeus** Kieffer.

Baeus (partim), Bugnion & Popoff, *Rev. Suisse Zool.,* XVIII, p. 731 (1910), [err. nom. Haliday 1833].

Aneurobaeus, Kieffer, in Adnrè, *Spec. Hymen. Eur.*, XI, pp. 87, 91 (1912), [Genotype: *A. apterus* (Bugn. & Pop.)]; *Das Tierreich, Lief.* XLVIII, p. 134 (1926).

apterus (Bugnion & Popoff).

Baeus apterus, Bugnion & Popoff, *Rev. Suisse Zool.*, XVIII, p. 731, pl. v, fig. 1-9 ♀ ♂(1910).[1]

Aneurobaeus apterus, Kieffer, in Andrè, *Spec. Hymne. Eur.* XI, p. 91 (1912);[2] *Das Tierreich*, Lief. XLVIII, p. 134 ♀ ♂ (1926).[3]

[1, 2, 3]Ambalagnoda: Ceylon.

9. **Dissacolus** Kieffer.

Acolus (Partim), Dodd, *Proc. roy. Soc. Queenslands*, XXVI, p. 127 (1914).

Dissacolus, Kieffer, *Das Tierreich*, Lief. XLVIII, p. 155 (1926), [Genotype : *D. bidentatus* (Dodd)♀].

Lakshmani Mani.

Dissacolus lakshmani, Mani, *Indian J. Ent.*, I, p. 96 ♂ (1939),[1] [Host : Eggs of a spider on sugarcane leaf].

[1]Delhi.

Subfamily III. TELEASINAE.*

Scelioninae (partim), Dalla Torre, *Cat. Hymen.*, V, p. 494 (1898).

Teleasini, Ashmead, *Bull. U.S.N. Mus.*, XLV, p. 180 (1893).

Teleasinae, Ashmead, J. N. Y. Ent. Soc., XI, p. 89 (1903). Brues, Gen. Ins., fas. 80, p. 18 (1908). Kieffer, *Gen. Ins.*, fas. 80 B, p. 91 (1910); *Das Tierreich*, Lief. XLVIII. p. 173 (1926). Dodd, *Proc. Linn. Soc. N. S. Wales*, LV, pp. 41-91 (1930), [Revision of Australian Forms].

Subfamily IV. **SCELIONINAE.**

Scelionidae (Partim), Förster, *Hymen. Stud.*, II, P. 27 (1856).

Scelionini, Thomson, *Ofv. Vet.-Akad. Förh.*, XV, p 417 (1858). Ashmead, *Bull. U. S. N. Mus.*, No. 45, p. 208 (1893).

---

* No Indian form has so far been described.

Scelioninae, Ashmead, *Bull. U. S. N. Mus.*, No. 45, p. 136 (1893). Dalla Torre, Ashmead, *J. N. Y. ent. Soc.*, XI, p. 90 (1903). Brues, *Gen. Ins.*, Fas. 80. p. 26 (1908). Kieffer, *Gen. Ins.*, fas. 80 B, p. 62 (1910); *Das Tierreich*, Lief. XLVIII, p. 264 (1926).

Key to genera

1.	Antennæ with 10 segments	2
	Anetannæ with 12 segments	3
2.	mesonotum not dentate, at the most with two longitudinal furrows; postmarginal vein short or undeveloped	*Scelio* ♂.
3.	Body flat, specially the head and thorax depressed, scape of antenna flattened and greatly expanded distally	*Platyscelio.*
	Body not as described above	4
4.	Upper part of face with one to three transverse curved lamellæ	*Sparasion.*
	Upper part of face without such a lamellæ	5
5.	Scutellum with two thorns or teeth or margins behind ; fore wings without pterostigma, median segment with one thorn or tooth in the middle; Parapsidal furrows obsolete; marginal vein very much shorter than stigmal vein	*Hadronotoides.*
	Southern neither toothed nor spined nor margined as described abvoe; at least fore wings with submarginal adn often marginal and stigmal veins; fore wings not narrowed basally	6
6.	Metanotum almost as long as scutellum or longer median segment lamellate; marginal vein of fore wings nearly punctiform	7
	Metanotum not as described abvoe	8
7.	Parapsidel furrows well developed with a distinct longitudinal line between them, metanotum rounded behind; fore wings with basal, medial cubital and discoidal veins, postmarginal vein absent.	*Oxyscelio.*
8.	Mesonotum with longitudinal line or furrow or with two anteriorly converging lines between the	9

	parapsidal furrows or only the scutellum with a medain longitudinal line	
	Mesonotum without such lines or furrows or even the parapsidal furrows	12
9.	Mesonotum with three longitudianl furrow	10
	Mesonotum with two longitudinal furrows	11
10.	Postmarginal vein longer than stigmal vein; metanotum with one or two teeth; eyes not pubescent; a crina connects the tegula with fore coxa	*Hoploteleia*
11.	metanotum normal, mesonotum with a median carina or without such a carina when the scutellum one median carina; subcostal vein arched and widely separated from anterior margin of wing, marginal vein punctiform	*Dicroteleia*
12.	Subcostal vien reaching the anterior margin of wing; segments of abdomen normal	13
13.	Marginal vein often replaced by pterostigma, postmarginal vein absent; adbominal tergits transverse	14
	Wings without pterostigma, postmarginal vein developed or absent	17
14.	Metanotum with one lamella or thorn or also with two teeth	15
	metanotum unarmed	16
15.	Metanotum with one transverse bilobed lamella	*Lepidoscelio*
16.	Eyes not pubescent, antennæ with a 6-9-jointed club; abdomen sessile; prothorax elongate	*Scelio* ♀
17.	Sixth abdominal tergite of female not much smaller than fifth; fore wings without a distinct vein-enclosed median cell	18
18.	metanotum or the middle of median segment with two teeth	19
	Metanotum or the middle of median segment with one often trilobed tooth or two-jointed apical process	22
	Metanotum or the middle or median segment unarmed	23

19. parapsidal furrows absent or indicated only posteriorly; scultellum rounded posteriorly	20
20. Postmarginal vein absent; first and second abdominal tergites transvese	*Dicroscelio*
Postmarginal vein longer than stigmal vein	21
21. First abdominal tergite short ; parapsidal furrows developed posteriorly; median segment with one tooth at the posterior angles	*Sceliacantha.*
22. Postmarginal vein well developed ; eyes pubescent; first abdominal tergite with horn in female; parapsidal furrows absent	*Probaryconus.*
23. Postmarginal vein at the most as long as the stigmal vein	24
24. parapsidal furrows absent, at the most indicated by a faint spur only	25
Parapsidal furrows distinct	31
25. Abdomen short, at the most as long as the rest of body, all tergites strongly transverse	26
Abdomen longer than the rest of body or at leat equal, when all the tergites are not transvese	28
26. Marginal vein shorter than postmarginal, often puncti-form	27
27. Body broad, second abdominal tergite longer than third Bod/slender, second abdominal tergite shorter than third.	*Hadronotus.* *Psiloteleia.*
28. Eyes not pubescent ; first abdominal tergite with horn in the female; marginal vein shorter than stigmal; second abdominal tergite shorter than third	*Pegoteleia.*
Eyes pubescent	29
29. Marginal vein punctiform or shorter than stigmal; first abdominal tergite elongated; postmarginal vein not distinctly longer than stigmal	30
30. Abdomen with one horn anteriorly in the female	*Baryconus.*
Abdomen arched dorsally and ventrally Rhacoteleia	*Rhacoteleia.*

10. **Sparasion** Latreille.

Sparasion, Latreille, *Hist. Crust. ins.,* III, p. 315 (1802), [Genotype: *S. frontalis Latr.]; ibidem,* XIII, p. 230 (1805) ; *Gen. Crust. Ins.* IV. p. 34 (1809). Ashmead, *J. N. Y. ent. Soc.,* XI, p. 93 (19030. Kieffer, *Gen Ins.,* fas. 80 B, p. 69 (1900); *Das Tierreich,* Lief. XLVIII, p. 282 (1926).

Ceraphron (Partim), Jurine, *Nouv. Meth. Class. Hymen.,* p. 303 (1807).

Scelio, (partim), Spinola, *Ann. Mus. Hist. Nat.,* XVII, p. 150 (1811).

Oxyurus, Lamarck, *Hist. Nat. Anim. s. Vert.,* XIV, p. 129 (1817).

Sparaison, Brues, *Gen. Ins.,* fas. 80, pp. 29, 41, 58 (1908).

Key to species

With parapsidal furrows, though indicated only posteriorly but deep	*S.albopilosellus.*
Without parapsidal furrows	*S. formosu.*

albopilosellus Cameron.

Sparasion albopilosellus, Cameron, *J. Bombay Nat. Hist. Soc.,* XVII, p. 98♂ (1906).[1] Kieffer, *Das Tierreich,* Lief. XLVIII, p. 292 (1926).[2] — [1,2]Quetta.

formosus Kieffer.

Sparasion formosum, Kieffer, *Bull. Soc. ent. Fr.,* P. 311♂ (1910);[1] *Gen. Ins.,* fas. 80 B, p. 69 (1910).[2] — [1, 2, 3]Malacca.

Sparasion formosus Kieffer, *Das Tierreich,* Lief. XLVIII, p 294 ♂ (1926).[3]

11. **Scelio** Latreille.

SYSTEMATIC.

Scelio, Latreille, *Hist. Nat. Crust. Ins.,* XIII, p. 227 (1805), [Genotype : *S. rugosulus* Latr. ♂]. Förster, *Hymen. Stud.,* II. p. 102 (1856). Dalla Torre, *Cat. Hymen.,* V, P. 494 (1898). Ashmead J. N. Y. ent., Soc., XI, p. 93

(1903). Brues, *Gen. Ins.*, fas. 80 B. p. 72 (1910) ; *Das Tierreich,* Lief. XLVIII. p. 308 (1926). Dodd, *Proc. Roy. Soc. Queensland,* XXXVIII, pp. 127-175 (1926), [Australian forms].

Cynipsillum (partim), Lamarck, *Hist. Nat. Anim. s. Vert.,* IV, p. 158 (1817).

Serlion, Say, *Contr. Maclur Lye. Arts & Sci. Philad.,* I, p. 80 (1828).

Bethylus (partim) Blanchard, in Cuvier. Règne Anim., (ed. 4) Ins., II, p. 116 (1849).

Aleria, Marshall, *Ent. Month. Mag.,* X. p. 208 (1874).

Caloptenobia Riley, *Rep. U. S. Ent. Comm. I,* p. 306 (1878).

Acerota, Provancher, Addit. Hymen Quebec, *p. 184 (1887).*

BIOLOGICAL

Scelio, Nikolski, V. V., *State Inst. Expt. Agron., Trans. Dept. Appl. Ent.* XII (2) pp. 1-330 (1924), [Bionomics as eggparasite of *Locusta migratoria].* LUK. Yano-vich, F. K., *Protect plants Ukraine,* 1925 (5-6) pp. 51-53 (1925), [Bionomics as egg-parasite of *Locusta migratoria* in Ukraine]. Oglobin, A.A., Bull. ent. Res., XVII (4) pp. 393-404 (1927), [*Scelio, uvarovi* Ogl., parasitic on eggs of *Locusta migratoria* phace danica in Ukraine *and S. nikolskyi Ogl., parasitic on eggs of L., migratoria* in Turkestan; correction to Kieffer's generic description of Scelio, Latr., that the maxillary palpi are 3-articulate]. Predtechenkskii, I. A., *Comment. Inst., astrachen. Defens. Plant.,* II (1) pp. 1-115 (1928), [Bionomics as egg-parasite of *L. migratoria in Lower* Volga region; in Russian with an English summary]; *Izv. otd. prikl. Ent.,* III (2) pp. 113-119 (1928), [*S. uvarovi* Ogl., as an egg-parasite of *L. migratoria* in Central Russia]. D (AMPF), A., *Biol. mens. Defensa agric Sec. Agric. Form Mexico,* III, (1-4) pp. 1-18 (1929),* [*Scelio fuscipennis* Ashm., as an egg-parasite of the Mexican migratory locust *Schistocerca paranensis]. Ferrière, Bull. ent. Res.,* XXI, pp. 33-44. (1930), [*S. zolotarevskyi Ferr.* as an egg-parasite of *L. migratoria migratorioides* from

* Not seen in original.

Madagascar]. Yazuikov Zakhvatkin, A.A., *Bull. ent. Res.*, XXII, pp. 385-391 (1931), [Bionomics of *S.nikolskyi* as an egg-parasite of locusts in Turkestan]. Williams, F. X., *Rep. Comm. Expt. Sta. Hawaii, Sugar Pl. Ass.*, 1935, pp. 1-17 (1936), [*S. pembertoni liberated againt Oxya chinensis* on sugarcane in hawaii]. Wong (Chi-Yu), *Ent. Phytop.*, III (21) pp. 418-421 (1935), [*S. uvarovi* as an egg-parasite of *L. Migratoria* in China.

acte Walker.

Scelio acte, Walker, *Ann. Mag. Nat. Hist.*, XVII, p. 185♀ (1846).[1] Dalla Torre, *Cat. Hymen.*, V, p. 494 (1898). Brues, *Gen. Ins.*, fas. 80, p. 43 (1908). Kieffer, *Das Tierreich,* Lief XLVIII, p. 322 (1926). — [1]Bombay.

calcuttaensis Mani.

Scelio calcuttaensis, Mani, *Rec. Indian Mus.*, XXXVIII, p. 337♀. (1936).[1] — [1]Calcutta.

hieroglyphi Timberlake.

Scelio hieroglyphi, timberlake, *Proc. hawaii. ent. Soc.*, VIII, p. 157 (1932),[1] [Host : Eggs of *Hieroglyphus banian].* — [1]Coimbatore.

horai Mani.

Scelio horai, Mani, *Rec. Indian Mus.*, XXXVIII, p. 336♀ (1936).[1] — [1]Chakradharpur.

oxyae Timberlake.

Scelio oxyæ, timberlake, *Proc. Hawaii. ent. Soc.*, VIII, p. 158 (1932),[1] [Host: Eggs of *Oxya velox*]. Ramachandra Rao, Y., *Rep. Proc. IV Ent. Meet. Pusa,* pp. 41-42 (1921),[2] [Bionomics as an egg-parasite of *Oxya velox* in South India]. — [1,2]Coimbatore

Paivai Mani.

Scelio paivai, Mani, *Rec. Indian Mus.*, XXXVIII, p. 336♀ (1936).[1] — [1]On board ship off Coconada.

pembertoni Timberlake.

Scelio pembertoni, Timberlake, *Proc. Hawaii. ent. Soc.*, VIII, pp. 153-162 (1932).[1] Pemberton, *Hawaii. Plant.* — [1]Selangor: Malaya.

Rec., XXXVII (1) pp. 7-9 (1933), [Introduction from Malaya into Hawaii against *oxya velox*]; *Proc. Hawaii. ent. Soc.* VIII, p. 253-264, pl. xv-xvi, figs. 1-15 (1933), [Bionomics]. Williams, F. X., *Rep. Comm. Expt. Sta. hawaii. Sugar Pl. Ass.*, 1935, pp. 17-25 (1936), [Liberations in Hawaii against *Oxya velox* on sugarcane].

serdangensis Timberlake.

Scelio serdangensis, Timberlake, *Proc. Hawaii, ent. Soc.*, VIII, pp. 153-162 (1932).[1] Pemberton, *Hawaii plant. Rec.*, XXXVII (1) pp. 7-9 (1933), [Introduction into Hawaii from Malaya against *oxya velox*]. — [1]Serdang: Malaya.

12. Lepidoscelio Kieffer.

Lepidoscelio, Kieffer, *Ann. Soc. Sc. Bruxelles*, XXIX, p. 129 (1905), [Genotype: *L. fuscipennis* Kieff.♀][1] Brues, *Gen. Ins.*, fas. 80, p. 30 (1908). Kieffer, *Gen. Ins.*, fas. 80 B, p. 72 (1910); *Das Tierreich*, Lief. XLVIII, p. 348 (1926).

viatrix Brues.

Lepidoscelio viatrix. Brues, *Proc. U.S.N. Acad. Sci., Washington*, III (2) p. 139, fig. 1 (1917);[1] *Psyche*, XXIV, p. 196 (1917),[1] [Host : eggs of *Colemania sphenarioides* Bot.].

13. oxycelio Kieffer.

Oxyscelio Kieffer, Zeitschr. Hymen. Dipt., VII, p. 310 (1907), [Genotype: *O. Foveatus* Kieff. ♀]. brues, *Gen. Ins.*, fas. 80, p. 46 (1908). Kieffer, *Gen. Ins.*, fas. 80 B, p. 68 (1910); *Das Tierreich*, Lief. XLVIII, p. 361 (1926). Dodd, *Proc. Roy. Soc. Queenslad*, XLII. pp. 72, 14 (1931), [Characters, List of spp., Synonymy : = *Dicroteleia* Kieff. *Camptoteleia* Kieff. = *Xenoteleia* Kieff.]. — [1,2]S. India.

foveatus Kieffer.

oxyscelio foveatus, Kieffer, *Zeitschr. Hymen. Dipt.*, VII, p. 310♀ (1907).[1] Brues, *Gen. Ins., fas. 80*, p. 46 (1908) Kieffer, *Gen. Ins.*, fas 80 B, p. 68 (1910); *Das Tierreich*, Lief. XLVIII, p. 361♀ (1926). — [1]Java.

14. **Hoploteleia** Ashmead.

Baryconus (partim) (non Förster), Ashmead, *Ent. Amer.*, III, p. 118 (1887).

Hoploteleia, Ashmead, *Bull. U.S.N. Mus.*, XLV, p. 227 (1893), [Genotype : *H. floridana* (Ashm.)♀ ♂]. Dalla Torre, *Cat. Hyme.*, V, p. 499 (1898). Ashmead, J. N. Y. *ent. Soc.*, XI, p. 94 (1903). Brues, *Gen. Ins.*, fas. 80, p. 29 (1908). Kieffer, *Gen. Ins.*, fas. 80 B. p. 71 (1910); *Das Tierreich,* Lief. XLVIII, p. 365 (1926). Dodd, *Proc. Roy. Soc.*, XL, pp. 39-50 (1929), [Australian species].

gravelyi Mani.

Hoploteleia gravelyi, Mani, *Rec. Indian Mus.*, XXXVIII, p. 469♀ ♂, (1936);[1] *Indian J. Ent.*, I, p. 96 fig. 10 ♂ (1939),[2] [Host: Unknown grass-hopper eggs].

[1]Madras.
[2]Cuttack, Coimbatore.

15. **dicroteleia** Kieffer.

Macroteleia (partim), Ashmead, *j. Linn Soc. London*, XXV, p. 222 (1894).

Prosanteris (Partim), Kieffer, *Ann. Soc. Sci. bruxelles*, XXXII, p. 136 (1908). Bures, *Gen. ins.*, fas. 80, p. 51 (1908). Kieffer, *Gen. Ins.*, fas. 80 B, p. 88 (1910), [Subgen. under *Trichanteris].*

Dicroteleia, Kieffer, *Not. Leyden Mus.*, XXX, p. 92 (1908), [Genotype : *D. rugosa* Kieff.]; *Gen. Ins.*, fas. 80 B, 68 (1910), [subgen. under *oxyscelio]; Das Tierreich,* Lief. XLVIII, p. 387 (1926).

rugosa Kieffer.

Dicroteleia rugosa, Kieffer, *Not. Leyden Mus.*, XXX, p. 92 ♂ (1908);[1] *Das Tierreich,* Lief. XLVIII, p. 388♀ (1926).

Oxyscelio (Dicroteleia) rugosa, Kieffer, *Gen. Ins.*, fas. 80 B, p. 68 (1910).

[1]Java.

16. **Sceliacantha** Dodd.

Sceliacantha, Dodd, *Arch. Naturgeich.*, LXXIX, A (6) p. 173 (1913). [Genotype: *S. quadrispinosa* Dodd ♂]. Kieffer, *Das* Tierreich, Lief. XLVIII, p. 412 (1926).

subplana Dodd.

Sceliacantha subplana, Dodd *Trans. ent. Soc. London,* p. 336 (1920).[1] — [1]Ceylon.

17. Psiloteleia, Kieffer.

Psiloteleia, Kieffer, in Voeltzkow, *Reis. in. Ostafrika, 1903-1905, wissensch, Ergebn.,* II, p. 53 (1910), [Genotype: *B. Striatigena* Kieff.♂]; *Gen. Ins.,* fas. 80 B, 82. (1910); *Das tierreich. Lief.* XLVIII. p. 451 (1926).

montana (Kieffer).

Hadronotus montanus. kieffer. *Ann. Soc. Sci. Bruxelles,* XXXII. p. 45.♂ (1908)[1], Brues, *Gen. Insc.,* fas. p. 51 (1948). Kieffer, *Gen. Ins., Fas* B p. 84 (1910). — [1]Matheran.

Psiloteleia montana, Kieffer. *Das.* Tierreich, Lief. XLVIII, p. 452 ♂ (1926).

18. Hadronotus, Forster.

Teleas (partim), Nees. Hymen. Ichneumraffin. Mongor., II. p. 290 (1834).

Hadronotus Förster, *Hymen. Stud., II.* 101, 105 (1856), [Genotype: *H. laticeps (Först) Kieff.*♀].* Ashmead, Bull. U. S. N. Mus., XLV. p. 229 (1893).

Dalla Torre, *Cat. Hymen.*V, p. 498 (1898). Ashmead, *J. N. Y. ent. Soc.,* XI, p. 92 (1903). Brues, *Gen. Ins., fas.* 80, pp. 37, 51 (1908). Kieffer, *Gen. Ins.,* fas. 80 B, p. 81 (1910); *Das Tierreich,* Lief. XLVIII, p. 453 (1926).

Telenomus (partim), Ashmead, *Bull. U. S. Depart, Agric., Ent.,* No. 14, p. 23 (1887).

Telenomoides, Dodd, *Trans. roy. Soc. S. Australia,* XXXVIII, p. 129 (1914).

Hadronotellus Keiffer, *Ent. Meded.,* XI, p. 341 (1917).

Hadronotus, Nixon, *Stylops,* III (1), pp. 1-5 (1934), [Javanese spp., *H. oophagus* Nix. from a Coreid, and H. *leptocorisae* Nix. from *Leptocorisa acuta].*

---

* Brues designates *H. exsculptus* Först. ♀as type instead of *H. laticeps* (Först. but it was not described until 1908 (kieffer. *Ann. Soc. Sc. Brux.,* XXXII, p. 144).

dasyni Nixon.

Hadronotus dasyni, Nixon, *Stylops,* III, p. 2 (1934).[1] — [1]Java.

homoceri Nixon.

Hadronotus homoceri, Nixon, *Stylops,* III, p. 4 (1934).[1] — [1]Java.

Javensis Dodd.

Hadronotus javensis, Dodd, *Arch. Naturgech.,* LXXX, A, (5), p. 162, ♀(1914).[1] Kieffer, *Das Tierrerich,* Lief. XLVIII, p. 460 (1926), [bred from eggs of Tineid moth (?) (Probably a pentatomid bug) on sugarcane]. — [1]Java.

19. Hadronotoides Dodd.

Hadronotus (partim), Dodd, *Trans. roy. Soc. S. Australia,* XXXVII, pp. 154, 178 (1913).

Hadronotoides, Dodd, *Arch. Naturgech.,* LXXIX, A(6), P. 171 (1913) [Genotype : *H. pentatomus* (Dodd) ♀ ♂]. Keiffer, *Das Tierreich,* Lief. XLVIII, p. 474 (1926).

rugostriatus Dodd.

Hadronotoides rugostriatus, Dodd, *Trans. ent. Soc. London,* p. 365 (1920).[1] — [1]Ceylon.

20. Rhacoteleia Cameron.

Rhacoteleia, Cameron, *Spol. Zeyl.,* III. p. 72 (1905), [Genotype : *R. pilosa* Cam. ♀ ♂]. Kieffer, *Gen. Ins., fas. 80 B, p. 85* (1910); *Das.* Tierreich, Lief. XLVIII, p. 476 (1926).

pilosa Cameron.

Rhacoteleia pilosa, Cameron, *Spol. Zeyl.,* III, p. 73♀ ♂ (1905).[1] Kieffer, *Gen. Ins.,* fas. 80 B, p. 85 (1910); *Das Tierreich,* Lief. XLVIII, p. 476♀ ♂(1926). — [1]Peradeniya: Ceylon.

21. Baryconus Forster.

Baryconus, Förster, *Hymen. stud.,* II, pp. 101, 104 (1856), [Genotype: B. ruficeps Kieff]♀. Dalla Torre, *Cat. Hymen.* V, P. 501 (1898). Ashmead, J.N. Y. ent. *Soc.,* XI, p. 91 (1903). Brues, *Gen. Ind.,* fas. 80, p. 31

(1908). Kieffer. *Gen. Ins.*, fas. 80 B. p. 84 (1910); *Das Tierreich.* Lief. XLVIII, p. 480 (1926).

Pictus Dodd.

Baryconus pictus Dodd, *Trans. end. Soc. London*, p. 335 (1919).[1] — [1]Ceylon.

22. **Probaryconus** Kieffer.

Baryconus subgen. *Probaryconus.*, Kieffer *Ann. Soc. Bruxelles*, XXXII, pp. 118. 165 (1908); *Gen. Ins.*, fas. 80 B. p. 84 (1910), pl. iii. figs. 16, 26.

Probaryconus, Kieffer, in Andre. Spe. Hymen. Eur., XI, p. 323 (1914), [Genotype : P. *Spinosus* (Kieff.)♀]; *Das Tierreich* Lief. XLVIII, p. 484 (1926).

indicus (Kieffer).*

Baryconus indicus, Kieffer, Zeitschr. *Hymen. Dipt.*, VII, p. 311. ♀(1907).[1] Brues, *Gen. Ins.*, fas. 80, p. 31 (1968). — [1]Semargon : Java.

Baryconus (Probaryconus) indicus, Kieffer *Gen. ins.*, fas. 80 B. p. 84 (1910).

Probaryconus indicus. Keiffer. Das Ţirreich, Lief. XLVIII, p. 485 ♀ (1926).

23. **Pegoteleia** Kieffer.

Caloteleia (partim) (nec. Westwood) Ashmead, *Bull. U.S.N. Mus.*, No. XLV, p. 212 (1893).

Ceratoteleia (partim), Kieffer, *Ann. Soc. Sc. Bruxelles*, XXXII, p. 121 (1908). Dodd, *Trns. roy. Soc. S. Australia* XXXVII, p. 140 (1913).

Baryconus (partim), Kieffer, *Bull. Soc. ent. France*, p. 294 (1910).

Pegoteleia, Kieffer, *Das Tierreich*, Lief. XLVIII, p. 510 (1926), [Genotype : *P. caloptera* (Kieff.)].

rufipes (Cameron)

Caloteleia rufipes, Cameron, *Indian For. Rec.* IV, p. 13, ♀, (1913)[1]. — [1]Dehra Dun.

---

* Since the second proof of this catalogue left my hands, I have seen specimens of this species bred from the eggs of an unknown grass-hopper at Poona (Note added in final proof).

Pegoteleia rufipes, Kieffer, *Das Tierreich,* Lief. XLVIII, p. 511 ♀ (1926).

24. **Platyscelio** Kieffer.

Platyscelio, Kieffer, *Ann. Mus. civ. stor. Nat. Genova,* XLII. p. 11 (1905), [Genotype : *P. pulchricornis* Kieff. ♀]. Brues, *Gen. Ins.* fas. 80, p. 40 (1908). Kieffer, *Gen. Ins.*, fas. 80 B, p. 66 (1910) *Das Tier reich,* Lief. XLVIII, p. 553 (1926). Brues, *Psyche,* XXIX, p. 21 (1922), [Notes].

punctatus Kieffer.

Platyscelio punctatus, Kieffer, *Insecta,* III, p. 321, ♂, (1913)[1]; *Das Tierreich,* Lief. XLVIII, p. 555, (1926). Mani, *Rec. Indian Mus.*, XXXVIII p. 337 ♂ (1936).[2]

[1]Luzon: Philippines.
[2]Puri : Orissa.

Subfamily V. PLATYGASTERINAE.

Cynipsera (Partim), Latreille, *Gen. Crust. Ins.*, IV, p. 31 (1809).

Platygastres, (partim), Latreille, *Ent. Mag.*, I. p. 269 (1833).

Platygasteroidae, Förster, *Hymen. Stud.*, II, pp. 20, 28, 106 (1856). Ashmead, *Bull. U.S.N. Mus.*, XLV pp. 24, 27, 247 (1893).

Sparasionidae (partim), Dahlbom, *Öfv. Akad. Förh.*, XIV, p. 290 (1858).

Platygasterini, Thomson, *Öfv. Akad. Förh.*, XVI, p. 69 (1859).

Platygasterinae, Dalla Torre, *Cat. Hymen.*, V, p. 469 (1898).

Platygasteridae, Marshall, *Cat. Brit. Hymen., OXyura,* p. 17 (1873). Ashmead, *J. N. Y. ent. Soc.*, XI, p. 94 (1903), [tables of genera]. Brues, in Viereck's *Hymen. Conn. Bull. Conecticut St. Geol. Nat. Hist. Surv.*, No. 22, p. 531 (1916), [Connecticut spp.]; *Bull. Mus. Compar. Zool. Harward Coll.*, LXXIII, pp. 494, 501 (1932). Fouts, *Bull. Bishop Mus., Honolulu,* CXIV, pp. 327-331, fig. 10 (1935), [Marquesas spp.].

platygasterinae, AShmead, *Bull. U. S. N. Mus.*, XLV, p. 24, 27, 247 (1893). Kieffer, *Centralbl. Bakt.parasit. Infektionaskrank.*, (2) XLVI, pp. 547-592 (1916), [Notes on the Platygasterinae and their life-histories]; *Broteria* (Ser. Zool.), XV, pp. 68-69 (1921), [a species parasitic on a cynipid]; *Ann. Soc. cient. Argent., XCI*, pp. 36-41, figs. 2 (1921); *ibid.*, XCIV, pp. 205-208, figs. 3 (1922), [Argentine spp.] Fouts *Proc. U.S.N. Mus.*, LXIII, art. 150, pp. 1-145 (1924), [revision of N. American forms]. Kieffer, *Das Tierreich*, Lief. XLVIII, p. 557 (1926), [tables and descriptions of genera and species of the world]. Gahan, *Misc. Pub. U.S. Depart. Agric. Washington*, No. 174, pp. 1-147, figs. 32 (1933), [Morphology and biology of American *Platygaster* spp.].

Key to genera.

1.	Fore wings with one closed basal cell or with a distinct, distally knobbed subcosta, which does not reach the anterior margin of wings	2
	Fore wings without a distinct vein, often with only a spur of subcosta, not thickened apically; tarsi 5-jointed	5
2.	Tarsi 5-jointed; fore wings with a subcosta not reaching the anterior margin, without cubital, radial or discoidal; often also without medial or basal	3
3.	Scape normal not broadened medially; abdomen in the female with or without horn	4
4.	Antennae with ten segments in both the sexes, without vertiscillate hairs; subcosta straight; fore tibial spur trifurcate, not combed; first abdominal tergite with a horn often reaching forward to the head; maxillary palpus biarticulate; Lateral ocelli separated from the front ocellus wider than from the eye border; labial palpus uniarticulate	*Inostemma.*
5.	Scape of antennae without a lamellate expansion ; scutellum unarmed or with only one tooth on the hind margin; abominal segments normal (the last three segments in the female not tubular), first abdominal tergite in the male shorter than the second; scutellum much shorter than mesonotum, without a nipple	6
6.	Scutellum elongated behind or with one thorn or process or hocker	7

	Scutellum distinctly semicircular, flat or arched; not long, without processes	9
7.	mesonotum long with four longitudinal furrows, the two median furrows converging behind; abdomen long, spindle-shaped.	*Anectadius.*
	Mesonotum at the most with two longitudinal furrows, abdomen in the female not elongate	8
8.	Antennae wit ten segments; parapsidal furrows deep, parallel behind; club absent in the female, the two terminal antennal segments distinctly separate from each other	*Xestonotidea.*
9.	Abdomen with 6 or 7 targites; antennae in the female with eight segments	10
	Abdomen with 6 or 7 tergites; antennae in the female and male with ten sogmnts	11
10.	Antennal club uniarticulate; fore tibial spur bifurcate and combed; scutellum arched	*Amitus.*
11.	Scutellum arched, head without a strong process between the antennae but with only the usual weak carina; two terminal antennal segments distinctly cut, four to nine segments not all equal; eyes not pubescent; maxillary palpus biarticulate, labial palpus uniarticulate	12
12.	Lebial palpus long	13
	Labial palpus short, transverse	15
13.	Abdomen in the female not elongate and narrow, with at the most only the two posterior-most tergites long	14
14.	Soutellum posteriorly with one short, thick hair bundle pointed above and backwards	*Tricharcis.*
	Scutellum without such hair-bundle; rounded behind, with often one hairy groove before ; body coarsely sculptured	*Polygontus.*
15.	The two posterior abdominal tergites transverse or at the most the last is elongate	*Platygaster.*

25. **Inostemma** Haliday.

Psilus (partim), Jurine, *Nouv. Mèth. Hymen.*, p. 318 (1807).

Platygaster (partim), Latreille, *Gen. Crust. Ins.*, IV, p. 31 (1809).

Inostemma, Haliday, *Ent. Mag.*, I, p. 270 (1833), [Genotype : *I. boscii* (Jur.) ♀]. Dalla Torre, *Cat. Hymen.*, V, p. 490 (1898). Brues, *Bull. Conn. St. Geol. Nat. Hist. Surv.*, No. 22, p. 532 (1916). Kieffer, *Das Tierreich*, Lief. XLVIII, p. 579 (1926).

indica Mani, sp. nov.

Inostemma indica Mani. This species was bred by me (10-5-1931), from the larvae of *Neolasioptera cephalandrae* Mani (Itonididae: Diptera), Producing galls on stem of *Coccinia indica.*[1] Abdominal horn almost equal to abdomen; second tergite almost equal to the remaining tergites combined, its greatest width more than its length; propodeum laterally with its margin 4-dentate; metatarsus longer than suceeding three tarsal segments combined. Type ♀ dissected, on slide in Imperial Pusa Collection, New Delhi.

[1]Tanjore: S. India.

26. **Anectadius** Kieffer.

Anectadius, Kieffer, *Ann. Soc. sc. Bruxelles*, XXIX, p. 187 (1905), [Genotype :*A. Striolatus* Kieff. ♀ ♂]; *Das Tierreich*, Lief. XlVIII, p. 607 (1926).

Key to species.

Frons and occiput transversely striped ; at least the first two admominal tergites are longitudinally striated	*A. Striolatus.*
Frons and occiput smooth, abdomen not striate	*A. bengalensis.*

bengalensis Kieffer.

Anectadius bengalensis, Kieffer, *Ann. Soc. sc. Bruxelles*, XXIX, p. 189 ♀ (1905);[1] Das Tierreich, Lief. XLVIII, p. 609 ♀ (1926),[2] [bred from midges in galls of *Artemisia*].

[1,2]Kurseong.

striolatus Kieffer.

Anectadius striolatus, Kieffer, *Ann. Soc. sc. Bruxelles,* XXIX, p. 187 ♀ ♂ (1905); [1] *Das Tierrecih,* Lief XLVIII, p. 608 ♀ ♂ (1926),[2] [Host : *Lasioptera textor Kieff*]. — [1,2]Kurseong.

27. **Xestonotidea** Gahan.

Xestonotus, Förster, *Hymen., Stud.,* II, pp. 107, 112 (1856), [preoccupied by *Xestonotus* Le Conte, *Trans. Amer. Pilos. Soc.,* X, p. 383 (1853)]. Dalla Torre, *Cat. Hymen.* V, p. 489 (1898), Ashmead, *J. N. Y. ent. Soc.* XI, p. 97 (1903).

Xestonotidea, Gahan *Proc. U.S.N. Mus., LVI,* p. 524 (1920), [Genotype : *X. andriciphilus*] (Ashm.)? ♀].

Axestonotus, Kieffer, *Das Tierreich,* Lief. XLVIII, p. 625 (1926)., [nom. nov. proposed without konwledge of Gahan's previous name].

foersteri Gahan.

Xestonotidea foersteri, Gahan, *Proc.* U. S. N. Mus., LVI, p. 524 (1920).[1] — [1]Coimbatore.

28. **Polygnotus** Förster.

Platygaster (partim), Nees, *Ichneum. affin. Monogr.,* II, p. 301, ♀(1834).

Rhacodia, Herrich-Schäffer, *Faun. Insect. German.,* P. 157, pl. xxiii, fig. 1, ♀ ♂(1838).

Polygnotus, Förster, *Hymen. Stud,* P. 108 (1856), [Genotype: *P. striolatus* (Nees) ♀]. Dalla Torre, *Cat. Hymen.,* V, p. 477 (1898). Ashmead, *J.N.Y. ent. Soc.,* XI, p. 98 (1903). Kieffer, *Das Tierreich,* Lief. XLVIII, p. 732 (1926).

sp. Gahan.

Polygnotus Sp., Gahan, *Proc. U.S.N. Mus.,* LVI, 523 (1920).[1] Ramakrishna, *Bull. ent. Res.,* XVIII, p. 76 (1927),[2] [Host : *Pachydiplosis Oryzae* Mani]. — [1]Pudukotah. [2]Tanjore.

29. **Trichacis** Förster.

Platygaster (partim), Walker, *Ent. Mag.,* III, p. 238 (1835).

Trichacis, Förster, *Hymen. Stud.,* II, pp. 108, 115 (1856), [Genotype : *T. remulus* (Walk.)♀ ♂]. Dalla Torre, *Cat. Hymen.,* V, p. 481 (1898). Ashmead, *J. N. Y. ent. Soc.,* XI, p. 98 (1903). Brues, *Bull. Conn. St. Geol. Nat. Hist. Surv.,* No. 22, p. 534 (1916). Kieffer, *Das Tierreich,* Lief. XLVIII, p. 710 (1926).

Trichasis, Thomson, *Öfv. Akad. Förh.,* XVI, p. 78 (1859).

indica Kieffer.

Trichacis Indica, Kieffer, *Berlin. ent. Zeit.,* LI, p. 265, ♂ (1907);[1] *Das Tierreich,* Lief. XLVIII, p. 716 (1926).[2] — [1,2]Java.

30. **Amitus** Haldeman.

Amitus, Haldeman, *Amer. J. Sci.,* (2) IX, p. 108 (1850), [Genotype : *A. Aleurodinis* Hald.♀ ♂]. Kieffer, *Das Tierreich,* Lief. XLVIII, p. 697 (1926).

Zacrita, Förster, *Verh. Ver. Rheinlande,* XXXV, p. 46 (1878).

Alaptus (partim), Cresson, *Trans. Amer. ent. Soc.* Suppl., P. 250 (1887).

aleurolobi Mani.

Amitus aleurolobi, Mani, *Indian J. Ent. I,* p. 97, ♂ (1939), [Host : nymphs of Bihar. *Aleurolobus barodensis*]. — [1]Pusa, Majhuatia Bihar.

31. **Platygaster** Latreille.

Scelio (partim), Latreille, *Hist. Crust. Ins.,* XIII, p. 227 (1805).

Psilus, Spinola, *Ins. Ligur.,* II, p. 235 (1808).

platygaster, Latreille, *Gen. Crust. Ins.,* IV, p. 31 (1809), [Genotype : *P. ruficornis* (Latr.)]. Dalla Torre, Cat. Hymen., V, P. 470 (1898), Brues, *Bull. Conn. St. Geol. Nat. Hist. Surv.,* No. 22 P. 540 (1916). Kieffer, *Das Tierreich,* Lief. XL VIII, p. 793 (1926).

Key to species.

Legs uniformly yellow coloured; scurellum strongly transverse and arched; wings with long marginal cilia; body brownish — *P. oryzae.*

Legs reddish, black in parts; scutellum somewhat elongated; fourth antennal segment of the female twice as long as thick, longer than fifth; club absent — *P. tibialis.*

oryzae Cameron.

Platygaster oryzae, Cameron, *Mem. Man. chester Lit. Philos. Soc.,* (4) IV, p. 182, ♀ (1891).[1] Dalla Torre, *Cat. Hymen.,* V, p. 475 (1898). Kieffer, *Das Tiereich,* Lief. XLVIII, p. 833 (1926),[2] [Host : *Pachydiplosis oryzae* Mani]. — [1,2]India. Tanjore, Coimbatore, Nagpur, Bengal, Burma.

tibialis Kieffer.

Platygaster tibialis, Kieffer, *Ann. Soc. Sci. Bruxelles,* XXIX, p. 191, ♀ (1905);[1] *Das Tierreich,* Lief. XLVIII, p. 833 (1926),[2] [Host : *Lasioptera textor* Kieff.] — [1, 2]Kurseong.

Family II. CALLICERATIDAE*

Ceraphronidae, Haliday, *hymen. Synop.,* p. ii (1939). Ashmead, *J. N. Y. ent. Soc.,* XI, p. 33 (1903). Kieffer, *Gen. Ins.,* fas. 94, pp. 1-27, pls. 2 (1909).

Ceraphrontinae, Dalla Torre, *Cat. Hymen.,* V, p. 522 (1898).

Calliceratidae, Kieffer, *Das Tierreich,* Lief. 42, pp. 63-238 (1914), [Monographic work]. Waterston, *Bull. ent. Res.,* XIV, (1) pp. 103-118 (1923), [*Calliceras dictynna* described as parasitic on *Stephanoderes hampei* in Uganda, Afrika]. Leefmans, S., *Meded. Koffiebessen-boekboekFonds,* 9, pp. 191-201 (1924), [Considered not really to be parasitic on the coffee scolytid]. Szelènyi, *Ann. Mus., hungar.,* XXX, pp. 50-66 (1936), [Palaearctic forms].

Key to subfamilies.

Pterostigma replaced by the marginal vein in fore wing; antennae with 11 segments in the male, seldom with 10 segments; in the female with 10 or seldom 9 sgements; with 11 segments only in the case of two genera, when the second abdominal tergite — *Calliceratinae*

* The change in the family name is necessitated by the fact that *Ceraphron* Jurine is invalid according to the International rules of Zoological Nomenclature, as Jurine did not publish a description of the genotype *Calliceras* Nees, the next oldest genus is accepted instead and hence the family name Calliceratidae.

has four longitudinal striations, the median one converging Calliccratinae behind.

Pterostigma well developed and large; antennae with 11 segments both in the male and female; second abdominal tergite at the most finely and densely covered with a circlet of hairs anteriorly. *Megaspilinae.*

Subfamily I. MEGASPILINAE.

Megaspilinae, Ashmead, *J. N. Y. ent. Soc.,* XI, p. 33 (1903). Kieffer, *Das Tierreich,* Lief. 42, p. 138 (1914).

Key to genera.

Mesonotum with one longitudinal furrow or without; flagellum of male serrate *Atritomellus.*

Mesonotum with three longitudinal furrows; metanotum unarmed; wings finely pubescent, marginal cilia present, pterostigma somewhat elongated; scape without teeth, flagellum serrately toothed or remose *Lygocerus.*

32. **Atritomellus** Kieffor.

Ceraphron (partim) Nees, Hymen. Monogr., II. 275 (1834).

Lygocerus (partim), Marshall, *Ent. Month. Mag.* V, p. 158 (1868).

Atritomus, Förster, *Verh. Ver. Rheinlande, XXXV,* p. 56 (1878), (pre-occupied Ritter 1877).

Dendrocerus (partim), Kieffer, in Andrè, *Spec. Hymen. Eur.,* X, p. 18 (1907).

Atritomellus, Kieffer, *Das Tierreich,* Lief. 42, p. 141 (1914), [Genotype : *A cocophagus* Först.♀ ♂]

Indicus Mani.

Atritomellus indicus, Mani, *Indian, J. Ent.,* I, p. 98 fig. 11,♀ ♂ (1939)[1], [Host : Nymphs of *Nimboa basipunctata* Withcombe]. [1]Delhi ; Karnal: Punjab.

33. **Lygocerus** Förster.

Calliceras, Westwood, *Intr. Classif. Ins.,* II, syn. p. 77 (1840), (non Nees].

Lygocerus, Förster, *Hymen. Stud.*, II. pp. 97, 99 (1856), [Genotype: *L. ramicornis* (Bohem.)♀ ♂]. Ashmead, *J. N. Y. ent. Soc.*, XI, p. 34 (1903). Kieffer, *Das Tierreich*, Lief. 42, p. 145 (1914) Haviland, *Quart, J. micr. Soc.*, LXV, pp. 101-127 (1920), [Development]; *Proc. Cambridge Philos. Soc.* XIX, 293-295 (1920), [early stages, life-history notes and hyperparasitism]. Regnier, R., *Rev. Bot. app. & Agric. Colon*, III, (19), 169-185 (1923), [*L. Syrphidarium* Kieff., parasitic on *Syrphus balteatus* and *S. ribessi* in France]. Ishii, T., *Oyo-dobuts. Zasshi*, III, (5) pp. 295-300 (1931), [larval parasite of *Homalotylus flaminius* in Japan]. Ward, K.M.J. Dept. Agric. *Victoria*, XXXII (2, 3 & 5), pp. 97-104, 134-145 258-268 (1934), [*L. niger as parasitic on Myzus persicae in* Victoria, Australia]. Pether Bridge, F. R. & Mellor, J.E.M., *Ann. appl. Biol.*, XIII(2), 329-341 (1936), [*L. testaceimanus* Kieff. parasitic on *Brevicoryne brassicae* in Englands]. Ishii, T., *Agric. & Hort.*, XII (1), pp. 60-70, (1937), [In japanese. Bionomics as parasitic of *Homalotylus flaminius* in Japan].

rufipes (Thomson).

Ceraphron, rufipes, Thomson, *Öfv. Kongl. Vet.-Akad. Förhand.*, XV, p. 293 ⚥ (1858).

Lygocerus rufipes, Dalla Torre, *Cat. Hymen.*, V, p. 534 (1898). Kieffer, Das Tierreich, Lief. XLII, p. 153 ♀ ♂ (1914).[1] Malaqui, A. & Moitie, A., *Comp. Rend. Soc. Biol. Paris*, LXXVI (16), pp. 803-805 (1914), [Parasitic on *Aphiseuonymi*, the black beet-aphis, in France]. Mani, *Indian J. Ent.*, III, p. 26 ♀ (1941),[2] [Host: Cocoons of a chrysopid on sugarcane].

[1]Sweden.
[2]Jullunder: Punjab.

Subfamily II. CALLICERATINAE.

Ceraphroninae, Ashmead, *J. N. Y. ent. Soc.*, XI, p. 34 (1903).

Calliceratinae, Kieffer, *Das Tierreich*, Lief. XLIV, 69 (11914).

34. **Calliceras** Nees.

Ceraprhon, Jurine, in Ms. in *Panzer, Krit. Revis.*, II, p. 135 (1806) ; Nouv. *Mèth. Class. Hymen.*, P. 303 (1807), [not Jurine 1806 Ms.]. Dalla Torre, Cat. *Hymen.*, V,

p. 524 (1898), Ashmead, *J. N. Y. ent. Soc.*, XI, p. 35 (1903). Kieffer, *Gen. Ins.*, 94, p. 17 (1909).

Calliceras, Nees, *Hymen. Monogr.* II, p. 278 (1834), [genotype: *Calliceras sulcatus* (Nees)]. Kieffer, *Das Tierreich*, Lief. XLII, p. 70 (1914). Ferrière, *Stylops*, II (4-5), pp. 86-96, 97-108 (1933), [*Calliceras fijiensis* as a parasite of *Apanteles tirathabae*, a parasite of *Tirathaba trichogramma* in Fijiu; *c. manilae* parasitic on *A. tirathabae* in Java].

athanasii (Girault)*

Ceraphron athanasii, Girault in Rama krishna, *Bull. ent. Res.*, XVIII, p. 76 (1927),[1] [Host : eggs of *Plusia agramma* Guen.]. [1]Coimbatore.

Diapriinae, Dalla Torre, *Cat. Hymen.*, V, p. 431 (1898). Kieffer, *Gen. Ins.*, fas. 124, pp. 1-175 (1911) ; *Das Tierreich*, Lief. XLIV, p. 3 (1916).

manfilae (Ashmead).

Ceraphron manilae, Ashmead, *Proc. U.S.N. Mus.*, XXVIII, p. 135 ♀ (9134).[1] [1,2]Manila: Philippines

Calliceras manilae, Kieffer, *Das Tierreich*, Lief. XLII, p. 104 ♀ (1914)[2]. Ferrièr, *Stylops*, II (5), p. 105 (1933),[3] [Host: *Apanteles machaeralis* at Dehra Dun]. [3]Dehra Dun: India; Java.

Family III. **DIAPRIIDÆ.**

Proctotrupii (partim), *Latreille, Gen. Ins. Crust.*, IV, p. 34 (1833).

Diapriae, Haliday, Ent. Mag., I, pp. 269, 274 (1839).

Diapriadae, Haliday, *Hymen. Dipt. syn.*p. 3 (1840).

Diapriides, Westwood, *Intr. Classif. Ins.* Syn, p. 75 (1857).

Diaprioidae, Förster, Hymen. Stud., II, pp. 20, 28 121, (1856).

Diapridae, Haliday, *Nat. Hist. Rev.*, IV, p. 167 (1858).

Diapriidae + Belytidae, Ashmead, *Bull. U.S.N. Mus.*, XLV, pp. 27, 384 (1893); *Proc. U.S.N. Mus.*, XXIII, p. 198

---

* The original description of this species is not traced in the literature available in this country.

(1900): *J.N.Y. ent. Soc.*, X, p. 244 (1902); XI, p. 28 (1903). Kieffer, *Gen. Ins.* Fas. 107, pp. 1-47 (1910); fas. 124, pp. 1-75 (1911).

Diapriidae, Kieffer, *Das Tierreich,* Lief. XLIV, pp. 1-627 (1916), [Monographic revision of world genera and species].

Psilidae Dahlbom, *Öfv. Kongl. Vet.-Akad. Förhand.*, XIV, p. 289 (1958).

Key to subfamilies.

Antennae 10-11 jointed in ♀, 13-14 jointed in ♂; fore wings without a well developed venation; hind wings without a distinct cell; wings often undeveloped — *Diapriinae.*

Antennae 15-jointed in ♀, seldom only 14-jointed; fore wings usually with a well developed venation, hind wings with one distinct closed basal cell, seldom without cell or apterous or with reduced wings when the antennae are 15-jointed. — *Belytinae.*

Subfamily **DIAPRIINAE.**

Diaprioidae, Förster, Hymen. Stud. II. pp. 20, 121 (1956).

Diapriini, Thomson, *Öfv. Kongl. Vet.-Akad. Förhand.*, XV, p. 359 (1858).

Diapridae, Marshall. *Cat. Brit. Mus., Hymen., Oxyura,* p. 10, (1873).

Diapriinae + Spilomicrinae, Ashmead, *Bull. U.S.N. Mus., XLV,* pp. 27, 384 (1893); *J.N.Y. ent. Soc.* XI, pp. 28, 30 (1903).

Diapriinae, Dalla Torre, *Cat. Hymen.*, V, p. 431 (1898). Kieffer, *Gen. Ins.*, fas. 124, pp. 1-175 (1911); *Das Tierreich,* Lief. XLIV, p. 3 (1916).

Key to genera.

1. Females — 2

Male — 34

2. Second abdominal tergite not longer than the third: Antennae with 13 segments. Pronotum, neck and legs wiht skinny lamellae	*Loboscelidia.*
Second abdominal tergite much longer than the third	3
3. Wings reduced or absolete.	4
Wings normally developed	12
4. Antennae with 13 segments	5
Antenane with 12 segments	6
5. Scutellum distinct, with one or two grooves in fronte; posterior angles of propodeum not spiny; abdomen rounded posteriorly and falttened above anteriorly, the antrior margin of the second tergite is higher than petiole in level.	*Spilomierus.*
Abdomen pointed posteriorly and laterally compressed; second abdominal tergite in level with the petiole	*Paramecius.*
6. Body smoth and shiny	7
7. Scutellum not distinctly separated from mesonotum, without grooves	8
Scutellum distinctly separated from mesonotum, with one or two grooves	9
8. Antennal club 3 or 4 jointed	*Phaenopria.*
9. Mandibles like an elongated beak	10
Mandibles not as above	11
10. Mandible turned obliquely behind; vertex with numerous teeth on the anterior margin	*Galesus.*
11. Head transverse, not flattened above; antennal club 3 jointed; pronotum much shorter than mesonotum, the latter-without parapsidal furrows	*Loxoptera.*
12. Fore wings straight or truncated distally	13
Fore wings rounded distally, without lobe	14
13. Antennae with 12 segments; mandibles wedge-shaped and elongated behind; wings without a distinct venation	*Galesus.*

14	Antennae with 13 segments, the terminal segment not forming a club; disral end of sceutellum without frenum, at the most with two grooves in front; lateral lobe of mesonotum arched or flat	15
	Antenane with 12 segments.	22
15.	Mesonotum with 7 longitudinal carinae	*Pleuropria.*
	Mesonotum without any such carinae	16
16.	Head coarsely punctate, often striate	*Odontopria.*
	Head smooth, nonpunctate, without striae, at the most with only a crinate cheek	17
17.	Submarginal vein not exceeding the basel third of fore wing	18
	Submarginal vein exceeding the basal of fore wing	19
18.	Antennae almost filiform or only exceptionally thickened	*Spilomicrus.*
19.	Prothorax short; cheek without carina; scutellum with one or two anterior grooves, lateral ones absent; marginal vein of fore wing quadrate and thick or elongate; abdomen behind not truncate	20
20.	Posterior end of abdomen rounded and flattened from above; second abdominal tergite at a higher level than petiole; length of marginal vein one to two time its breadth	*Spilomicrus.*
	Posterior end abdomen pointed and laterally compressed; second abdominal tergite on a level with the petiole; length of marginal vein three or four times its breadth	21
21.	Mesonotum with parapsidal furrows	*Paramesius.*
22.	Mandibles prolonged beak-like either perpendicular on the longitudinal axis of the head or oblique and covering the prosternum; face abnormal	23.
	Mandible not beak-like; face normal	24
23.	Wings without distinct venation; parapsidal furrows deep and complete; scutellum mostly with six grooves; antennae mostly without a distinct club	*Galesus.*

24.	Scutellum unarmed; propodeum unarmed or at the most with a sharp-pointed median longitudinal carina; fore wings at least with subcostal and marginal veins, the former reaching the margin of the wing and not or a little exceeding the bassal third of the length of wing; petiole distinctly cylinderical and much thinner than the second abdominal tergite; parasidal furrows absent; thorax smoth and shiny	25
25.	Fore wings with a basal vein	26
	Fore wings without a basal vein	27
26.	Head unarmed (not spiny) and the antennae with a 3-jointed club	*Loxoptera.*
27.	Scutellum without a groove in front	28
	Scutellum with one or two grooves in front	29
28.	Scape normal; antennal club 3 or 4-jointed	*Phaenopria.*
29.	Scape normal; terminal antennal segment not inordinately thickened; marginal vein obvedge-shaped; eyes mostly naked	30
30.	Scutellum carinate	31
	Scutellum noncarinate	33
31.	Head not pentangular	32
32.	Anterior margin of the second abdominal tergite angularly incised and projecting; antennal club exceptionally shaped, 6-jointed	*Diapria.*
	Anterior margin of second abdominal tergite not incised and projecting; antennal clud 3 to 6-jointed, prominent	*Ashmeadopria.*
33.	Head rounded or transverse; antennal club 3-5 jointed	*Trichopria.*
		35
34.	Antennae with 13 segments	40
	Antennae with 14 sgements	36
35.	Fourth antennal segment dentately produced; mouth not beak like; scutellum with one or two grooves; fore wings rounded apically; propodeum mostly without thorn; antennal scape spindess; face not excavated	

36.	Head with large punctures, face above antennae with 1-3 carinae; mesonotum with 4 longitudinal furrows	*Odontopria.*
	Head nonpunctate; mesonotum at the most with only two longitudinal furrows	37
37.	Third antennal segment not half so long as the fourth; second abdominal tergite in level with petiole, from which it is not distinctly separated	38
	Third antennal segment about as long as or longer than the fourth; second abdominal tergite very distinctly separated from the petiole or with 1-3 furrows in front.	39
38.	parasidal furrows complete	*Paramesius.*
39.	Second abdominal tergite at a higher level than the petiole	*Spilomicrus.*
40.	Third antennal segment dentateley produced	41
	Fourth antennal segment dentately produced	42
41.	Mandibles not beak-like and projecting; parapsidal furrows absent; flagellum with whorls of setao	*Ashmeadopria.*
42.	Scutellum unarmed	43
43.	Wings obsolete or vestigial	44
	Wings normally developed	47
44.	Scape normal, not flattened and broad	45
45.	Scutellum without grooves in front	*Phaenopria.*
	Scutellum with one or two grooves infront	46
46.	Parapsidal furrows absent	*Loxoptera.*
47.	Scutellum without grooves in front	48
	Scutellum with one or two groove in front	49
48.	Petiole of abdomen at the most somewhat longer than thick, body compact	*Phaenopria.*
49.	Subcosta reaching anterior margin of wing or not reaching it.	50
50.	Face clongate behind, mandible beak-like covering prosternum; wings without distinct venation	*Galesus.*

	Face not elongate mandible normal; parapsidal furrows absent; propodeum unarmed or at the most with a sharp-pointed median carina; subcosta not reaching the basal third of the anterior margin of wing; face non-carinate	51
51.	Fore wings with a basal vein; flagellum without whorl of haris and also witout long hairs	52
	Fore wings without a basal vein	53
52.	Third antennal segment not distinctly shorter than the fourth; mesonotum elongate, marginal vein long, obwedge-shaped	*Loxoptera.*
53.	Thorax smooth and shiny, propodeum with one carina or of a leathery texture	54
54.	Scutellum carinate	55
	Secutellum noncarinate	57
55.	Head not pentangular; femora and tibiae club-shaped; wings without stigmal vein	56
56.	Flagellum with long, erect, scattered haris; anterior margin fo second abdominal tergite incised and projecting	*Diapria.*
	Flagellm with whorls of hairs; anterior margin of second abdominal tergite not incised and not projecting	*Ashmeadopria.*
57.	Scape cylindrical, somewhat thinner basally than spically; flagellum with whorls of hairs or scattered, erect, long hairs or only fine pubescence	*Trichopria.*

35. **Loboscelidia** Westwood.*

Loboscelidia, Westwood, *Thesaur, ent. oxon.,* p. 171 (1874), [Genotype : *L. rufescens Westw.* ♀] Dalla Torre, *Cat. Hymen.,* V, p. 431. (1898). Kieffer, *Das tierreich,* Lei. XLIV, p. 14 (1916). Fouts, Philip *J. Sc.,* XX, p. 619 (1922), [key to species]

Loboscelidoidea, Rye in *Zool. Rec.,* XI, p. 365 (1876)

* The student should consult Fouts (loc. cit.) for key to species.

antennata Fouts.

Loboscelidia antennata, Fouts, *Philip. J. Sc.,* XX p. 622 (1922).[1] — [1]Singapore.

carinata Fourts.

Loboscelidia carinata, Fouts, *Philip, J. Sc.,* XX, p. 626 (1922).[1] — [1]Singapore.

collaris Fouts.

Loboscelidia collaris, Fouts, *Philip. J. Sc.,* XX, p. 627 (1922).[1] — [1]Singapore.

36. **Phaenopria** Ahsmead.

Phaenopria, Ashmead, *Bull. U.S.N. Mus.,* XLV, pp. 436 (1893), [Genotype: *P. minutissima Ashm.* ♀]. Dalla Torre, *Cat. Hymen.,* V, p. 433 (1898). Ashmead, J.N.Y. *ent. Soc.* XI, pp. 32, 33 (1903). Kieffer, *Gen. Ins.,* fas. 124 p. 49 (1911); *Das Tierreich,* Lief. XLIV, p. 52 (1916)

cavernicola Kieffer.

Phaenopria cavernicola, Kieffer, *Berlin. ent., Zeit.,* LI, p. 261 ♂ (1907); *Gen. Ins.,* fas. 124, p. 49 (1911); *Das Tierreich,* Lief. XLIV, p. 57 ♂ (1916).[1] — [1]Java.

37. Diapria Latreille.*

Ichneumon (partim), Fabricius, Syst. Ent., p. 343 (1775).

Chalcis (partim) Fabricius, Ent. Syst. suppl., p. 242 (1778).

Diapria, Latreille, *Prècis des Caractères,* II, p. 110 (1796), [Genotype: *D. conica* (Fab.)]. Dalla Torre, *Cat. Hymen.,* V, p. 435 (1898). Ashmead, *J.N.y. ent. Soc.,* XI, pp. 31, 32 (1903), Kieffer, *Gen. Ins.,* fas. 124, p. 58 (1911); *Das Tierreich,* Lief. XLIV, p. 70 (1916).

Psilus, Jurine, *Nouv. Mèth. classif. Hymen.,* p. 319 (1807).

Oxyurus (partim), Lamarck, *Hist. Nat. Anim. Vert.,* (ed.2). IV, p. 342 (1835).

Tropidopria, Ahsmead, *Bull. U.S.N. Mus.,* XLV, p. 416 (1893).

---

* Key to the five Indian species of *Diapria* Latr. cannot be constructed from the meagre description of the species by Motchulsky and Dalla Torre; the Species have not also been recognised after their description by the protologists.

affinis Motschulsky.

Diapria affinis, Morschulsky, *Bull. Imp. Soc. Nat. Moscou,* XXXVI, (2) p. 46♀ (1863).[1] Dalla Torre, *Cat. Hymen.,* V. p. 435 (1898). Kieffer, *Gen. Ins.,* fas. 124, p. 59 (191); *Das Tierreich,* Lief. XLIV, p. 75 ♀ (1916). — [1]Mt. patannas : Ceylon.

impressieollis Motschulsky.

Diapria impressicollis, Motschulsky, *Bull. Imp. Soc. Nat. Moscou,* XXXVI, (2), p. 45 (1863)[1]. Dalla Torre, *Cat. Hymen., V.,* p. 435 (1238).* Kieffer, *Gen. Ins.* fas. 124, p. 59 (1911);* *Das. tierreich,* Lief. XLIV, p. 75 (1916). — [1]Mt.patannas: Ceylon.

motschulskyi Dalla Torre.

Diapria longicornis, Motschulsky, *Bull. Soc. Imp. Nat. Moscou,* XXXVI (2), p. 45 (1863),[1] [Pre-occupied Thomson]. — [1]Mt. Nuwara Eliya: Ceylon.

Diapria motschulskyi Dalla Torre, *Wien. ent. Zeit.,* IX, p. 97 (1890); *Cat. Hymen.* v, p. 437 (1898). Kieffer, *Gen. Ins.,* fas. 124, p. 59 (1911); *Das Tierreich,* Lief. XLIV, p. 75 (1916).

walkeri Dalla Torreh.

Diapria apicalis Walker, Ann. Mag. Nat. Hist. (3), VI, p. 359 ♀ (1860), [preoccupied Say].

Diapria walkeri, Dalla Torre, *Wien. ent. Zeit.,* IX, p. 97 (1890); *Cat.* Hymen., V, p. 439 (1898).[1] Kieffer, *Gen. Ins.,* fas. 124, p. 59 (1911); *Das tierriech* Lief. XLIV, p. 75 ♀ (1916). — [1]Ceylon.

38. Ashmeadopria Kieffer.

Diapria (partim), Latreille, *Hist. Crust. Ins.,* XIII, p. 231 (1805).

Psilus (Partim), Jurine, *Nouv. Mèth Hymen.,* p. 319 (1807).

Platymischus, Provancher, *Addit., Hymer Quebec,* 182 (1887).

Tropidopria (partim), Ashemead, *Bull. U. S. N. Mus.,* XLV, p. 416 (1893).

---

* *impressicornie* of Dalla Torre and Kieffer is probably a misprint for *impressicornie* Motschulsky.

Diapria (Tropidopria), Kieffer, in Andrè *Spec. Hymen.* Eur., X, p. 951 (1911).

Ashmeadopria, Kieffer, *Gen. Ins.,* fas. 124, p. 59 (1911), [Genotype: *A. verticillata* (Latr.) ♀]; *Das Tierreich,* Lief. XLIV, p. 119 (1916). Voinovskaya-Kriger. T., *Izu. prikl. Ent.,* IV (1), pp. 185-188 (1929), [Bionomics as larval parasite of *Oscinella frit* in Leningrad. In Russion with a summary in German].*

jacobsoni (Kieffer)

Tropidopria jacobsoni, Kieffer, *Zeit. Hymen. Dipt.* VII, p 311 ♀ (1907).[1]

[1]Semarong : Java.

Ashmeadopria jacobsoni Kieffer *Gen. Ins.,* fas. 124, p. 60 (1911); *Das Tierreich,* Lief XLIV, p. 136 ♀ (1916).

39. **Trichopria** Ashmead.

Psilus (partim), Say, *Boston J. Nat. Hist.,* I, p. 284 (1836).

Diapria (partim), Thomson, *Öfv. Kongl. Vet.-Akad. Förhand.,* XV, p. 366 (1858).

Aneurhynchus (partim), Ashmead, *Canad. Ent.* XIX, p. 105 (1887).

Platymiscus (partim) Ashmead, *Canad. Ent.,* XIX. p. 195 (1887).

Galesus (partim), Ashmead, *Canad. Ent.,* XI, p. 195 (1887).

Loxoptera (partim), Ashmead, *Canad. Ent.,* XX, p. 53 (1888).

Ceratopria, Ashmead, *Bull. U. S. N. Mus.,* XLV, p. 428 (1893).

Trichopria, Ashmead, Bull. U.S.N. Mus., XLV, p. 407, 431 432, (1893), [Genotype : *T, pentaplasta** Asham ♀ ♂]. Dalla Torre *Cat. Hymen.,* V, p. 434 (1898). Ashmead, *J.* N. Y. *ent. Soc.,* XI, p. 31 (1903). Kieffer, *Gen. Ins.,* fas 124, p. 62 (1911); *Das Tierreich,* Lief. XLIV, p. 77 (1916).

planopria, Kieffer, *Bull. Soc. Metz.,* XXV, p. 19 (1908).

---

* Not seen in original.

* The genotype *T. pentaplasta,* is referred to the subgens *Ceratopria* Ashm.. by Kieffer in *Das Tierreich.*

Trichopira (Orthopria), Kieffer, in Andrè *Spec. Hymen. Eur.,* X, p. 984 (1910).

Key to subgenera.

Flagellar segments with short fine prostrate heirs in ♂	*Trichopria.*
Flagellar segments with whorls of long erect hairs in ♂	*Planopria.*

Key to species.

1.	Antennal club 3-jointed; body red; wings exceeding nearly by half its length the abdomen	*T. (T.) rufa.*
	Antennal club 5-jointed	2.
2.	Propodcum with only one median longitudinal carine; red; rest of thorax black	*T. (T.) Indica.*
	Thorax entirely black	*T. (T.) tachnidiarum.*
	Scutellum with two grooves in front	*T. (P.) versicolor*
	Scutellum with one groove in front	*T. (P.) syrphidarum.*

39a. **Trichopria (Trichopria)** Ashmead.

indica Kieffer.

Trichopra (Trichopria) indica, Kieffer, *Bull. soc. Nat. Metz.,* XXV, p. 106 ♀ (1905); *Das Tierreich,* Lief. XLIV, p. 91 ♀ (1916).[1] — [1]Matheran : India.

Trichopria indica, Kieffer, *Gen. Ins.,* fas. 124, p. 65 (1911).

feuerboni Bischoff.

Trichopria feuerboni, Bischoff, *Arch. Hydrobiol. Suppl.* IX, p. 746 (1932).[1] — [1]Java.

rufa (Kieffer).

Diapria rufa, Kieffer *Berlin. ent. Zeit.,* LI, p. 261 ♀ (1907).[1] — [1,2,3]Tjibodas: Java.

Trichopria rufa, Kieffer, *Gen. Ins.,* fas. 124, p. 65 (1911).[2]

Trichopria (Trichopria) rufa, Kieffer, *Das Tierreich,* Lief. XLIV, p. 90 ♀ (1916).[3]

tachinidarum Ferrière.

Trichopria tachinidarum, Ferrière, Stylops, II, p. 104 (1933),[1] [Host: Tachinid parasite of *Tirathaba* sp]. [1]Buitenzorg : Java.

39b. Trichopria (Planopria)

Diapria (partim), Nees, *Hymen. Ichaneum. Först. Monogr.*, II, p. 322 (1834). Ashmead, *Bull. U.S.N. Mus.*, XLV, pp. 407, 420 (1908). Kieffer.

Planopria, Kieffer, *Bull. Soc. Nat. Metz.*, XXV, p. 19 (1908).

Trichopria (Orthopria), Kieffer, in Andrè *Spec. Hymen. Eur.*, X, p. 984 (1911); *Gen. Ins.*, fas. 124, p. 63 (1911).

Trichopria (Planopria), Kieffer, *Das Tiereich,* Lief. XLIV, p. 106 (1916).

syrphidarum (Kieffer).

Diapria syrphidarum, Kieffer, *Zeit. Hymen. Dipt.*, VII, p. 312 ♀ ♂ (1907).[1] [1]Semarong : Java.

Trichopria Syrphidarum, Kieffer, *Gen. Ins.* fas. 124, p. 66 (1911).[20].

Trichopria (Planopria) syrphidarum. Kieffer, *Das Tierreich,* Lief. XLIV, p. 113. (1916),[3] [Host : Larva of *Eristalissp*].

versicolor (Kieffer).

Diapria versicolor, Kieffer, *Berlin. ent. Zeit.*, LI, p. 262 ♂ (1907).[1] [1,2]Tjibodas : Java.

Trichopria versicolor, Kieffer, *Gen. Ins.*, fas. 124, p. 66 (1911).[2]

Trichopria (Planopria) versicolor, Kieffer, *Das Tierreich,* Lief. XLIV, p. 113 ♂ (1916).[3] [3]Java.

40. Loxoptera Förster.

Diapria (partim), Nees, *Hymen. Ichneum. Monogr.*, II, p 328 (1834).

Rhacodia (?) (Partim), Herrich-Schäffer, *Faun. Ins. Germ.*, 157 t. 23. f. 2 (1838).

Loxoptera, Förster, *Hymen. Stud.,* II, pp. 122, 123, 126 (1856), [Genotype: *L. dispar* (Nees) ♀]. Dalla Torre, *Cat. Hymen.,* V, p. 440 (1898). Ashmead, *J. N. Y. ent Soc.,* XI, p. 31 (1903). Kieffer, *Gen. Ins.,* fas. 124, p. 55 (1911); *Das Tierreich,* Lief. XLIV, p. 167 (1916).

Diapria (Basalys) (Partim), Haliday, *Nat. Hist. Rev.,* IV, p. 171 (1857).

Basalys (partim), Thomson, *Öfv. Kongl. Vet.-Akad* Förhand., XV, p. 368 (1858).

indica Kieffer.

Loxoptera indica, Kieffer, *Berlin. ent. Zeit.,* LI, p. 262 ♂(1907);[1] *Gen. Ins.,* fas. 124, p. 56 (1911);[2] *Das Tierreich,* Lief. XLIV, p. 185 ♂ (1916).[3] — [1, 3]Java. [2]India.

41. **Galesus** Curtis.

Psilus, Panzer, *Faun. Ins. Germ.,* VII, part 83, t. 11 (1801).

Oxyurus (partim), Lamarck, *Hist. Anim. Vert.,* IV, p. 128 (1817).

Galesus, (Haliday),* Curtis, *Brit, Ent.,* VIII, p. 341 (1831), [Genotype: *G. fuscipennis Curtis* ♂]. Dalla Torre, *Cat. Hymen.,* V, p. 442 (1898). Kieffer, *Gen Ins.,* fas. 124, p. 39 (1911); *Das Tierrich* lief. XLIV, p. 200 (1916).

Coptera, Say, *Boston J. Nat. Hist.,* I, p .280 (1836).

Anisoptera, Herrich-Schäffer, *Nomen. Ent.,* II, pp. 57, 69 (1840).

Entomacis, Provancher, *Addit. Faun. Canada, Hymen.,* P. 404 (1888).

Schizogalesus, Kieffer, in Andrè *Spec. Hymen. Eur.,* X, p. 833 (1910); *Gen. Ins.,* fas. 124, p. 40 (1911).

Key to species.

Face between antennae and ocelli with two teeth which become short carinae behind and do not reach the ocelli	*G. Magrettii.*
Face with longitudinal carinae	*G. taprobanicus.*

* Haliday manuscript name.

magrettii Kieffer.

Galasus magrettii, Kieffer, *Ann. Soc. sc. Bruxelles,* XXIX, p. 138 ♂ (1905);[1] *Das Tierreich,* Lief. XLIV, p. 222 ♂ (1916).[2] — [1, 2, 3]Carim Cheba : Burma.

Galesus (Galesus) magrettii, Kieffer, *Gen. Ins.,* fas. 124, p. 41 (1911).[3]

taprobanicus (Dalla Torre).

Diapria nigricornis, Motschulsky, *Bull. Soc. Imp. Nat. Moscou,* XXXVI (2), p. 45 ♂ (1863),[1] [name preoccupied Thomson]. Dalla Torre, *Cat. Hymen.,* V, p. 438 (1898). — [1]Mt. Nuwara Elliya : Ceylon.

Diapria taprobanica, Dalla Torre, *Cat. Hymen.,* V, p. 438 ♀ (1898).

Galesus (Galesus) taprobanicus, Kieffer, *Gen. Ins.*fas. 124, p. 42 (1911).

Galesus taprobanicus, Kieffer, *Das Tierreich,* Lief. XLIV, p. 224 ♀ (1916).

42. **Pleuropria** Kieffer.

Pleuropria, Kieffer, *Ann. Soc. sc. Bruxelles,* XXIX, p. 136 (1905); [Genotype : *P. maculipennis* Kieff. ♀]; *Gen. Ins.,* fas. 124, p. 35 (1911); *Das Tierreich,* Lief. XLIV, p. 27 (1916).

maculipennis Kieffer.

Pleuropria maculipennis Kieffer, *Ann. Soc. Sc. Bruxelles,* XXIX, p. 137 ♀ (1905);[1] *Gen. Ins.,* fas. 124, p. 35 (1911); *Das Tierreich,* Lief. XLIV, p. 275 ♀(1916). — [1]Carim Cheba : Burma.

43. **Odontopria** Kieffer.

Odontopria, Kieffer, *Ann. Mus. Civ. Stor. Nat. Genova,* XLII, p. 24 (1905), [Genotype: *O. cornuta* Kieff. ♀]; *Ge. Ins.,* fas. 124, p. 29 (1911); *Das Tierreich,* Lief. XLIV, p. 275 (1916).

Spilomicrus (partim), Kieffer, *Not. Leyden Mus.,* XXX, p. 93 (1908).

cribrata (Keiffer).

Spilomicrus cribrata, Kieffer, *Not. Leyden Mus.,* XXX, p. 93 ♂ (1908).[1] — [1, 2, 3]Java.

Odontopria cribrata, Kieffer, *Gen. Ins.,* fas. 124, p. 29 (1911);[2] *Das Tierreich,* Lief. XLIV, p. 277 ♂ (1916).[3]

nigripes Kieffer.

Odontopria nigripes, Kieffer, *Berlin. Ent. Zeit.,* LI, p. 259♂ (1907);[1] *Gen. Ins.,* fas. 124, p. 29 (1911);[2] *Das Tierreich,* Lief. XLIV, p. 277♂ (1916).[3] — [1, 2, 3]Java.

44. **Spilomicrus** Westwood.

Spilomicrus, Westwood, *Philos. Mag.,* (30) I, p. 129 (1832), [Genotype: *S. stigmaticalis* Westwood ♀]. Dalla Torre, *Cat. Hymen.,* V, p. 446 (1898). Ashmead, *J. N. Y. ent. Soc.* XI, P. P. 29, 30, (1903) Kieffer. *Gen. Ins.,* fas. 124, p. 22 (1911); *Das Tierreich,* Lief. XLIV, p. 283 (1916).

Chlidonia (partim), Herrich-Schäffer, *Faun. Ins. Germ.,* 157, t. 24 (1838).

Diapria (Spilomicrus), Haldiay, *Nat. Hist. Rev.,* IV, P, p. 171 (1857).

Paramesius (partim), Provancher, *Addit. Hymen. Quebec.,* p. 405 (1888).

Ismarus, Ashmead, *Bull. Colorado Biol. Ass.,* I, p. 11 (1890).

maculipennis Kieffer.

Spilomicrus Maculipennis, Keffer, *Berlin ent. Zeit.,* LI, p. 260 ♂ (1907); [1] *Gen. Ins.,* fas. 124, p. 24 (1911);[2] *Das Tierreich,* Lief. XLIV, p. 298 ♂ (1916).[3] — [1, 2, 3]Tjibodas : Java.

45. **Paramesius** Westwood.

Cerapsilon, Curtis, *Guide Brit. Mus. Ins.,* p. 127 (1829), [without description].

Paramesius, Westwood, *Philos. Mag.,* (3), I, p. 129 (1832), [Genotype : *P. rufipes* Westwood ♂]. Dalla Torre, *Cat. Hymen.,*V, p. 447 (1898). Ashmead *J.N.Y. ent. Soc.,* XI, pp. 29, 30 (1903). Kieffer, *Gen. Ins.,* fas. 124, p. 20 (1911); *Das Tierreich,* Lief. XLIV, p. 308 (1916).

Diapria (partim), Nees, *Hymen. Ichneum. Affin. Monogr.*, II, 335 (1834).

Psilus (partim), Say, *Boston J. Nat. Hist.*, I, p. 282 (1836).

Rhacodia, (Partim) and *Chlidonia* (partim). Herrich-Schaffer, *Faun. Ins. Germ.*, 157, t. 23, 24 (1838).

incompletus Kieffer.

Paramesius incompletus, Kieffer, *Berlin, ent. Zeit.*, LI, p. 259 ♀ ♂ (1907);[1] *Gen.Ins.*, fas. 124, p. 22 (1911);[2] *Das Tierreich,* Lief. XLIV, p. 319 ♀ ♂ (1916).[3] — [1,2,3]Tjibodas : Java.

Family IV. **SERPHIDAE**

Proctotrupii, (partim), Latreille, *Gen. Crust. Ins.*, IV, p. 33 (1809).

Proctotrupes, Haliday, *Ent. Mag.*, I, p. 273 (1833).

Codrini (partim), Nees, *Hymen. Ichneum. offin. Monogr.*, II, p. 313 (1834).

Proctotrypini, Nees, *Hymen. Ichneum. affin. Monogr.* II, p. 346 (1834).

Proctotrypoidae, Förster, *Hymen. Stud.*, II, pp. 20, 99 (1856).

Codrinidae, Dahlbom, *Öfv. Kongl. Vet.-Akad. Förhand.*, XIV, p. 289 (1858).

Proctotrypinae, Ashmead, *Bull. U.S.N. Mus.*, XLV, p. 331 (1893). Dalla Torre, *Cat. Hymen.*, V, p. 461 (1898).

Proctotrypidae, Marshall, *Cat. Brit. Hymen. Oxyura,* p. 1 (1873). Ashmead, *J.N.Y. ent. Soc.*, X, p. 243 (1902). Morley, *Entomologist,* LV, p. 2 (1922), [Key to genera from Gr. Britain]; *Ent. Month. Mag.*, LIX, pp. 142, 184, 228, 263 (1923), [Synopsis of British species].

Serphinae, Kieffer in Andrè *Spec.Hymen. Eur.*, X, p. 269 (1907).

Serphidae, Kieffer, *Gen. Ins.*, fas. 95, pp. 1-10 (1909); *Das Tierreich,* Lief. XLII, pp. 1-62 (1914), [monographic account of world forms]. Eastham, *Parasitology,* XXI, (1-2), pp. 1-21, Pls. i-iii, fig. 1 (1929), [List of hosts, literature, etc.].

Subfamily I. SERPHINAE

Serphinae, Kieffer, *Das Tierreich,* Lief. XLII, p. 1 (1914).

46. Serphus Schrank.

Serphus, Schrank, *Schrift. Berlin. Gesell. Natur. Fr.,* I. p. 307 (1780), [Genotype : *S. brachypterus* Schr. ♀]. Kieffer, *Das Tierreich,* Lief. XLII, p. 7 (1914).

Proctotrupes, Latreille, *Prècis Caract. Ins.,* p. 108 (1796).

gravidator (Linnæus).

Ichneumon gravidator, Linnæus, Syst. Nat., ed. 10, p. 565 (1758).[1] — [1, 2]Whole of Europe.

Banchus gravidator, Fabricius, *Syst. Pez.,* p. 128 (1804).

Codrus gravidator, Jurine, *Nouv. Mèth. Hymen.,* I, p. 309 (1807).

Proctotrupes gravidator, Lepeletier, *Encycl. Mèth.* X, p. 208 (1825).

Serphus gravidator, Kieffer, in Andrè *Spec.Hymen. Eur.,* X, p. 297 ♀ ♂ (1908);[1] *Das Tierreich,* Lief. XLII, p. 11 ♀ ♂ (1914);[2] [*Hosts : S. gravidator gravidator* Parasitic on *Boletophila fusca*]. Gardner, J. C. M., *Indian For. Rec.,* XIV, p. 107 (1929);[3] [Host : *S. gravidator partipes* Dodd Parasitic on larvæ of *Nebria cameroni*].* — [3]Dehra Dun.

sp. Jardine.

Serphus sp., Jardine, *Ceylon Administration Rep. For 1917, Dept. Agric.,* (1918), [Parasitic on *Homona coffearia*].[1] — [1]Ceylon.

Subfamily II. HELORINAE

Heloridae, Föster, *Hymen. Stud.,* II, p. 142 (1856). Ashmead, *J. N. Y. ent. Soc.,* X, p. 242 (1902).

Helorini, Thomson, *Öfv. Kongl. Vet.-Akad. Förhand.,* XV, p. 379 (1858).

---

* The author thanks Dr. C.F.C. Beeson, Forest Entomologist, Dehra Dun, or drawing his attention to this reference (Note added in second proof).

Helorinae, Ashmead, *Bull. U. S. N. Mus.,* XLV, p. 334 (1893). Dalla Torre, *Cat. Hymen.,* V, p. 468 (1898). Kieffer, *Das Tierreich,* Lief. XLII, p. 57 (1914), [Monograph].

47. **Helorus** Latreille.

Sphex, (partim), Panzer, *Faun. Ins. Germ.,* V, (52), t. 23 (1798).

Helorus, Latreille, *Hist. Crust. Ins.,* III, p. 309 (1802), [Genotype : *H. ater (Nees)*]. Kieffer, *Das Tierreich,* Lief.XLII, p. 59 (1914).

Oxyurus (partim), Lamarck, *Hist. Anim. sns. Vert.,* IV, p. 130 (1817).

Copelus, Provancher, *Natural. Canad.,* XII, p. 207 (1881).

striolatus Cameron.

Helorus striolatus, Cameron, *J. Bombay Nat. Hist. Soc.* XVII, p. 99 (1906).[1]

[1]Qeutta.

PART XXVII – ISOPTERA ORDER. ISOPTERA.

Isopteres Brulle, *Exped.Sc. Morie. 3 (Zoologie),* Paris: 66 (1832).

Isoptera Hagen, *Linn. Ent.*, X : 1—144, 270-325 (1855); *Ibid.*, XII: 1—342 (1858); *Ibid.*, XIV: 73—128 (1860). Comstock, *Manual Stud. Ins.*, Ithaca: 95 (1895). Handlirsch, *Sitzungsb. Akad. wiss. Wien*, CXII: 729 (1903); Desneux, *Genera Insect.*, XXV: 1—54 (1904). Silvestri, *Fauna Südwest Aust.*, II (17): 279–314 (1909). Holmgren, . *Svenska Vetenska Aka. Handl.*, XLVI (6): 1—88, pls. 1–6, 6 figs. (1911); *Ibid.*, XLVIII (4): 1–166, pls. 1—4, 88 figs. (1912). Banks and Snypder, *Bull. U.S. nat. Mus.*, 108: I—228, pls. 1—35, 70 figs. (1920). Light, *philipp. J. Sci.*, XIX (1): 23—61, pls. 1—6, m 3 figs. (1921). emerson, *Zoologica N. Y.*, VI (4); 291—459, Pls. A. B and figs. 24—94 (1925). Sjöstedt, *K. svenska Vetenska Akad. Handl.*, (3), III (1) : I—419, pls. 1—16, 83 figs. (1925). Emperson, *Bull.Amer. Mus. nat Hist.*, LVII: 401—574 (1928). Light, *Termites and Termite Control* by Kofoid,*et al.*, Berkeley, California: 111—112 (1913). Hill, *Termites from the Australian Region,* Melbourne: 6 (1942).

The first attempted asystematic monograph was that of Hagen (1855—60). Later Desneux (1904) recognized three subfamilies under this group, i.e. Mastotermitinae Desn., Calotermitinae Desn., and Termitinae Desn., which together constituted the single family Termitidae latreille. Silvestri (1909) elevated the Mastotermitinaeto family rank while Desneux's second and third subfamilies remained to form the family Termitidae. The next notable system of classification was that of Holmgren (1911), which formed the basis of all the succeeding classifications. Unfortunately, the names which Holmgren used for three of the families

(Protermitidae, Mesotermitidae and metatermitidae) were not based on type genera and thus did not conform to the accepted rules of nomenclature and were changed by Banks and Snyder (1920), who, however, recombined Mesotermitidae and Metatermitidae of Holmgren in their termitidae. These two families were again separated by Light (1921) along with lines laid down by Holmgren, and given the names Rhinotermnitidae and termitidae. These were accepted by Emerson in 1925 and 1928. Sjöstedt (1925) separated Hodotermitinae from Kalotermitidae of previous workers, and raised it to family rank; and this change was accepted by Emerson (1928). Thus five families are recognized at present, viz. (1) Mastoteritidae, (2) Hodotermitidae, (3) Kalotermitidae, (4) Rhinotermitidae (5) Termitidae.

Key to Families.*

1. Tarsi 5 jointed in all castes; hind wings with a well developed anal. lobe; genitalia well developed; fontanel absent in all castes; ocelli present	Mastotermitidae(Not found in India).
Tarsi 4-jointed or rarely, imperfectly 5-jointed in all castes; hind wings without anal lobes; genitalia lacking or vestigial	2
2. Fontanel and oceliabsent; pronotum saddleshaped and narrower than the head in al castes; anterior wing scales short, not overlapping the posterior wing scales	Hodotermitidae.
Fontanel absent in all castes; ocelli usually present; pronotum flat, usually broader than head in all castes; anterior wing scales large, overlapping the posterior wing scales; worker caste absent.	Kalotermitidae.
Fontanel present; mandibles of so-called *Leucotermes* or *Serritermes* type (Figs. 1 and 2); anterior wing scales large (excent in *Psammotermes*): wing often reticulate and withouthairs; pronotum of worker and soldier flat, without anterior lobe.	Rhinotermitidae.

* This key is copied from Emerson, *Bull. Amer. Mus. nat. Hist,* LVII: 401-574 (1928) Kofoid, *et al, Termites and Termite Control,* University of California, Berkeley: 111-112 (1934).

Fontanel present; mandibles of so-called *Leuctoermes* or *Serritermes* type; anterior wing scales never large; wing only slightly reticulate, if at all; wing membranc and margin more or less hairy; pronotum of soldier and worker narrow with uplifted median anterior lobe. Termitidae.

FAMILY 1. HODOTERMITIDAE

Hodotermitini Desneus, Genera Insect., XXVI: 15 (1904).

Protermitidae (partim) Holmgren, *Trans. Linn. Soc. Lond. Zool.*, 2, XIV: 137 (1910).

Hodotermitinae Holmgren, K. *svenska vetenskaAkad. Handle.*, XLVI (6): 1—88 (1911); *Ibid.*, XLVIII, (4): 1—166 (1912); *mem.Dep. Agric.India Ent.*, *V* (3): 138—139 (1917). Hegh, *Les Termites,* Bruxelles: 681 (1922).

Hodotermitidae Sjöstedt, *K. svenska Vetensk Akad. Handl.*, (3), III (1): 1–419 (1925). emerson, *Bull. Amer. Mus. nat. Hist. LVII:* 401—574 (1928). Light, in Kofoid's *Termites and Termite Control,* Berkeley, California: 111 (1934).

Hodotermes Hagen, *Monat. Akad. wiss berlin:* 480 (1853); *Linn. Ent.*, XII: 81—85 (1858); *Reise nach Mossambique, Insekten:* 59 (1862). Haviland, *J. linn. soc. Zool.*, XXVI: 371—372 (1898). Sjöstedt, *K. Svenska VeternskAkad. Handl.*, XXXIV: 21 (1900). Desneux, *Genera Insect.*, XXV:L 15—17 (1904).

Hodotermes (Anacanthotermes) Jacobson, *Ann. Mus. zool. Acad. St. Petersb. K. svenska Vetenska Akad. Handl.*, XLVI, (6): 1—88 (1911); *Mem. Dep. Agric. India Ent.*, V (3): 138 (1917).

1. königi Holmgren

SYSTEMATICS.

Hodotermes königi Holmgren, *Mem, Dep. Agric. India Ent.*, V (3): 138—139 (1917)[1]. Margabandu, *J. Bombay nat. Hist. Soc.*, XXXVII (3): 701 (1934).

[1]'Coimbatore.

2. **macrocephalus** Desneux.

SYSTEMATICS.

Hodotermes macrocephalus Desneux, *Ann. Soc. ent. Belg.*, XLIX: 344—348 (1905)[1].

Hodotermes (Anacanthotermes) macrocephalus Holmgren, *K. svenska VetenskaAkad. Handl.*, XLIII (1), fig. 10 (4), (1908); *ibid.* XLVI (6): 43, pl. 1, fig. 10, (1911). [1,2,4,6]Sind.

BIOLOGICAL

Lefroy, Indian Insect Life, Calcutta, I: 120 (1909)[2], Bugnion, *Mus. Hist. nat.* Paris, XX, (4): 175 (1914)[3]. Holmgren *Mem. Dep. Agric. India Ent.*, V (3): 139 (1917)[4]. Magrabandu, *J., Bombay nat. Hist. Soc.* XXXVII (3): 701 (1934). Besson, Forest Insects, Dehra Dun: 541 (1941)[5]. [3]Ceylon, [5]Punjab.

(Hosts: Primarily grass; occasionally Heliotropium spp. and transplants of *Prosopsis* spp.).

3. **tristis** Cockerell.

SYSTEMATICS

Hodotermes tristis Cockerell, *Ann. ent. Soc. Amer.*, X: 329 (1917), (Fossil in Burmese Amber of Miocene Blue clays). Burma.

4. **viarum** König

SYSTEMATICS

Termes viarum König, *Schr. berl. naturf. Freunde*, IV (1): 24, pl. 1, figs. 12—14 (1779)[1]. [1,2]Tranqueber (Madras)

Hodotermes viarum Hagen, *Linn. Ent.*, XII: 90—93 (1858)[2]. Desneux, *Genera Insect.*, XXV: 17 (1904). Fletcher, *Agric. J. India*, VII (3): 224 (1912); *Proc. 4th ent. Mtgs. Pusa:* 330, pl. 52, figs. 12—14 (1921)[3], (Translation of König's paper of South Indian Termites). [3,5,6,7]Coimbatore (S. India).

BIOLOGICAL

Termes viarum Bugnion, *Bull. Mus. Hist. nat. Paris*, XX (4): 175 (1914)[4]. [4]Ceylon

Hodotermes viarum Delmeelo, F., *Proc. 3rd ent. Mtgs. Pusa,* III: 1010—1019 (1919)[5].

Fletcher, Ibid., I: 314 (1919)6; *Bull. Dep. Agric. India,* 100: 246 (1921)[7].

Margabandu, *J. Bombay nat. Hist. Soc.* XXXVII (3): 701 (1934).

(Host: Grass).

FAMILY II. KALOTERMITIDAE

Termitina (partim) Stephens, *Ill. Brit. Ent.,* VI: 97 (1836). Hagen, *Linn. Ent.,* XII: 4—31 (1858).

Calotermitinae Froggatt, proc. Linn. Soc.N.S.W., XXI (4): 516 (1896). Desneux *Genera Insect.,* XXV: 11—12 (1904); *Ann. Soc. ent. belg.,* XLVIII: 284 (1904). Holmgren, K. *Svenska Vetensk Akad. Handl.,* XLVI (6): 33—34 (1911); *Ibdi.,* XLVIII (4), (1912); *Mem. Dep. Agric. India Ent.,* V (3): 138—139 (1917). Hegh, *Les Termites,* Bruxelles: 681 (1922).

Glyptotermitinae (partim) Silverstri, *Redia,* I: 17 (1903).

Protermitidae Holmgren, *Zool. Anz.,* XXXV: 284—286 (1910); *in Escherich Termitenleben auf Ceylan,* Gustav Fischer, Publisher, Jena: 183—189 (1911).

Kalotermitidae Banks nad Snydper, *Bull. U.S. nat. Mus.,* 108: 16 (1920). Light, *Philipp. J. Sci.* XVIII: 245 (1921). Sjöstedt, *K. svenska VetenskAkad. Handle.,* (3), II (3): 22 (1925). Light, *Termites and Termite Control* By Kofoid *et al.,* Berkeley, California: 112 (1934).

Key to subfamilies

Imago Ocelli absent; antennae with over20 segments; wings with a well developed radius complex; tarsi t-jointed; cerci 5—7 jointed *Soldier:* Antennaewith over 20 segments ; cerci 5–7 segmented ; tarsi 5 jointed; tibia with spines before apex	Termopsinae.
Iamgo: ocelli present; antennae with les than 20 semgnets; wings with a simple radius and a radius complex consisting of a radius and a radial sector; tarsi 4-jointed; cerci 2—5 jointed. *Soldier:* Antennae with less than 20 segments; cerci 2—5 jointed; tarsi 4—jointed; tibia at most with 3 apical spines.	Kalotermitinae.

Subfamily I.TERMOPSINAE

Temitina (partim). Stephenes, *Ill. Brit. Ent., VI:* 97 (1836). Hagen, *Linn. Ent.,* XII: 4—31 (1858).

Termopsis Desneux, *Genera Insect.,* XXV: 12 (1904).

Hodotermitinae (Partim) Holmgren, *Zool. Anz.,* XXXV: 284—286 (1910).

Termopsinae Holmgren, *K. svenska VtenskaAkad. Handl.* XLVI (6); 35 (1911).

Key to genera*

Imago : Facetted eyes kidney shapped: antennae 23-segmented; pronotum considerably small; forewing scale cut straight without forming an arch; cerci with up to 8 joints. *Soldier:* Mandibular teeth very powerful; pronotum convex in front; cercio very long, 7-segmented. *Worker:* Cerci long, 6—7 segemented; pronotum convex in front ... *Archotemopsis,*

Imago : Facettedeyes oval; antennae up t 27- sgemented; pronotum considerably small; forewing-scales forming strong arch; cerci 5-jointed. *Soldier:* Mandibular teeth little powreful; pronotum straight in front; cerci 4-segmented. *Worker:* Cerci 4-jointed pronotum straight in front ... *Termopsis.*

Termopsis (Archotermopsis) Desneux, *Genera Insect.,* XXV: 13—14 (1904).

Archotermopsis Holmgren, *Zool. anz.,* XXXV: 284—286 (1910); *K. svenska Vetenska Akad. Handl.,* XLVI (6); 35—37 (1911).

wroughtoni Desneux.

SYSTEMATICS

Termopsisradcliff Radcliffe, *Indian For. Mag.,* p. 412 *(1904).*

Termopsis wroughtoni Desneux, *J. Bombay nat. Hist. Soc.* XV (3): 445—446 (1904); *Ann. Soc. ent. Belg.,* XLVIII: 280—285 (1904).

Termpsis (Archotermposis) wroughtoni Desneux, *Genera Insect.,* XXV: 14, pl. I, fig. 3, 3*a* (1904); *J. Bombay*

* Key is translated from Holmgren, *K. svenska VetenskAkad. Handl.,* XLVI (6): 35-36 (1911).

nat. Hist. Soc., XVII: 293—298, (1906). Lefory, *Indian insect Life*, Calcutta, I: 119, fig. 48 (1909).

Archotermopsis wroughtoni Holmgren, *K. Svenska VetenskaAkad. Handl.*, XLVI (6): 36—37, pl. 1, fig. 6, pl. 2, fig. 1, pl. 3, fig. 1, and text fig. 3 (1911). Fletcher,, *Agric. J. India* VII (3): 222—223 (1912). IMMS, *PhIl. Trans.* B, 209: 75—80, Pls. 3—10 (1919). Hegh, *LesTermites*, Bruxelles: 45 (1922). IMMs, *A General Text Book of Entomology*, London: 266, figs. 272, 273 (1942). Beeson, Forest insects, Dehra Dun : 539—540, fig. 151, no. 1 (1941). — Kashmir to Kumaon 4,000 to 9,000 ft. in the Himalayas,

Biological

Desneux, *J.Bombay nat. Hist. Soc.*, XVII: 293—298 (1906). Cleveland, *Biol. Bull. mar. biol. Lab.*, XLVI (4—5): 191—193 (1924). Magabandu, *J. Bombay nat. Hist. Soc.* XXXVII (3): 714 (1934). Beeson, *Indian For. Rec. Ent.*, IV (2): 53 (1941); *Forrest insets*, Dehra Dun: 539—540) (1941).

(Hosts: Decayed stumps and fallen trees of *Abies pindrow, Cedrus deodara, Pinus excelsa and P. longifolia*).

GENUS II.TERMOPSIS HEER.

Termpsis Heer, *Insekten fauna der Tertiargebille von Oeningen*, II: 23 (1849). Hagen, Monat. Ber. Akad. wiss. Berlin: 480 (1853); *Linn.Ent.*, XII: 74 (1858); *Peter's Reisem Mossambiquei Insekten*, II: 59 (1862). Haviland, J. *Linn. Soc. (Zool)*, XXVI: 372 (1898). Desneux, Genera Insect., XXV: 12-14 (1904).

1. **Swinhoei** Cockerell

SYSTEMATICS

Termopsis swinhoei Cockerell, *Ann. Etn. Soc. Amer.*, IX: 138 (1916), (Fossil in Burmese Amber of Miocene Blue Clays). — Burma.

Subfamily II. KALOTERMITINAE

Termitina (Partin) Stephens, Ill. Brit. *Ent.*, VI: 97 (1836). Hagen, *Linn Ent.*, XII: 4—31 (1858).

Calotermitinae (partim) Frotggatt, *Proc. Linn. Soc. N.S.W.*, XXI (4): 516 (1896).

Glyptotermitinae (partim) Froggatt, *Pro. Linn. Soc. N.S.W.*, XXI (4): 518 (1896). Silvestri, *Redia,* I: 17 (1903).

Calotermitini Desneux, *Ann. Soc. ent. Belg.*, XLVIII : 285 (1904); *Genera Insect.*, XXV: 20 (1904).

Hodotermitini (partim) Desneux, *Genera Insect,* XXV: 14-15 (1904).

Protermitidae (Partim) Holomgren, Zool. anz., XXXV: 284-286 (1919).

Calotermitinae Holmgren, *KI. Svenska Vetensk Akad. Handl.*, XLVI (6): (1920). Emerson, *Zoologica N. Y.*, VI (4): 316 (1925).

Kalotermes Hagen, *Monat. Ber. Akad. wise Berlin:* 480 (1853).

Calotermes, Hagen, *Linn. Ent.*, XII: 33 (1858). Froggat, *Proc. Linn. Soc. N.S.W.*, XXI 94): 517, 521 (1896). Wasmann, *Abh. senckenb. naturf. Ges.*, XI (1): 150. Havilnad, *J. linnn. Soc. (Zool).* XXVI (169): 373 (1898). Sjöstedt, *K. Svenska VetenskAkad. Handl., XXXIV: 35* (1900). Silvestri, Redia, I: 20 (1903). Desneux, *Genera Insect.*XXI: 20 (1904). Froggatt, *Agric. Gaz. N.S.W., XVI:* 650 (1905)

Key to subgenera*

(basedon characters of the imago)

1.	Radial sector with numerous branches to the anteric wing Radial sector without branches	2 *Glyptotermes.*
2.	Media does not join radial sector	3
	Mediajoins the radial sector nar or beyond the centre of the wing	4
3.	Media not heavily chitinized and lies parallel to and intermediate between the radial sector and cubitus	Kalotermes Hagen s. str.
	Media heavily chitinized and lies parallel to and near the radial sector	*neotermes.*
4.	Smaller forms; head 0.8 mm. wide larger fors; head 1.0 mm. or more in width.	*Cryptotermes. Plasocryptotermes.*

* Keys are copied from Light, *Philipp. J. Sci.* XLII (1): 16-19 (1930).

Key to subsgenera*
(Based on characters of the soldier)

1.	Head reddish or reddish-brown mostly large form with head more htan 1.5 mm. wide.	2
	Head smoky brown to balck, never distinctly reddish; small to very small forms with head less than 1.5 mm. wide.	3
2.	Pronotum larger than half its breadth; femora more or less dilated.	*Kalotermes* Hagen s.str.
	Pronotumshorter than half its breadth; femora little, if any dilated	*Neotermes.*
	Pronotum shorter than half its breasth; femora little, if any, dilated	4
3.	Head shorter, truncated, with an anterior concavity	*Glyptoterms.*
4.	Head high, rough; anterior concavity dee: mandibles small, weakly toothed	*Cryptoterms.*
	Head low, smooth, broad and flat; anterior concavity shallow; mandibles fairly large, distinctly toothed	*Planocryptotermes.*

Subgenus. 1 **Kalotgermes** Hagen. S. str.

Calotermes (Calotermes) Holmgren, K. Svenska Vetenska Akad. *Handle.,* XLVI (6): 53-54 (1911); *Ibid.,* L (2): 37 (1913). Lill, Termites from Australian Regione, Melbourne: 38 (1942).

Kalotermes (Kalotermes s. str.) Banks and Snyder, *Bull.* U.S. *nat.* Mus., 108: 16-17 (1920). Fuller, *S. Afr. J. nat. Hist.,* III (1): 16 (1921). Light, *Philipp. J. Sci.* XIX (1): 29 (1921). Snyder, *Proc. U.S. nat. Mus., LXI:* 1 (1922). Emerson, *Zoologica, N. Y.,* VI (4): 316 (1925). Light, *proc. Calif. Acad. Sci.,* XXI (20): 236 (1935).

Calotermes Sjöstedt, K. *svenska VeteskaAkad. Handl.,* (3), I (3): 22-23 (1925). Light, *Philipp. J.* Sci., XLII (1): 15-16 (1930). Kemner, *K. Svenska Veternsk Akad. Handle.,* (3), XIII (4): 29 (1934).

1. beesoni Gardner

SYSTEMATICS

Kalotermes beesoni Gardner, *Indian J. Ent.*, IV (1-2): 103 (1944). Haldwani (U.P.).

2. pintoi Kemner.

SYSTEMATICS

kalotermes pintoi Kemner, *Ent., Tidskr.*, LIII: 143 (1932)[1]. Hutson, *Adm. Rep. Dir. Agric. Ceylon, 1930: 73 (1931).* Pinto, *Indian J. Ent., III (1): 80 (1941).* (Host: Dead tree stump). [1]Ceylon.

Subgenus **2. Cryptotermes** (Banks).

Cryptotermes Banks, *Ent. News*, XVII, 336 (1906), Bull. *U.S. nat. Mus.*, 108: 16, 35 (1920). Light, *Philipp. J. Sci.*, XIX (1): 36 (1921). Sjöstedt, *K. svenska VetenskaAkad. Handle.*, (3), III (1): 28 (1925). Emerson, *Zoologica N.Y.*, VI (4): 326 (1925). Kemner *K. Svenska Vetenska Akad. Handl.*, (3), XIII (4): 44 (1934).

Calotermes (Procryptotermes) Holmgren, *Trans. Linn. Soc. Lond. Zool.*, pt. 1, (2), XIV: 139 (1910).

Allotermes Wasmann, *Voeltzknow, Reise Ostafrika in den Jahre 1903-5,* III: 121 (1910).

Calotermes (Cryptotermes) Holmgren, *K. svenska VetenskaAkad.* XLVI (6): 53 (1911).

Kalotermes (Cryptotermes) Snyder, *Proc. U.S. Nat. Mus.*, LXI: 12 (1922). Light, *Proc. Calif. Aad. Sci.*, (4), XXI (2): 242 (1935). Hill, *Termites from the Australian Rejoin,* Melbourne: 65-66 (1942).

1. Bengalensis Snyder.

SYSTEMATICS

Cryptotermes bengalensis Snyder, *Indian For. Rec. Ent.*, XX (11): 4-6 (1934)[1]. Beeson, *Forest Insects*, Dehradun: 540 (1941). [1]Sundarbans (Bengal).

(Hosts: Wood of *Heriteria fomes* and *Erythrina indica)*

2. **domesticus** Haviland.

SYSTEMTICS

Calotermes domesticus Haviland, *J. linn. Soc. (Zool),* XXVI: 374, pl. 23, figs. 3-6 (1898)[1]. Desneux, *Genera Insect.,* XXV: 23 (1904). [1,4]Singapore (Malaya) & Sarawaka (Borneo).

Calotemes (Cryptotermes) domesticus Holmgren *K. Svenska Vetenska Akad. Hanadl.,* XLVI (6): 55, pl. 2, fig. 9 (1911). Hegh, *Les Termites,* Bruxeelus: 48 (1922)[2]. [2,3,4]Ceylon.

BIOLOGICAL

Green, Spolia zeylan., (X) (33) : 8 (1913)[3]. Jepson, *Bull. Dep. Agric. Ceylon,* 85 : 7 (1929). Magabandu, *J. Bombay nat. Hist. soc.,* XXXVII (3): 702 (1934). Hutson *et al., Tech. Rep. Dep. AGric. Ceylon, 1928;* 7-8 (1929). Beeson, *Foresst Insects,* Dehradun: 540 (1941)[4].

(Host: Wood work of building).

3. **perforans** Kemner.

SYSTEMATICS

Cyptolermes perforans Kemner, *Ent. Tidskr.,* LIII : 133-155 (1932).

BIOLOGICAL

Hutson, *Adm. Rep. Dir. Agric. Ceylon,* 1930: 73-74 (1931) Jepson, *Ibid., 1934: 117 (1935).* Pinto, *Indian J. Ent.,* III (1): p. 83 (1941). Ceylon

[Hosts: Timber and dead branch of *Eugenia cumini (jambolina)].*

Subgenus 3. **Planocryptotermas** Light.

Planocryptotermes Light *Philipp, J. Sci.,* XIX (1): 36 (1921).

Kalotermes (*Planocryptotermes*) Light, *Philipp. J. Sci.* XLVII (1): 15 (1930). Kemner, *K. svenska Vetenska Akad.Handl.,* (3), XIII (4): 12, 30, 47 (1934).

1. prius Kemner.

SYSTEMATICS

Planocryptotermes primus Kamner, *Ent., Tidskr.* LIII: 133-155 (1932).

BIOLOGICAL

Hutson, *Adm. Rep. Dir. Agric. Ceylon,* 1930: 73-74 (1931). Jespon, *Ibid., 1934:* 117-120 (1935). Summary of the work of Div. of Ent. Dep. Agric. Ceylon in 1935, *Trop. Agriculturist,* LXXXVI (6): 345 (1936). Jepson, *Adm. Rep. Dir. Agric Ceylon, 1935: 50 (1936).* Pinto, *indian J. Ent.,* III (1): 85 (1941). Ceylon

(Host: Wood).

Subgenus 4. Glyptotermes Froggatt.

Glyptotermes Froggatti, *Proc. Linn. Soc. N.S.W.,* XXI: 518-543 pl. 4 (1896); *Agric. Gaz. N.S.W., XVI : 654 (1905); Fmrs' Bull. N.S.W. Dep. Agric.,* 60 (2nd ed.): 25 (1915). Snyder, *Proc. U.S. nat. Mus.,* LXIV: 1, 9 (1924). Sjöstedt, *K. Svenska vetenska Akad. Handl.,* (3), III (1): 22-23, 42 (1925). Light, Philipp. *J. Sci.,* XLII (1): 15-16, 18 (1930). Kemner, *K. Svenska VetenskkAkad. Handl.,* (3), XIII (4): 38 (1934).

MI9ICAR

Calotermes (partim) Desneux, *Genera Insect.,* XXV: 20 (1904).

Calotermes (Glyptotermes) Holmgren, *Trans. Linn. Soc. Lond. Zool.,* (2), XIV (1): 140 (1910); *K. Svenska Vetenska Akad.* XLVI (6): 53 55 (1911); *Ibid.,* L (2): 37, 49 (1913).

Kalotermes (Glyptotermes) snyder, *J. Wash. Acad. Sci.* XV (7): 152, 157 (1925); *Proc. ent. Soc. Wash.* XXXV (5): 57 (1933); *Indian For. Rec. Ent.,* XX (11): 24, 27 (1934). Hill, *Termites from the Australian Region,* Melbourne: 85 (1942)

1. almorensis Gardner.

SYSTEMATICS

Glyptotermes almorensis Gardner, *Indian J. Ent.,* IV (1-2): 104-105, fig. 1 (1944).[1] [1]Bajwara (Almora, U.P.)

[Host: Dead wood of *Rhododendron azarea (arborescens)*].

2. **Buttlerupeni** Holmgren.

SYSTEMATICS

Calotermes (Glyptotermes) buttlerupeni Holmgren, Zool. Zb., XXXVI: 233 (1914). — Ceylon.

3. **ceylonics** Holmgren.

SYSTEMATICS

Calotermes (Glyptotermes) ceylonicus Holmgren, *in Escherish Termitendleben auf Ceylan*, Gustav fischer, Publisher, Jena: 189 (1911). Green, *Spolia zeylan*, IX (33): 8 (1913). Holmgen, *Ibid.*, XVIII: 279-280 (1913). Hegh, *Les Termites*, Bruxelles; 48 (1922). John, *Treubia*, VI (3-4): 380 (1925)). — Ceylon.

BIOLOGICAL

Bugnion, *Bull. Mus. hist. nat. Paris*, XX (4): 176-177 (1914). John, Treubia, VI (3-4): 380 (1925). Hutson *et al.* Tech. rep. Dep. *Agric. Ceylon, 1928:* 4-5 (1929). Hutson, *Falio, Peradeniya:* 1-2 (1930). Jepson, *Trop. Agricullurist*, LXXV (3): 148, 151, 155, p. II (1930). Jepson *et al.*, *Tech. Rep. Dep. Agric. Ceylon, 1929:* 7 (1930). Hutson, *Adm., Rep. Dir. Agric. Ceylon, 1930:* 73 (1931). Margabandu. *J. Bombay, nat. hist. Soc.* XXXVII (3): 702 (1934). Jepson, *Adm. Rep. Dir. Agric. Ceylon, 1934:* 117 (1935). beeson, Forest Insects, Dehr dun: 541 (1941). Pinto, *Indian J. Ent.*, III (1): 86 (1941).

[Hoss: Decayed logs of *hevea* (rubber), wood of *Acacia* and Artocarpus *integrifolia*].

4. **coorgensis** Holmgren.

SYSTEMATICS.

Calotermes (Glyptotrmes) coorgensis, Holmgren, *Mem. Dep. Agric. India Ent.*, V (3): 139-140 (1917). Margabandu *J. Bombay nat. Hist. Soc.*, XXXVI (3): 702 (1934). — Sidapur 3500 (ft. (Coorg).

(Host: Wood of red cedar).

5. **Dilatatus** Bugnion.

SYSTEMATICS

Calotermes (Glyptotermes) dilatatus Bugnion *Mem., Soc. zool. Fr.,* XXIII: 137-143 (1910). Holmgren, *Spolia Vetensk Akad. Handl.,* L(2) : 56 (1913). Green, *Spolia zeylan.,* IX (33): 8 (1913. Bugnion, *Bull Mus. His. Nat. Paris,* XX (4): 176, , pl. IV, figs. *Hist. nat. paris,* XX (4): 176, pl. IV, figs. 1-4 (1914). John, *Treubia,* VI (3-4): 380 (1925). Hegh, *Les Termites,* Bruxelles: 48 (1922). Kemner, *Bull. ent. Res.,* XVI (4) : 383-384 (1926). Ceylon.

BIOLOGICAL

Hutson, *Trop. Agriculturist,* LIX: 83-87 (1923): *Leafl. Dep. Agric. Ceylon,* 24: 2-3, Pl. I (1923). Jepson, *Yearb.* Dep. *Agric. Ceylon, 1924* : 7-8, pl. I (1924). John, *Treubia,* VI (3-4): 380 (1925). Anonymous, *Trop. Agriculturist,* LXVII (2): 65-66 (1926). Jepson, *Quart.J. Indian Tea Ass.,* pp. 134-142 (1926). Rutherford, *Planter's Note Book,* 8th ed., Colombo, Ceylon: 234- 532 (1926). Jepson, *Yearb. Dep. Agric. Ceylon. 1928:* 3-7 (1929). Jepson, *Bull. Dep. Agric. ceylone,* 86: 2-6 (1929); *Trop. Agriculturist,* LXXV (5): 307-311 (1929); *Folio, paradeniya:* 1-2 (1930); *Trop. Agriculturist,* LXXV (3) 147-148, pl. I and II (1930); *Ibid.,* (4): 191-195 (1930). Jepson *et al., Tech. Rep. Dep. Agric. Ceylon, 1929:* 6-10 (1930). Hutson, *Adm. Rep. Dir. Agric. Ceylon, 1930:* 73 (1931); *Ibid.,* 1931: 114-1145 (1932); *Ibid., 1933:* 118-122 (1934).magabandu, *J. Bombay nat. Hist. Soc.,* XXXVII (3) : 702 (1934). Jepson, *Adm. Rep. Dir. Agric. Ceylon, 1934:* 117-119 (1935). Summary of the work of the Div. of Ent. Dep. of Agric. Ceylon, 1935 *Trop. Agriculturist,* LXXXVI (6): 345 (1936). Jepson, *Adm. Rep. Dir. Agric. Ceylon, 1935*: 51 (1936). Rau, S. Ananda, Rep. Tea Dep. *U.P.A.S.I., Madras, 1935-1936:* 35-45 (1936)[1]. King, *Tea Quart.,* XXX (4): 166 (1937). Beeson, *Forest Insects,* Dehra Dun: 541 (1941). [1]Madras.

[Hosts: *Aleurites* Spp., *Artocarpus integrifolia* (Jak), *Casuarina equisetifolia, Coffea robusta* (coffee), *Cedrela toona, Crotalaria anagyroides,Caryophyllus aromaticus* (clove), *Camellia (Thea) sinensis* (tea), *Clerodendron nutans, Erythrina lithosperma (dadap), Ficus tsiela, Grevillea* spp, *Eriobotrya japonica (loquat), Hevea brasiliensis* (rubber), *Ligustrum walkeri, Magnolia grandiflora,, Musa sapinetum (paradisiaca)* (plantain), *Moringa pterygosperma (oleifeira)* (horse radish tree), *Delonix (Poinciana) regia, Psidium guajava* (guava), *Theobroma cacao; Wormia triquetra*].

6. jepsoni Kemner.

SYSTEMATICS

Glyptotermes jepsoni Kemner, *Ent. Tidskr.*, LIII: 133-155 (1932).

BIOLOGICAL

Hutson, *Tech. Rep. Dep. Agric. Ceylon, 1928:* 4-5 (1929). Jepson, *Ibid.*, 1929: 6 (1930). Pinto, *Indian J. Ent.*, III (1): 87 (1941)[1]. — [1]Ceylon.

(Hosts: Preference for dead and decaying wood but even attacks living tissue of *Doona zey lanica, Albizzia moluccana, Cedrela toona* and *Syzygium gardneri).*

7. **minutus** Kemner.

SYSTEMACTICS

Glyptotermes minutes Kemner, *Ent.Tidskr.*, LIII: 133-155 (1932).

BIOLOGICAL

Jepson, *Tech. Rep. De. Agric. Ceylon, 1929: 6 (1930). Pinto, Indian J. Ent.*, III (1): 87-88 (1941). — Ceylon.

(Hosts: Wood of *Pithecolobium saman* and living tree of *Cupressus knightiana*).

Subgenus 5. **Neotermes** Holmgren.

Calotermes (Neotermes) Holmgren., *K. Svenska VetenskAkad. Handl.* XLVI (6): 53-54 (1911); *Ibid.*, L (2): 36 (1913).

Neotermes Silvestri, *Ann. Mus. Civ. Storia Nat Genova,* (3), III (1): 33 (1912). Banks, *Bull. U.S. nat. Mus.*, 108: 16, 32, 78 (1920). Mjöber, *Ark. Zoml.*, XII, 15 : 11 (1920). Sjöstedt, *K. Svenska VetenskaAkad. Handl.*, (3), III (1): 22, 33 (1925). Light, *Philipp. J. Sci.*, XLII (1) : 15-16 (1930). Kemner *K. Svenska VetenskAkad. Handl.* (3)., XIII, (4): 31 (1934).

Kalotermes (Neotermes) Emerson, *Zoologica N. Y.*, VI (4: 316) (1925); *Bull. Amer. Mus..nat. Hist.* LVII : 421 (1928). Snyder, *Indian For. Rec. Ent.* XX (11) : 24 (1934). Hilli, *Termites from the Australian Region*, Melbourne: 44-45 (1942).

1. **andamanensis** Snyder.

SYSTEMATICS

Kalotermes (Neotermes) andamanensis Snyder proc.d U.S. *nat. Mus.*, LXXXII, ARt. 16 (2957) : 2 (1933). Magrabandu, *J. Bombay: nat. Hist. Soc.* XXXVIII (1): 208-209 (1935). — Andaman Islands.

2. **assmuthi** Holmgren.

SYSTEMATICS

Calotermes (Neotermes) assmuthi Holmgren, *J. Bombay nat. Hist. Soc.* XXII (2) : 102-103 (1913)[1]; *K. svenska VetenskaAkad. Handl.*, L (2) : 45 (1913). — [1]Bangalore.

BIOLOGICAL.

Assmuth, *J. Bombay nat. Hist. Soc.* XXII (2) : 374-382 (1913). Holmgren, *K.svenska VentenskAkad. Handl.* L(2): 45 (1913). Magrabandu *J. Bombay nat. Hist. Soc.* XXXVII (3): 702 (1934). *Hist. Soc.* XXXVII (3): 702 (1934). Beeson, *Forest Insects*, Dehra Dun : 541 (1914).

(Hosts: Logs of wood lying outside and rarely the wood work of buildings).

3. **Bosei** Snyder.

SYSTEMATICS

Kalotermes (Neotermes) bosei Snyder, *Proc. U.S.nat. Mus.* LXXXII, Art. 16 (2957): 2-3 (1933)[1]; *Indian For. Rec. Ent.*, XX (11): 2-4 (1934)[2]. Magrabandu, *j. Bombay nat. Hist. Soc.*, XXXVIII (1): 208 : 209 (1935). — [1]Mathranwala (Dehra Dun). [2]Buxa (Bengal).

Beeson, *Forest Insects*, Dehra Dun: 541 (1941).

(Hosts: Wood of *Ficus* spp. and *Salix* spp.)

4. **fletcheri** Holmgren.

SYSTEMATICS

Calotermes (Neotermes) fletcheri Holmgrein, *Mem. Dep. Agric. Ent., V* (3): 139 (1917)[1]. Margabandu, *J. Bombay nat. Hist. Soc.,* XXXVII (3) : 702 (1934). [1]Coimbatore.

5. **gardneri** Snyder.

SYSTEMATICS

Kaloterms (Neotermes) gardneri snyder, Proc. U.S. nat. Mus., LXXXII. Art. 16, (2957) : 3-4 (1933)[1]. Margabandu, *J. Bombay nat. Hist. Soc.,* XXXVIII (1): 208-209 (1935)[2]. Beeson, *Forest Insects,* Dehra Dun: 541 (1941). [1,2]Dehra Dun (U.P.)

Host: Dead branch of living tree of *Mangifera indica).*

6. **greeni** (Desneux).

SYSTEMATICS

Calotermes greeni Desneux, *Ann. Soc. ent. Belg.,* LI: 388-389 (1907). Bugnion, *Mem. Soc. zool. Fr.,* XXIII: 125-137, fig. 4, and Pl. III, figs. 1-12 (1910). Holgremen, *in Escherish Termitenlebenauf ceylan,* Gustav Fischer, Publisher, Jena: 189 (1911). Bugnion, *Bull. Mus. Hist. nat. Paris,* XX (4): 175-176, 179, pl. III, figs. 1-3 (1914). Ceylon.

Calotermes (Neotermes) greeni Holmgren, *K.svenska VetenskaAkad. Handl.,* XLVI (6): 54 (1911). Green, *Spolia, zeylan.,* IX (3): 8 (1913). Kemner, Bull. ent. Res., XVI (4): 382-382 (1926).

Neotermes greeni King, *Tea Quart.,* X (3) : 166 (1937). Beeson, *Forest Insects,* Debra Dun; 541 (1941).

BIOLOGICAL

Bugnion, *Bull. Mus. Hist. nat. paris,* XX (4) : 175-176, pl. III, figs. 1-3 (1914). DE Mellro, *proc. 3rd ent. Mtgs. Pusa,* III : 1010, 1019, 1022 (1919). Hegh, *Les Termites, Bruxelles.* 48 (1922). Hutson, *Leafl. Dep. Agric. Ceylon,* 24: 2-3, pl. I (1923); *Trop. Agriculuturist,* LIX: 83-87 (1923); *Ibid.,* LXIII : 91-92 (1924); *Yearb. Dep. Agric. Ceylon, 1924:* 7-8, pl. I (1924). John, *Treubia,* VI (3-4): 380 (1925). Anonymous, Trop. *Agriculturist,* LXVII (2): 65-66

(1926). jepson, *Quart. J. Indian Tea Ass., pp; 134-142 (1926), Trop. Agriculturist,* LXVII (2) : 67-79 (1926). Rutherford, *Planter's Note Book, 8th ed. Colombo, Ceylon* : 234, 532 (1926). Jepson, *Yearb. Dep. Agric. Ceylon, 1927* : 20-21 (1927). Hutson, *et all., Tech. Rep. Dep. Agric. Ceylon, 1928* : 3-5 (1929). Hutson, *Falio, peradeniya* : 1-2, (1930). Jepson, *Trop./ Agriculturist,* LXXV (3) : 147-148, 151 (1930); *Ibid.,* (4) : 191-195 (1930); *Tech. Repp. Dep. Agric. Ceylon, 1929* : 6-7 (1930). Hutson, *Adm. Rep. Dir. Agric. Ceylon,* 1930 : 73 (1931); *Ibid.,* 1931: 114 (1932); *Ibid.,* 1932: 125 (1933). Jepson, *Ibid., 1933:* 118 (1934). Mar-Gabandu, *J. Bombay nat. hist. Soc.,* XXXVII (3) : 702 (1934). Jepson, *Adm. Rep. Dir. Agric. Ceylon, 1934:* 117-119 (1935). King, *Tea Quark.,* X (3) : 166 (1937). Beeson, *Forest Insects,* Dehra Dun: 541 (1941). Pinto, *Indian J. Ent.,* III (1) : 80-81 (1941).

[Hosts : Aleurites spp., Artoearpus *integrifolia, araucaria cookii, Anacardium occiden-tale* (cashewe nut), *Cassia multijua, Camellia sinensis* (tea), *Casuarina equisetijolia, Cedrela Crotalaria anagyroides, Cupressus knightiana, Croton, aromaticus* (croton oil plant), *Derris* spp., *Erythrina lithosperma (dadap), Elaecoarpus seerratus* (wild olive), *Ficus elastica, Grevillea robusta, Hevea brasiliensis* (rubber), *Michelia champaca, Macaranga tomentosa, Liguistrum walkeri, Litsea, sebifera, nephelium lappaceum, ormosia dasycarpa, Pithecolobium saman, Pittosporum viridiflorum, Spondia lutea, Shorea robusta, Solanum macranthum, Sterculea* spp. and *Pithecolobium* spp.]

7. militaris (Desneux).

SYSTEMATICS

Calotermes militaris Desneux, *Ann. Soc. ent. Belg.,* XLVIII: 146 (1904); *Genera Insect.,* XXV: 24 (1904). Bugnion, *Mem., Soc. Zool. Fr.,* XXIII: 137 (1910). Holmgren, *in Esherich Termitenleben af Ceylan,* Gustav Fischer, publisher, Jena: 189, pl. II, fig. A (1911). Green *Spolia zeylan.,* IX (33): 7 (1913). Holmgren, *Ibid., VIII* (32): 277-278 (1913). Jepson, *Trop. Agriculturist,* LXXVII (2): 67-79, pl. I (1926). Kemner, *Bull. ent. Res.,* XVI (4): 380-381 (1926). Magrabandu, *J. Bombay nat. Hist. Soc.,* XXXVII (3): 702-703 (1934). Ceylon.

Neotermes militaris Beeson, *Forest Insects,* Dehra Dun: 541, fig. 151, no. 2 (1941). Pinto, *Indian J. Ent.,* III (1): 81, figs. 17, 18 (1941).

BIOLOGICAL

Green, *Trop. Agriculturist,* XXVIII: 181-182 (1907). Dobel, *Spolia, Zeylan.,* VII: 74, 80 (1907). Bugnion, *Bull. Mus. Hist. nat. Paris.,* XX (4): 175 (1914). Ruther-ford, *Trop. Agriculturist,* XLII: 305; (1914). De Mello, *Proc. 3rd ent. Mlgs. Pusa,* III: 1017 (1919). Hutson, *Trop. Agriculturist,* LIII (5): 341 (1919); *Planters' Chron. Coimbatore,* XV (12): 207 (1920); *Trop Agriculturist* ,LVII (3): 194 (1921). Hegh, *Les Termites,* Bruxelles: 48 (1922). Hutson, *leafl. Dep. Agric. Ceylon:* 7-8, pl. I (1924). Anonymous, *Trop. Agriculturist* LXVII, (2): 65-66 (1926). Jepson, *quart. J. Indian Tea Ass., pp.* 134-142 *(1926); Trop. Agriculturist,* LXVII (2): 67-69, pl. I (1926). Rutherford, *Planter' Note Book, 8th ed.,* Colombo, Ceylon: 234, 528, 532 (1926). Jepson, *yearb. Dep. Agri, Ceylon, 1927:* 20-21 (1927). Hutson, *Tech. Rep. Dep. Agric. Ceylon, 1928: 2-7* (1929). Jepson, *Bull. Dep. Agric. Ceylon,* 86: 2-6 (1929); *Trop. Agriculturist,* LXXII (5): 307-311 (1929); *Ibid.,* LXXV (3): 155 (1930); *Ibid.,* (4): 191-195 (1930); *Tech., Rep. Dep. Agric. Ceylon, 1929:* 6-10 (1930). Hutson, *Adm. Rep. Dep. AGric. Ceylon, 1930: 72-73 (1931);* Ibid., 1931: 114-115 (1932); Ibid., 1932; 124-127 (1933). Jepson, *Ibid.,* 1933: 118-122 (1934). Margabandu, *J. Bombay nat. Hist. Soc.* XXXVII (3): 702-703 (1934). Jepson, *Adm. Rep. Dri. Agric. Ceylon, 1934:* 117-119 (1935). Summary of the work of the Div. of Ent. Dep. Agric. Ceylon, 1935, Trop. *Agriculturist,* LXXXVI (6): 345 (1936). Jepson, *Am. Rep. Dep. Agric. Ceylone, 1935:* 50-51 (1936). King, *Tea Quart.* X, (3): 160-166, pl. I, (1937); Ibid., (4): 19.5.205, 4 pls. (1937). Beeson *Forest Insects,* Dehra Dun: 541, fig. 151, no. 2 (1941). Pinto, *Indian J. Ent.* III (I): 81, figs. 17 and 18 (1941).

[Hots: *Acacia decurrens, Aleurites* spp. *Albizzia moluccana, Camellia, sinensis,* (tea), *Cedrela toona, Crotalaria, anagyroides, Calophyllum tomentosum, Cinnamomum camphora* (camphor), *Cryptomeria, japonica* (Japanese cedar), *Casuarina equisetifolia, Erythrina lithosperma (dadap), Doona zeylanica, Eucalyptus* spp., *Eugenia gardneri, Grevillea, robusta,Linociera purpourea, Litsea zeylanica, Semacarpus coriacea, Symplocos spicata* and *Wednlandia notoniana*].

Var. unidentatus Kemner.

Calotermes (Neotermes) militaris var. *unidentatus* Kemner, *Bull. ent. Res.*, XVI, (4): 381-382 (1926). XXVII (3): 703 (1934). Pinto, *India J. Ent.*, III (1): 81 (1941). Ceylon.

[Host: *Camellia sinensis,*the tea plant). [Thisspecies was decribed by Kemner from a single soldier among specimens of *Calotermes (Neotermes) militaris* and it may be regarded as a freak as suggested by Pinto].

Family III. RHINOTERMITIDAE

Mesotermitidae Holomgren, *K. senska VetensAkad. Handl.*, XLVI, (6): 62-84 (1911); *in Escherich Termitenleben auf Ceylan,* Gustav Fischer, Publisher, Jena: 206, 209-210 (1911); *Mem. Dep. Agric. India Ent.*, V (3): 140-144 (1917).

Termitina (partim) Banks and Snyder, *Bull. U.S. ant. Mus.* 108: 75-76 (1920).

Rhinotermitidae Light, *Philipp. J. Sci.*, XVIII: 245 (1921); *Ibid.*, XIX: 25 (1921). (Emerson, *Zoologica* N. Y., VI (4): 340 (1925). Light, *Termites and Termit Control,* by Kofoid *et al.*, Berkeley, Calitfornia: 112 (1934). Hill, *Termites from the Australian Region,* Melbourne: 6 (1942).

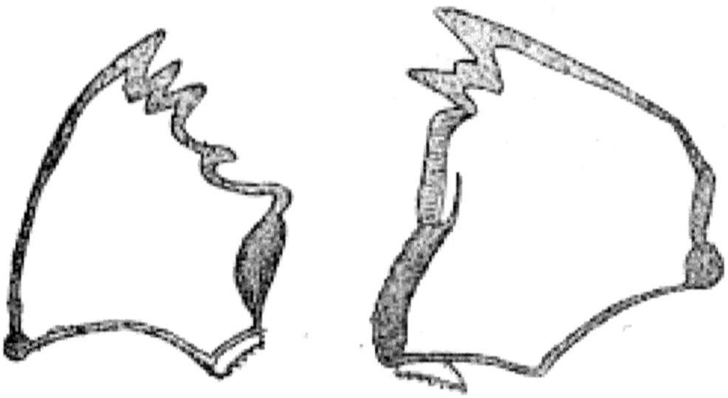

FIG. 1.

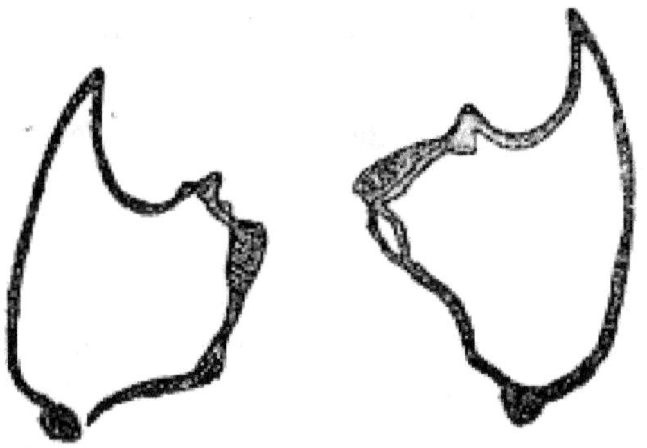

Fig. 2

*Holmgren, *K. svenska Vetensk Aked. Handl.*, XLVI (6): 48, 82 Text figs. 5 and 6 (1911).

*Key to subfamilies***

(I) Imago

1.	Mandibles of *Serritermes* type armour (Fig. 2)	*Serritermitinae* (Not found in India)
	Mandibles of *Leucotermes* type armour (Fig. 1).	2
2.	Fore wing scales not larger than the hinder	*Psammotermitinae* (Not found in India)
	Fore wing scales large	3
3.	Head parallel sided, oval Head contracted, circularto 3-edged	*Leucotermitinae.* 4
4.	Clypeus strongly enlarged, often drawn out like nature form ; forehead with a groove; fontanel open, with fontanel gland.	*Rhinotermitinae*
	Clypeusnot strongly enlarged, not nasute forming; forehead without groove	5
5.	Head round, 3-edged, posteriorly straight cut; fontanel with fontanel gland; pronotum very short, having a fissure anteriorly; wings thick, very finely haired and very weakly reticulate.	*Termitogetonina*
	Head almost round to circular; fontanel with or without gland or plate; pronotum proportionately broad; wings more or less reticulate	6

* This key is translated from Holmgren *K. svenska VetenskAkad. Handl.*, XLVI (6): 62.63, and 64-84 (1911). While redrawing the key the description of subfamily Stylotermitinae as given by Holmgren, *Mem. Dep. Agric. India Ent.*, V (3) ; 141 (1917), is also taken into account.

6.	Head almost circular; fontanel with fontanel plate; wings more or less reticulate.	*Coptotermitinae.*
	Head almost circular; fontanel with fontanel gland; wings strongly reticulate	*Stylotermitinae.*

(II) Soldier

1.	Clypeus relatively long, crosswise oval; mandibles along the outside of the base considerably strongly concave powerfully serrated or toothed.	*Serritermitinae.* (Not found in India)
	Clypeus short; mandibles along the outside of the base never strongly concavely bent, never distinctly sorrated or toothed	2
2.	Head parallel sides; laborum with or without hyaline tip	3
	Head narrowed towards infront.	5
3.	Only one soldier form; labrum without hyaline tip; pronotum semicircular, concave anteriorily; styli absent; tibiae with 3-apical spines.	*Stylotermitinae.*
	Labrum with relatively big hyaline tip, styli present; two soldier forms	4
4.	Smaller soldier form little unlike the bigger one; mandibles powerful with big teeth, left one wit more teeth than the right; antennae of large soldier strongly narrowed out; pronotum trapezoid, considerably deeply cut infront; styli small; foretibiae with 3 powerful short apical spines, medium and hinder with 2-extremely thick short spines.	*Psammotermitinae* (Not found in India)
	Often 2nd soldier form a little varying from the big; mandibles small considerably straight, left more freqeuntly with a rudimentary tooth; antennae never strongly narrowed; pronotum concave fore and behind; tibiaewith 3-apical spines.	*Leucotermitinae.*
5.	Left mandible with two teth and right with one big tooth; forehead and labrum with a groove mostly with too strong unlike soldier forms	*Rhinotermitinae.*
	Mandibles without dentition or with rudimentary tteth; one soldier form	6

6. Forehead with a groove or formed into a tube: posterior end of head convex in the form ofarch; with or without facetted eyes; antennae 13-176 segmented; pronotum flat, infront without median process; stylip present	*Coptotermitinae.*
Forehead without bot grooveand tube; posterior end of head concave; without facetted eyes; antennae 14-15 segmented; pronotum flat, infront drawn out inaprocess; stylia absent.	*Termitogetoninea*

(III) Worker

1. Cylpeus relatively short; mandibles of *Leucotermes* type armour (Fig. 1).	2
Clypeus long; madibles of *Serritermess* type armour (Fig.2)	*Serritermitinae.* (Not found in India)
2. Pronotum flat	3
Pronotum saddle-shaped	*Rhinotermitinae.*
3. Pronotum not produced forward; head oval or circular, not 3-edged; styli present or absent	4
Pronotum infront produced forward; head 3-edged, rounded off; styli absent	*Termitogetoninae*
4. Head circular; styli absent; meso and meta notum bent posteriorly like the stumps of a wing.	*Stylotermitinae.*
Head broad, oval; styli present; meso and metanotum not bent posteriorly like the stumps of a wing	5
5. Fore tibiae with 3, hind tibiae with 2 apical spines; first tooth of mandible dustinctly longer than the second.	*Psammotermitinae* (Not found in India)
All tibiae with 3-apical spines	6
6. First tooth of left mandible little longer than the second. First tooth of left mandible distinctly bigger than the second	*Leucotermitinae.* *Coptotermitinae.*

Subfamily I. **COPTOTERMITINAE**

Coptotermitinae Holmgren, *K. svenska VetenskAkad. Handl.*, XLVI (6): 71 (1911); *Mem. Dep. Agric. India Ent.*, V (3): 144 (1917). Emerson, *Zoologica N. Y.*, VI (4): 344 (1925)

*Key to genera**

Imago: Clypeus large, considerably raised up; atennae 19-22 segmented ; wings not usually thickly hair, stronlgy reticulate; median springing from the cubitus or missing. *Soldier:* Without frontal tube; fontanel in normal situation; with facetted eyes. *Worker:* Clypeus big; often with facetted eyes.	*Prorhinotermes.*
Imago: Clypeus small, usually flat; antennae 19-23 segmented; wings usually thickly hairy, membrane rough; median of the fore wing arises freely from the wing basc. *Soldier:* With a short frontal tube; fontanel adjusted towards infront facetted eyes missing. *Worker:* Clypeus small; without facetted eyes	*Coptotermes.*

Genus i. **Coptotermes** (Wasmann).

Termes (Coptotermes) Wasmann, *Ann. Mus. Civ. Storia Nat. Genova;* XVI (36): 629 (1896); *Abh. senckenb. naturf. Ges.,* XI (1): 151 (1897). Desneux, *Genera Insect.,* XXV: 29, 32, 34 (1904).

Arrhinotermes Wasmann, *Zool. Jb., Abt. I,* XVII (1): 103 (1902). (heimi as type).

Coptotermes Silvestri, *Redia,* I: 46 (1903). Froggatt, *Agric. Gaz. N.S.W.,* XVI: 760 (1905). Holmgren, *K. Svenska VetenskaAkad. Handl.,* XLIV (3): 28-29, 110, 169 (1909): *Ibid.,* XLVI (6); 73 (1911); *Ibid.,* L(2): 71 (1913). Light, *Philipp. J. Sci.,* XIX (1): 48-49 (1921). Oshima, *Capita zool.,* XI (3): 5 (1923). Sjöstedt, *K. svenska VetenskAkad. Handl.,* (3), III (1): 55 (1925). Emerson, *Zoologica N. Y.,* VI (4): 344 (1925). Light, *Univ. Caif. Publ. Ent.,* IV (5); 75-88 (1927). Emerson, *Bull. Amer. Mus.nat. Hist.,* LVII, Art. 7: 423 (1928), Kemner, *K. svenska VetenskaAkad. Handl.,* (3), XIII (4): 57 (1934). Hill, *Termites from the Australian Region,* Melbourne: 140-141 (1942).

1. **ceylonicus** Holmgren.

SYSTEMATICS.

Coptotermes ceylonicus Holmgren, *in* Escherish Termitenleben auf Ceylan, Gustav Fischer,

* Key is translated from Holmgren, *K. svenska Venteskakad. Handl.,* XLVI (6) 71-72 (1911).

Publisher, Jena: 192-193, pl. 3, fig. D (1911); *J. Bombay nat. Hist. Soc.* XXI, (3): 777 (1912); *K. svenska VetenskAkad. Handl.*, L (2): 76 (1913). Hegh, *Les Termites,* Bruxelles: 48 (1922). John, *Treubia,* VI (3-4): 381 (1925). Kemner, *Bull. ent. Res.*, XVI (4): 384 (1926).

BIOLOGICAL.

Green, *Rep. Ent. Dep. Agric. Ceylon, 1911-12,* (4): 2-5 (1912); *Spholia zeylan.*, IX (33): 9 (1913). Bugnion, *Bull. Mus. Hist. nat. Paris.*, XX (4): 177-179, pl. V, figs. 1-3 (1914), John, *Treubia,* VI (3-4): 381 (1914). Hutson, *Tech. Rep. Dep. Agric. Ceylon, 1928: 7-8 (1929).* Jepson, *Bull. Dep. Agric. Ceylon,* 85: ,7 pl. II (1929); *Ibid.*, 86: 2-3 (1929); *Trop. Agriculturist,* LXXII (5): 309 (1929). Hutosn, *Folio, peradeniya:* 1-2 (1930). Jepson, *Trop. Agriculturist,* LXXV (3): 151-152, pl. II (1930). Jepson, *Tech. Rep. Dep. Agric. Ceylon, 1929:* 6-12 (1930). Hutson, *Adm. Rep. Dir. Agric. Ceylon, 1930:* 73-74 (1931); *Ibid., 1931:* 114 (1932); *Ibid., 1932:* 124-128 (1933). Dover and Mathur, *India For. Rec. Ent.*, XX (7): 3 (1934). Jepson, *Adm. Rep. Dir. Agric. Ceylon, 1933:* 118-120 (1934). Magabandu, *J. Bombay nat. Hist. Soc.*, XXXVII (3): 703 (1934). Jepson, *Adm. Rep. Dir. Agric. Ceyloln, 1934:* 117-120 (1935). Beeson, *Indian For. Rec. Ent.*, IV (2): 69 ; *Forest Insects,* Dehra Dun: 542, fig. 151, no. 4 (1941)[1].

[1]Ceylon to Indo-China, and South India.

[Hosts: Logs, beams, stack timber packingcaes, doors windows in contact with soil and also living plants of *Camellia sinensis* (tea), *Hevea brasiliensis* (rubber), *Cocos nucifera* (coconut palm), *Ligustrum walkeri, Albizzia stipulata, Aleurites* spp., *Acacia* spp., *Tamarindus indica, madhuca (Bassia) longifolia, Gliricidia maculata, Theobroma cacao,* and grape fruit tree].

2. exiguus (Hagen).

SYSTEMATICS.

Termes exiguus Hagen, *Linn. Ent.*, XII: 208 (1858).

Eutermes exiguus Desneux, *Genera Insect.*, XXV: 42 (1904). Ceylon.

Coptotermes exiguus Jepson, *Trop. Agriculturist.* LXXV (3): 155 (1930); *Tech. Rep. Dep. Agric. Ceylon, 1929:* 6 (1930).

[Hosts: Logs, beams, stacktimber, packing cases, doors, windows in contact with soil and also living plants of *Camellia sinensis* (tea), and *Erythrina lithosperma (dadap)*].

3. **gestroi** (Wasmann).

SYSTEMATICS.

Termes gestroi Wasmann, *Ann. Mus. Civ. Storia Genova,* XVI: 628 (1896)[1]. Haviland *J. linn. Soc. Lond. (Zool.),* XXVI: 390-391 (1898). Pratt, *Bull. Dep. Agric. F.M.S.,* (1908)[2]. — [1, 4, 6]Burma. [1]Singapore (Malaya).

Coptoterme gestroi Desneux, *Genera Insect.,* XXV: 34 (1904). Holgren, *Mem. Dep. Agric. India Ent.,* V (3): 145 (1917)[3]. — [1, 4,]Sumatra.

Coptotemes curvignathus Holmgren, *KI. Svenska Vetensk Akad. Handl., L.* (2): 7 (1913)[4]. John, *Treubia,* VI (3-4): 381 (1925). Caresche, *C.R. Trav. Inst. Rech. Agron. For Indo China,* II: 195-212, pl. 6 (1937)[5]. — [2, 4, 6]Malaya. [3]Assam.

BIOLOGICAL.

Pratt, *Bull. Dep. Agric. F. M. S., I* (1908). Lefroy, *Indian Insect Life,* Calcutta, I: 119 (1909)6. Begnion, *Mem. Soc. Fr.,* XXIII: 100-123 (1900). Fletcher, *Agric. J. India,* VII, (3): 225 (1912)[7]; *Bull. agric. Res. Inst. Pusa,* (59): 39 (1916). John, *Treubia,* VI (3-4): 381 (1925). Margabandu, *J. Bombay nat. Hist. Soc.* XXXVII (3): 704 (1934). Beeson, *Forest Insects,* Dehra Dun: 542-543 (1941)[8]. — [5]Indo-China [7]Japan. [8]North India.

[Hosts: Serious pest of *Heva brasiliensis* (rubber); also found damaging *Albizzia procera, Artocarpus blumei, Bombax malabaricum, Canarium commune, Cocos nucifera, Eriodendron anfractuosum, Ficus elastica, Magnifera indica, Oroxylon indicum, Salix* spp., *Sapium bacca-tum, Semecarpus anacardium, Shorea robusta* and wooden structures].

4. **parvulus** Holmgren.

SYSTEMATICS.

Coptotermes parvulus Holmgren, *J. Bombay nat. Hist. Soc.*, XXII (1): 104 (1913)1. Assmuth, *Ibid.*, (2): 374, 378-382 (1913)[2]. Holmgren, *K. Svenska VetenskaAkad. Handl.*, L (2): 73 (1913).

[1]Anand,Vadtal (Bombay).
[2]Karachi (Sind).

BIOLOGICAL.

Assmuth, *J. Bombay nat. Hist. Soc.*, XXII (2): 373, 378-383 (1913). Annandale, *Rec. Indian Mus.*, XXV (2): 233 (1923)[3]. Silvestri, *Ibid.*, (2): 225 (1923)[4]. Magabandu, *J. Bombay nat. Hist. Soc.*, XXXVII (3): 704 (1934). Beeson, *forest Insects*, Dehra Dun: 544 (1941).

[3, 4]Barkuda Islands (Orissa)

(Hosts: Wood and wooden articles touching ground).

5. **travians** (Haviland).

SYSTEMATICS.

Ternes travians Haviland, *J. linn. Soc. (Zool.)*, XXVI: 391-392, pl. 23,figs. 19, 22 (1898)[1].

[1, 4]Singapore (Malaya)

Coptotermes travians Desneux, *GeneraInsect.*, XXV: 34 (1904). Bugnion, *Mem. Soc. zool. Fr.*, XXIII: 100-123 (1910)[2]. Holmgren, *J. Bombay nat. Hist. Soc.*, XXI (3): 777 (1912)[3]. Light, *Philipp. J. Sci.* XIX (1): 48-49 (1921). John, *Treubia*, VI (3-4): 381 (1925)[4].

[1]Sarawak (Borneo).
[2]Ceylon.
[3, 6]Bombay.

Arrhinotermesheimi Wasmann, *Zol. Jb., Abt. I*, XVII (1): 104, pl. 4, fig. la-If (1902)[5]. Desneux, *Genera Insect.*, XXV: 34 (1904). Light, *Philipp. J. Sci.*, XIX (1): 48-49 (1921). Magabandu, *J. Bombay nat. Hist. Soc.* XXXVII(3): 703 (1934).

[4]Sumatra, Borneo Java.
[5]India.

Coptotermes heimi Holmgren, *J. Bombay nat. Hist. Soc.* XXI (3): 777 (1912). Magrabandu, *Ibid.*, XXI (3): 777 (1912). Margabandu, *Ibid.*, XXXVII (3): 704 (1934).

[7]Calcutta, Mangalore, Ranchi

BIOLOGICAL.

Bugnion, *Mem. Soc. zool. Fr.*, XXIII: 100 123 (1910). Lefroy, *Indian Insect Life*, Calcutta, I: 119 (1909). Holmgren, *J. Bombay nat Hist. Soc.*, XXI (3): 777 (1912).Assmuth, *Ibid.*, XXII (2): 374, 378-383 (1913)[6]. Oshima, *Annot. zool. Jap.*, VIII: 555 (1914). Assmuth,

[8]Jullundur (Punjab), Pusa (Bihar), Noakhali and Chaumuhani (Bengal)[2], Bellary (Madras), South

J. Bombay nat. Hist. Soc., XXIII(4): 691 (1915)[7]. Holmgren, *Mem. Dep. Agric. india Ent.*, V (3): 145 (1917)[8]. De Mello, *Proc. 3rd ent. Mtgs. Pusa*, III: 1018 (1919). Light, *philipp. J. Sci.*XIX (1): 48-49 (1921). Annandale, *Rec. Indian Mus.*, XXV (2): 233, 250-251 (1923)[9]. Silvestri, *Ibid.*, (2): 225 (1923). John., *Treubia*, VI (3-4): 381 (1925). Dover, *India For.*, LVII (7): 342-345 (1931) Dover and Mathur, *Indian For. Rec. Ent.*, XX (7): 3 (1934). Magabandu, *J. Bombay sot. Hist. Soc.* XXXVII (3): 704 (1934). Beeson, *Forest Insects*, Dehra Dun: 543-544 (1941). — Canara and Margalli (Coorg). [9]Barkuda Islands (Orissa)

(Hosts: Wood, wooden articles and tree stamps.)

Genus II. **Prorbinotermes** Silvestri

Arrhinotermes Wasmann, *Zool. Jb., Abt. I, XVII* (1): 103, 106 (1902), (*oceanus* as type). Holmgren, *K. svenska Vetensk Akad. Handl.*, XLVI (6): 71-2 (1911); *Ibid.*, L (2): 66 (1913). Sjöstedt, *K. Svenska Vetenska Akad. Handl.*, (3), III (1): 55 (1925).

Coptoermes (Arrhinotermes) Desneux, *Genera Insect.*, XXV: 33 (1904).

Prorhinotermes Silvestri, *Fauna Südwest-Aust.*, II (17): 286 (1909). Holmgren, *K. svenska Vetenska Akad. Handl.*, XLIV (3): 28 (1909). Banks and Snyder, *Bull. U. S. nat. Mus.*, 108: 145 (1920).Light, *Philipp. J. Sci.*, XIX (1): 48-49 (1921).Emerson, *Zoologica N.Y.*, VI (4): 344 (1925). Snyder, *J. Wash. Acad. Sci.*, XV (17): 398 (1925). Light, *Philipp. J. Sci.*, XLII (1): 15 (1930). Hill, *Termites from the Australian Region*, Melbourne: 136-137 (1942).

Procoptotermes Holmgren, *K. Svenska VenteskaAkad. Handl.*, XLVI (6): 73 (1911).

Termitogetonella Oshima, *Philipp. J. Sci.*, XVII (5): 496-497, pl. 3 figs. 3, 4, p. 4. figs. 3, 4 (1920).

1. **Flavus** (Bungnion).

SYSTEMATICS.

Coptoermes flavus Bugnion, *Mem., Soc. zool. Fr.*, XXVIII: 100-1234 (1910).

Coptotermes (Arrhinotermes) flaves Bugnion, *Mem. Soc. zool. Fr., XXIV:* 97-106, 2 *plz.* (1911). — Ceylon.

Arrhinotermes flavus Green, *Spolia zeylan.*, IX (33): 9 (1913). Hegh, *Les Termites,* Bruxelles: 48 (1922). John, *Treubia,* VI (3-4): 381 (1925).

BIOLOGICAL

Bugnion, *Bull. Mus. Hist. nat. Paris.*, XX (4): 178 (1914). De Mello, *Proc. 3rd ent. Mtgs.* Pusa, III: 1019-1022 (1919). John, Teubia, VI (3-4): 381 (1925). Magabandu, *J. Bombay Nat. Hist. Soc.* XXXVII (3): 703 (1934).

(Host: stems of Mangrove).

Subfamily II. **HETEROTERMITINAE**

Heterotermitinae Froggatt, *Agric. Gaz. N. S. W.* XVI: 656 (1905). Emerson, *Bull. Amer. Mus. nat. Hist.* LVII: 423 (1928).

Lecotermitinae Holmgren, *K. svenska VetenskAkad. Handl.* XLVI (6): 68-69 (1911); *Mem. Dep. Agric. india Ent.*, V (3): 140 (1917). Emerson, Zoologica N. Y., VI (4): 340 (1925).

Key top genera*

Imago: Clypeus very short, nearly 4-times as broad as long, strongly bilobed infront; ocelli distinct distinct; wings clear, slender and without hairs,but more or less reticulate, their outer and hinder margins not ciliated; legs reaching beyond tip of the body. *Soldier:* Head more or less elongate; mandibles slender, slightly curved,without teth; labrum narrowed and pointed at tip; gula narrowed in middle, much widened infront; pronotum croader infront, front margin indented in the middle; a small kyaline ocellus like not present above the base of each notenna. *Reticulitermes.*

Amago: Clypeus 3-times as broad as long, bilobed infront; no celli; wings opaque, more or less hairy, not recticulate,margins eliated, legs short, reaching but little beyond tip of the body. *Soldier:* Head rather long; mandibles long, slightly curved, without teeth; labrum hardly pointed at tip; gula *Heterotermes.*

* Key is drawn by consulting the description of genera given by Banks and Snyder, *Bull. U. S. nat Mus.*, 108 : 42-43, 75-76 (1920).

not narrowed in middle and wider infron tthan behind; pronotum broader infront, front margin deeply indented in the middle; no hyaline cellus like dot above the base of any antenna.

Genus i. **heterotermes** Frogatt.

Heteroterms Froggatta, *Proc. Linn., Soc. N.S.W.* XXI (4): 518, 550 (1896). Desneux, Genera, *Insect.*, XXV: 32 (1904). Froggatt, *Agric. Gaz. N.S.W.* XVI: 656 (1905); *Frmrs; Bull. N.S.W. Dep. Agric.* 60 (2nd ed.): 26 (1915). Emerson, *Bull. Amer. Mus. nat. Hist.*, LVII: 423 (1928). Light, *Philipp. J. Sci./*, XLII (1): 18-19 (1930). Hill., *Proc. Roy. Soc. Vict.*, XLIV (2): 143 (1932). Light. Kemner, *K.svenska VetenskaAkad. Handl.*, (3), XIII (4): 67 (1934). Hill, *termites from the Australian Region*, Melbourne: 120-121 (1942).

Leucoterms Silvestri, *Bull. Mus. Zool. Anat. comp. Torino.*, XVI (389): 3 (1901); *Redia*, I: 37, fig. 33 (1903). Holmgren, *K. Svenska VetenskaAkad. Handl.*, XLVI (6): 62, 69 (1911). Banks, *Bull. U.S.nat. Mus.*, 108: 75 (1920). Light. *Philipp. J. Sci.* XIX (1): 57 (1921). Sjöstedt., *K. svenska VetenskaAkad. Handl.*, (3), III (1): 59 (1925). Emerson, *Zoologica, N.Y.*, VI (4): 340 (1925). John. Treubia, VI (3-4): 381 (1925). Silvestri, *Bull. Soc. ent. Fr.*, 10: 203 (1936).

Psalidotermes Silvestri, *Fauna Süwets-Aust.*, II (17): 296 (1909).

1. **Ceylonicus** (Holmgren).

SYSTEMATICS

Leucotermes ceylonicus Holmgren, *in Escherish Termitenleben auf Ceylan*, Gustav Fischer, Publisher, Jena: 190-192, pl. 2, figs. CI-C2 (1911). Green, *Spolia Zeylan*, IX (33): 8 (1913). hegh. *Les Termites*, Bruxelles: 48 (1922). John, *Treubia*, VI (3-4): 381 (1925). Ceylon

Heterotermes ceylonicus Beeson, *Forest Insects*, Dehra Dun: 544 (1914).

BIOLOGICAL

Bugnion, *Bull. Mus. Hist. nat. Paris*, XX (4): 177 (1914). John, *Treubia*, VI (3-4): 381 (1925). Hutson, *Tech.*

Rep. Dep. Agric. Ceylon, 1928: 7-8 (1929). Jepson, *Bull. Dep. Agric. Ceylon,* 85: 7-8 (1929). Margabandu, *J. Bombay nat. hist. Soc.,* XXXVII (3): 703 (1934). Beeson, *Forest Insects,* Dehra Dun: 544 (1941).

(Hosts: Wood works of buildings and stumps of *Grevillea* spp.).

2. **Indicola** (Wasmann).

SYSTEMATICS.

Leucotermes indicola Wasmann, Zool. *Jb., Abt. I,* XVII (1): 118, pl. 4, fig. 2 (1902)[1]. Desneux, *Genera Insect.,* XXV: 33 (1904). Holmgren, *J. Bombay nat. Hist. Soc.* XXI (3): 776 (1912)[2]; *Ibid.,* XXII (1): 103 (1913)[3]; *Mem. Dep. Agric. India Ent.,* V (3) 140 (1917)[4]. Hegh, *Les Termites,* Bruxelles: 48 (1922)[5].

[1]India.
[2,3]Bombay.
[4]Gurdaspur (Punjab)
[5]Ceylon.

Heterotermes indicola Beeson, *Indian For. Rec. Ent.,* IV (2): 68 (1941); *Forest Insects,* Dehra Dun: 544, fig. 151, no. 3 (1941)[6].

[6]Throughout India up to 7,000 ft. in Himalayas.

BIOLOGICAL

Lefroy, *Indian Insect Life,* Calcutta, I: 119: (1909). Fletcher, *Agric. J. India,* VII (3): 223-224 (1912). Assmuth, *J. Bombay nat. Hist. Soc.,* XXII (2): 278-389 (1913). Bugnion, *Bull. Mus. hist. Nat. Paris,* XX (4): 177 (1914). Assmuth, *J. Bombay nat. Hist. Soc.,* XXXIII (4): 690-692 (1915)[7]. DeMello, *Proc.#rd ent. Mtgs. Pusa,* III: 1011-1012 (1919). Dover, *Indian for.,* LVII (7): 3470 (1931). Margabandu, *J. Bombay nat. Hist. Soc.,* XXXVII (3): 703 (1934). Beeson, *Indian For., Rec. Ent.,* IV (2): 68 (1941)[8]; *Forest Insects, Dehra Dun:* 544 (1941). [Hosts: Wood and wooden structures, timbers such as *Cedrela toona, Pinus longifolia, Flacourtia ramontchi, Shorea robusta, Tectona grandis* and *Quercus* spp. (an imported one)].

[7]Calcutta, Mangalore and South Canara.
[8]Dehra Dun.

3. **malabaricus** Snyder.

SYSTEMATICS.

Heterotermes malabaricus Snyder, *Proc. U.S. nat. Mus.,* LXXXII, Art. 16 (2957): 4-6 (1933). Margabandu, *J. Bombay nat. Hist. Soc.,* XXXVIII (1): 208-209 (1935).

South Mangalore and Malabar coast.

Genus ii. **Reticulitermes** (Holmgren).

Leucotermes (Reticulitermes) Holmgren, *K. svenska VetenskAkad. Handl,* L (2): 103 (1913).

Reticulitermes Banks and Snyder, *Bull. U.S. nat. Nus.,* 108: 39, 42-43 (1920).

1. **assamensis** Gardner.

SYSTEMATICS.

Reticulitermes assamensis Gardner, *Indian J. Ent.,* IV (1-2): 105-106) (1944). Sadiya (Assam).

[Host: *Terminalaia myriocarpa).*

Subfamily III. **RHINOTERMITINAE**

Rhinotermitinae Froggatt, *Agric. Gaz. N. S. W.,* XLVI 6: 76-77 (1911). Banks and Snyder, *Bull. U. S. nat. Mus.* 108: 39 (1920). Emerson, *Zoologica N.Y.,* VI (4): 346 (1925).

Genus **Rhinotermes** (Hagen).

Termes (Rhinotermes) Hagen, *Linn. Ent.,* XII: 233 (1858). Haviland, *J. linn. Soc. (Zool.),* XXVI: 392 (1898).

Rhinotermes Froggatt, *Proc. Linn. Soc. N.S.W.,* XXI (4): 518, 539 (1896). Sjöstedtn, K. *svenska VetenskaAkad. Genera Insect.,* XXV: 26-28 (1904). Froggatt, *Agric. Gaz. N. S. W.,* XVI: 654 (1905). Holmgren, *K. svenska VetenskaAkad. Handle.,* XLV (6): 79, pl. 6, figs. 1-4 (1911); *Ibid.,* L (2): 80 (1913). HiLL, *Termites from the australian Region,* Melbourne: 175 (1942).

Subgenus **Schedorhintoermes** (Silvestri).

Schedorhinotermes Silvestri, *Fauna Südwest-Aust.,* II (17): 289 (1909). Sjöstedt, *K. Svenska VetenskaAkad. Handl.,* (3), III: 51 (1925). Light, *Philipp. J. Sci.,* XLII (1): 15 (1930). Kemner; *Lunds, Univer.Arbseratt.,* (2), XXVII (13): 23 (1931); *K. svenska VetenskaAkad. Handl.,* (3), XIII (4): 50 (1934).

Rhinotermes (Schedorhinotermes) Holmgren, *K. svenska Venteksa Akad. Handl.,* XLVI (6): 79, pl. 6 figs. 1-4 (1911): *Ibid.,* (2): 81-82 (1913). Oshima, *Annot. zool.*

Jap., VIII: 562 (1914). John, *Treubia*, VI (3-4): 384 (1925) Snyder, *J. Wash. Acad. Sci.*, XV (17): 404 (1925). Emerson, *Bull. Amer. Mus., nat. Hist.*, LVII: 425 (1928). Hill, Termites *from the Australian Region*, Melbourne: 176 (1942).

1. magnificus (Silvestri).

SYSTEMATICS

Schedorhinotermes magnificus Silvestri *Rec. Indian Mus.*, VIII (5): 425-426, 2 figs. (1914). — Misty Mollow, W. Dawna Hills, (Lower Burma).

Subfamily IV. **STYLOTERMITINAE**

Stylotermitinae Holmgren, *Mem. Dep. AGric. India Ent.*, V (3): 141 (1917).

Genus Stylotemess Holmgren.

Stylotermes Holmgren, *Mem. Dep. Agric. India Ent.* V (3): 141 (1917)

1. fletcheri Holmgren.

SYSTEMATICS

Stylotermes fletcheri Holmgren, *Mem. Dep. Agric. Indian Ent.*, V (3): 142-144 (1917)[1]. — [1, 2]Shevaroy Hills 4000 it. (S.India)

BIOLOGICAL.

Fletcher, *Proc. 3rd ent. Mtgs. Pusa*, I: *P. 314* (1919)[2]; *Bull. Dep. Agric. India*, 100: 246 (1921). Magabandu, *J. Bombay nat. Hist. Soc.* XXXVII (3): 704 (1934).

(Host: Wood of *Mangifera indica*, Mango.).

Subfamily V. **TERMITOGETONINAE**

Termitogetoninae Holmgren, *K. sveska VetenskAkad. Handl.*, XLVI (6): 75 (1911).

Genus **Termitogeton** (Desneux).

Termes (Termitogeton) Desneux, *Ann. Soc. ent. Belg.*, XLVIII: 374 (1904); *Genera Insect.*, XXV: 47 (1904).

Termitogeton Holsmgren, *K. svenska VetenskAkad. Handl* XLVI (6): 75-76 (1911). John, *Treubia,* VI (3-4): 384 (1925).

1. umbilicataus (Hagen).

SYSTEMATICS.

Termes umbilicatus Hagen, *Zool. Bot. Ver. Wien:* 472 (1858); *Linn. Ent.,* XIV: 100-113 (1860).

Termes (Termitogeton) umbilcatus Desneux, *Genera Insect.,* XXV: 47 (1904).

Termitogeton umbilicatus Holmgren., *in* Fischer, Publisher, Jena: 193-194, pl. 2, fig. K (1911). Green, *Spolia zeylan.,* IX (33): 9 (1913). Bugnion, *Ann. Soc. ent. Fr.,* LXXXIII: 39-47. pl. I (1914); *Spolia zeylan.,* IX: 149 (1914). Hegh, *Les Termites,* Bruxelles: 48 (1922). John, *Treubia,* VI (3-4): 384 (1925). Ceylon.

Termes planus Haviland, *J. Linn. Soc. (Zool.),* XXVI: 397, p. 24, fig. 30-33 (1898)[1]. Desneux, *Genera Insect.,* XXV: 37 (1904). Holmgren, *in Escherich Termitenleben auf Ceylan:* 193-194, pl. 2, fig. K (1911). [1]Sarawak(Borneo)

BIOLOGICAL

Bugnion, *Ann. Soc. ent. Fr.,* LXXXIII: 39-47, pl. I (1914); *Bull. Mus. Hist. nat. paris,* XX (4): 179, pl. VI, figs. 1-4, pl.VII, figs. 1-2 (1914). De Mello *Proc.3rd ent. Mtgs. Pusa,* III: 1019 (1919). John, Treubia, VI (3-4): 384 (1925). Margabandu, *J. Bombay nat. Hist. Soc.,* XXXVII (3): 704 (1934).

(Host: Wood in contact with soil.)

Family IV. **TERMITIDAE**

Termes Linnnaeus, *Syst. Nat., ed.,* 10 (1758); *Ibid.,* ed., 13, Gmelin, I: 2911 (1788-1793). Hagen, *Monat. Ber. Akad.wiss. Berlin*: 480 (1853); *Linn. Ent.* XII: 107-108 (1858); *Peter's Reise Mossambique, Insekten:* 86 (1862). Wasmann, *Abh. senekenb. naturf. Ges.,* XXI (1): 150 (1897). Haviland, J. *linn. Soc. (Zool),* XXVI: 377 (1898). Sjöstedt, *Mon. Term. Afr., 1900*: 67 (1900).

Termitina (partim) Froggatt, *Proc. Linn. Soc. N. S. W.,* XXI (4): 518 (1896). Desneux, *Ann. Soc. ent. Belg.,*

XLVIII: 285 (1904); *Genera Insect.*, XXV: 25-32 (1904). Banks and Snyder, *Bull. U.S. nat. Mus.*, 108: 54-75 (1920).

Metatermitidae Holmgren, *in Escherich Termitenleben auf Ceylan*, Gustav Fisher, Publisher, Jena: 207-209, 210-212 (1911); *K. svenska VetenskaAkad. Handl.*, XLVIII (4): 1-166 (1912); *Mem. Dep. Agric. India Ent.*,V (3): 145-171 (1917).

Termitidae Light, *Philipp. J. Sci.* XVIII: 245 (1921). Emerson, *Zoologica N.Y.*, VI (4): 357 (1925). Light, *Termites and Termite Controly by Kofoid, et al.*, Berkeley, California: 112 (1934). Hill, *Termites from the Australion Region*, Melbourne: 6 (1942).

*Key to genera and subgenera**

1. *Imago:* Big to medium big species with large clear clypeus ; left mandible with 2 mostly straight big poined teetch, a toothless than broad. *Soldier:* Head towards infront more or less narroed; clypeus projected,4-edged; pronotum weakly saddle shapped, usually considerably long and broad; antennae 15-17 segmented; left mandible at least with toothed base, or with a toothless pace up to the forwardly adjusted marginal toth 3

   Imago: Medium to small species; clypeus of varying length; left mandible with 2 pointed teeth, and an elongated short-indented toothelss space behind; labrumbroader thanlong; fore tibiae with 2 spines. *Soldier:* Head with frontaltube (Nose); mandibles rudimentary, (i.e. nasute soldiers) 6

   Imago: Head of very varying form; clypeus always considerably long:mandibles varied, Ist tooth often bigger than the end; for tibiaewith three spines. *Soldier:* Without fontanel tube; head more or less circular or ectangular and also may be 2

---

* The Key has been drawn after consulting the following papers:–

(2) Holmgren, in *Escherich Termitenleben auf Ceylan*, Gustav Fischer, Publisher, Jena: 206-212 (1911); *K. svenska VetenskaAkad.Handl.*, XLVIII (4) : 59-62 (1912).

(3) Silvestri, *Rec. Indian, Mus.*, XXV (2): 221 (1923).

(4) Snyder, *Indian For. Rec. Ent.*, XX: 28 (1934).

(5) Hill. *Termites from the Australian Region*, Melbourne: 204, 378, 419, 422 (1942).

narrowed anteriorly; antnnae not more than 14 - segment (when more sgmented both mandibles at the bas strongly bent, both with a tooth in the middle,or wound like ascre or rod like with a strong basal tooth on the left); pronotum short, saddle shaped: tibiae with 3 spines.

2. *Imago:* Head broadly oval; Ist tooth of mandible somewhat bigger than the 2nd; fontanel in the middle of the head of often prolonged anterioly. *Soldier:* Antennae of 12-13 segments; mandibles rod like with a strong basal tooth on the left or slightly to strongly asymctrical; labrum with the point spilt. 10

Imago: Head, if broadly oval, the fontanel mostly situated to wards the hinder side; if head circular, the fontanel situated in the centre of the head or a little to the rear; ist tooth of mandible somewhat bigger than the 2nd. *Soldier:* Mandibles at the base strongly bent like an arch, either short and strong or long and relatively weak, in the middle with a powerful tooth ; labrum short 11

Imago: Head long stretched oval; facetted eyes considerably small; alypeus as long as half the breath. *Soldier:* Antennae of 12-13 segments; both mandibles of equalsize, inset from the sides of the head, and at the base from the out side towards within strongly concavely bent, inner edgre often finely serrated. *Microcerotermes.*

3. *Soldier:* Small species; usually one soldier form, and this always smaller than worker; antennae with 16 semgnets at the most; mandibular teeth relatively small; labrum usually long; fore legs sub-fossoial. *Microtermes.*

Soldier: Large species with two soldier forms. Antennae 16-17 segmented 4

4. *Soldier:* Both formslarger than the workers; labrum with hyaline tip; 3rd semgent of antennae longer than 2nd; mandibles without prominent marginal teeth *macrotermes.*

Soldier: Both forms larger than the workers; labrum with hyaline tip; 3rd segment of antennae than 2nd; mandibles without prominent marginal teeth 5

Soldier: Labrum without hyaline tip; 3rd segment of antennae shorter thanor as long as the 2nd.	
5. *Soldier:* Left mandible usually with marginal tooth in the middle or in the outer third	*Termes (Ter L.)* hagen s.str.
Soldier: Left mandible with 4 notches at the base	*Termes (Hygmes)*
Soldier: Left mandible with an obtuse tooth at the basal part, infront of which a large excavation, whose lower angle acute and bounded by obtuse tooth above, shape of its apical part somewhat weakly arched	*Termes (Euscaioter)*
6. *Soldier* caste absent	*Anoplotere.*
Sodier caste absent	7
7. *Soldier:* head usually constricted behind antennae	8
Soldier: Head not constricted behind antennae	9
8. *Soldier:* Nasus cylindrical, long; mandibles not pointed	*Eutermes (subuliterm*
Soldier: Nasus elongate; mandibles pointed	*Eutermes (Trinerviter*
Soldier: Nasus short and thick ; mandibles pointed at apex one soldier form	*Eutermes (Grallatoter)*
Soldier: Nasus elongate, narrow; mandibles pointed at apex; 3rd segment of antennae longer than the 4th ; one soldier form.	*Eutermes (Hospitalis)*
9. *Soldier:* Nasus conical; mandibles pointed at apex; one soldier form	*Eutermes (Eutermes)* Holmgren
Soldier: Head circular; nasus cylindrical, short: mandibles with a point near base	*Eutermes (Rotunditter)*
Soldier: Head considerably elongated; nasus cylinderical, short	*Eutermes (Ceyloniterr)*
10. *Imago:* Fontanel in the middle of the head and pressed in; post-clypeus shorter than half its breadth; antennae 14-17 segmented; memo and meta-notum either pointed at the back or wide posteriorly, the former slightly grooved and the latter broadly crenated; wing membrane dark, not clothed with micraters.	*Capritermes.*

Soldier: Head not prolonged into a point; mandibles strongy asymetrically twisted; labrum long, either with extended lateral corners and rounded or inscised middle portion, or 3 short lobed; antennae wit 14-15 segments; fontanel when present free on anterior portion of head

Imago: Fontanel mostly cleft, often prolonged anteriorly; post-clypeus about as long as half its breadth; antennae 14-16 segmented; posterior margin of meso and meta-notum usually deeply incised medially; wing membrance dark, densely clothed with micrasters. *Soldier:* Head prolonged into a point infront; mandibles slightly a symetrical, lengthened, rod-like and with a strongly developed basal tooth on the left mandible; antennae 14-15 segmented; labrum either forked or with a lobed, prominent anterior margin, more or less grooved dorsally; fontanel larg, surrounded by hairs, placed vertically in the sloping part of the head below a more or less well developed prominence — *Mirotermes.*

11. *Soldier:* Head rectangular; fontanel large; antennae with 13 segments mandibles with 3 cornered tooth posterior to the middle — *Hamitermes (Synhamitermes).*

Soldier: Head longer than broad, fontanel present; antennae 12-13 segmented; mandibles fairly with a tooth in the middle — *Eremotermes.*

Soldier: Fontanel present or absent; antennae with 14 segments; mandibles short and usually strongly bowed. — 12

12. *Soldier:* Head thick and rectangular, fontanel absent; right mandible wiht a small rudimentary tooth — *Eurytermes.*

Soldier: Head rectangular; fontanel present; mandibles short and broad, wiht a broad tooth in the middle — *Hamitermes. (Euhamitermes).*

Soldier: Head circular; mandibles markedly incurved at apex, with marginal tooth near middle. — *Hamitermes (Globitermes).*

Soldier: Head narrowed anteriorly, much longer than broad, fontanel large; mandibles with a tooth in the middle — *Hamitermes. (Hamitermes* Silv.) .str.

Genus i. **Eremotermes** Silvestri.

Eremotermes Silvestri, *Bull. Lab. Zool. Gen. Agric. Portici,* VI: 108-109 (1912). HOlmgren, *Mem. Dep. Agric. India Ent., V* (3): 166 (1917).

seudomirotermes Holmgren, *K.Sevenska VEtneskaAkad. Handl.,* L(2): 232, pl. VIII, fig. 22 (1913).

Fletcheri Holmgren.

SYSTEMATICS.

Eremotermes fletcheri Holmgren, *Mem. Dep. Agric. India Ent.,* V (3): 167-168 (1917). Margabanu, *J. Bombay nat. Hist. Soc.,* XXXVII (3); 714 (1934). Coimbatore (S. India)

paradoxalis Holmgren.

SYSTEMATICS.

Eremotermes paradoxalis Holmgren, *K. Svenska Vetensk Akad. Handl.,* L(2): 232,Pl. VIII, Fig. 22 (1913); *J. Bombay nat. Hist. Soc.,* XXII (1); 115-116 (1913)[1]; *Mem. Dep. Agric. India Ent.,* V, (3): 166-167 (1917)[2]. [1](Bangalore. [2]Hoshangabad (M.P.) and Pusa (Bihar).

BIOLOGICAL.

Holmgren, *J. Bombay nat. Hist. Soc.,* XXII (1): 115-116 (1913); *Mem. Dep. Agric. India Ent.,* V (3): 166-167 (1917). Magrabandu, *J. Bombay nat. Hist. Soc.* XXXVII (3): 714 (1934).

(Host: Grass).

Genus ii. **Eurytermes Wasmann.**

Eurytermes Wasmann, *Zool. Jb., Abt I,* XVII (1): 124 (1902). Green, *Spolia, Zeylan.,* IX: 12 (1913), John, *Treubia,* VI (3-4): 407 (1925).

Termes (Eurytermes) Desneux, *Genera Insect.,* XXV: 12 (1904).

1. assmuthi Wamann.

SYSTEMATICS.

Eurytermes assmulthi Wasmann, *Zool. Jb., Abt I,* XVII (1): 124, p. 4, fig. 5 (1902). Desneux, *Genera Insect.,* XXV: [1, 3, 4]Ceylon. [2]Indian.

40 (1904). Holmgren, *in Escherich Termitenleben auf Ceylan,* Gustave Fischer, Publisher, Jena: 205 (1911)[1]. Silvestri, *Rec. Indian Mus.,* XXV: 228 (1923). — [5, 6, 7]Barkuda Islands (Orissa).

BIOLOGICAL.

*Lefroy, *Indian Insect Life,* Calcutta, I: 120 (1909)[2]. Green *Spolia zeylan.,* IX (33): 12 (1913)[3]. Bugnion, *Bull. Mus. Hist. nat. Paris,* XX (4): 189 (1914)[4]. Annandale, *Rec. Indian Mus.,* XXV (2): 233, 245-250 (1923)[5]. Bose, *Ibid.,* XXV: 153 (1923)[6]. Silvestri *Ibid.,* XXV: 228 (1923)[7]. Magabandu, *J. Bombay nat. Hist. Soc.,* XXXVII (3): 710-711 (1934).

[Hosts: Roots of *Hemigyrosa canescens* and *Ficus carica* (fig.) trees].

var. modestior Silvestri.

Eurytermes assmuthi var. *modestior* Silvestri, *Rec. Indian Mus.,* XXV (2): 228-229 (1923). Annandale, *Ibid.,* (2): 233-251 (1923). — Barkuda Islands (Orissa)

Margabandu, *J. Bombay nat. Hist. Soc.,* XXXVII (3): 711 (1934).

2. **cayionicus** Holmgren.

SYTEMATICS.

Eurytermes ceylonicus Holmgren, *K. svenska VetenskaAkad. Handl.,* XLVIII (4): 83 (1912); *Ibid.,* L (2): 222 (1913). Green, *Spolia zeylan.,* IX (33): 12 (1913). Bugnion, *Bull.Mus. Hist. nat. Paris,* XX (4): 184 (1914). John, *Treubia,* VI (3-4): 407 (1925). Magabandu, *J. Bombay nat. Hist. Soc.*XXXVII (3): 711 (1934). — Ceylon.

Genus iii. **Hamitermes** Silvestri.

Amitermes Silvestri, *Bull. Mus., Zool. Anat. comp. Toriono,* XVI (389): 4, (1901). Wasmann, *Zool. Jb., Abt. I,* XVII (1): 122 (1902). Desneux, *Genera Insect.,* XXV: 28 (1904). Banks and Snyder, *Bull. U.S. nat. Mus.,* 108: 5 (1920). Sjöstedt *K. Svenska VetenskaAkad. Handl., (3),* III (1): 181 (1925). Emerson, Zoologica,

* *Eutermes assmuthi* Wasm. is not known in literature as such; probably Leafray means *Eurythermes assemuthi* Wasmann.

N.Y., VI (3-4): 315, 430 (1925). Light, *Univ. Calif. Publ. Zool.*, V (9): 173 (1930); *Ibid.*, V (10); 215 (1930); *Ibid.*, V (17): 355 (1932); *Ibid.*, VI (6): 139 (1933).

Hamitermes Silvestri, *Redia*, I: 43, 217 (1903) (*Amitermes* replaced by *Hamitermes*): *Denkschr. med.—nat. Ges. Jena*, XIII: 77 (1908). Holmgren, *K. Svenska Vetenskad. Handl.*, XLVIII (4): 87-91 (1912); *Ibid.*, L (2): 27, 223 (1913). Mjöberg, *Ark. Zool.*, XII (15): 66 (1920). Fuller, *S. Afr. J. nat. Hist.*, III (1): 19 (1921); *Ibid.*, III (2): 122 (1922). Silvestri, *Zoologica N. Y.*, III (16): 309 (1923). John, *Treubia*, VI (3-4): 407 (1925). Hill, *Termites from the Australian Region*, Melbourne: 307-208 (1942).

Monodonterms Silvestri, *Fauna Siidwest Aust.*, II (17): 299 (1909).

Subgenus 1. **Eubamitermes** Holmgren.

Hamitermes (Euhamitermes) Holmgren, *K. svenska Vetenska Akad. Handl.*, XLVIII (4): 89 (1912); *Mem. Dep. Agric. India Ent.*, V (3): 165 (1917). Snyder, *Indian For. Rec. Ent.*, XX 28 (1934).

1. indicus Holmgren.

SYSTEMATICS.

Euhamitermes indicus Holmgren, *Mem. Dep. Agric. India, Ent.*, V (3): 165-166 (1917). Margabandu, *J. Bombay nat. Hist. Soc.*, XXXVII (3): 710 (1934). — Shevaroy Hills 4000 ft. (S. India).

(Host: Dead leaves in soil).

Subgenus 2. **Hamitermes Holmgren.**

Hamitermes (Hamitermes) Holmgren, *K. sevenska Vetensk Akad. Handl.*, XLVIII (4): 91 (1912); *Ibid.*, L (2): 225 (1913). Snyder, *Indian For. Rec. Ent.*, XX: 28 (1934).

1. belli (Desneux).

SYSTEMATICS.

Termes belli Desneux, *Ann. Soc. Ent. Belg.*, XLIX: 352-354 (1905). Lefroy, *Indian Insect Life*, Calcutta, I: 120 (1909). Margabandu, *J. Bombay nat. Hit. Soc.* XXXVII (3): 704 (1934). — Karachi (Sind).

Amitermes belli Beeson, *Forest Insects,* Dehra Dun: 545 (1941).

(Host: Roots of *Cistanche tubulosa*).

2. lighti Snyder.

SYSTEMATICS.

Amitermes lighti Snyder, *Proc. Wash.,* XLVI: 93 (1933). U.P. (India).

Subgenus 3. **Globitermes** Holmgren.

Hamitemes (Globitermes), Holmgren, *K. svenska VetenskAkad. Handl.,* XLVIII (4): 87 (1912); *Ibid.,* L (2): 228 (1913), Snyder, *Indian For. Rec. Ent.,* XX: 28 (1934).

1. audax Silvestri.

SYSTEMATICS.

Globitermes audar. Silvestri, *Rec. Indian Mus.,* VIII (5): 432-434, text fig. 8 (1914). Moulmein (Burma).

2. birmanicus Snyder.

SYSTEMATICS.

Globitermes birmanicus Synder, *Indian For. Rec. Ent.,* XX (11): 17-18, fig. 9 (1934). Beeson, *Forest Insects,* Dehra Dun: 547 (1941). Zibyaung W. Salween (Burma).

(Hosts: Stumps, logs and sap wood of *Tectona grandis*).

Subgenus 4. **Snynhamitermes** Holmgern

Hamitermes (Synhamitermes) Holmgren, *J. Bombay nat. Hist. Soc.,* XXI (3): 792 (1912); *K. svenska VetenskaAkad. Handl.,* XLVIII. (4): 91 (1912); *Ibid.,* L (2): 225 (1913). Snyder, *Indian For. Rec. Ent.,* XX: 28 (134).

1. ceylonicus Holmgren.

SYSTEMATICS.

Hamitermes (Synhamitermes) ceylonicus, Holmgren, *K. svenska VetenskaAkad. Handl.,* XLVIII (4): 91 (1912); *Ibid.,* L (2): 225, pl. VIII. fig. 21 (1913). Green, Ceylon.

Spolia zeylan. IX (33): 13 (1913). Bugnion, *Bull. Mus. Hist. nat. Paris,* 7 XX (4): 184 (1914). Hegh, *Les Termites,* Bruxelles: 48 (1922). John, *Treubia,* VI (3-4): 408 (1925). Mar. Gabandu, *J. Bomay nat. Hist. Soc.,* XXXVII (3): 710 (1934).

2. quadriceps (Wasmann).

SYSTEMATICS.

Amitermes quadriceps Wasmann, *Zool. Jb., Abt. I,* XVII (1): 123, pl. 4, fig. 5 (1902). — [1, 2]India.

Eutermes quadriceps Desneux, *Genera Insect.,* XXV: 44 (1904)[1]. Lefroy, *Indian Insect Life,* Calcutta, I: 120 (1909)[2]. — [3, 5]Ceylon. [4]Khandala (Bombay)

Hamitemes quadriceps Holmgren. *in Eschierich Termitenleben auf Ceylan,* Gustav Fischer, Publisher, Jena: 206-212, pl. 2, fig. I, (1911)[3], *J. Bombay nat. Hist. Soc.,* XXI (3): 792 (1912)[4].

Hamitermes (Synhamitermes) quadriceps Green, *Spolia zeylan,* IX (33): 13 (1913)[5].

BIOLOGICAL.

Assmuth, *J. Bomabay nat. Hist. Soc.,* XXII (2): 374-375 (1914). Bugnion, *Bull. Mus. Hist., nat. Paris,* XX (4): 184 (1914). Margabandu, *J. Bombay nat. Hist. Soc.* XXXVII (3): 710 (1934).

(Host: Dried blades of grass).

Genus iv. **Microcerotermes** Silvestri.

Microcerotermes Silvestri, *Bull. Zool. Anat. comp. Torin.,* VI (386): 3 (1901). Wasmann, *Zool, Jb. Abt I,* XVII (1): 119 (1902). Silvestri, Redia, I: 41 (1903) Desneux, *Genera Insect.* XXV: 29 (1904). Holmgren, *K. svenska, VetenskaAkad. Handl.,* XLVIII (4): 121 (1912)*. *Ibid.,* L (2): 29, 256 (1913). Sjöstedt, *Ibid.,* (3), III (1): 193 (1925). Emerson, *Zoologica N.Y.,* VI: 453 (1925). John, *Treubia,* VI (3-4): 418 (1925). Emerson, *Bull. Amer. Mus. nat. Hist.,* LVII: 543

* *Microtermes* is wrongly written by Holmgren here. He infact intended to *Microcerotermes.*

(1928). Kemner, *Lunds, Univ. Arbsberätt.*, (2), XXVII (13), (1931); *K. svenska VetenskAkad. Handl.*, (3) XIII (4): 158 (1934). *Hill, Termites from the Australian Region*, Melbourne: 421-422 (1942).

1. annandaei Silvestri.

SYSTEMATICS.

Microcerotermes annandalei Silvestri, *Rec. Indian Mus.*, XXV (2): 225-228, figs. 3, 4 (1923)[1]. Annandale, *Ibid.*, XXV (2): 233 (1923)[2]. Magabandu, *J. Bombay nat. Hist.Soc.*, XXXVII (3): 709 (1934). Beeson, *Forest Insects*, Dehra Dun: 549, fig. 151, no. 6 (1941). — [1, 2]Barkuda Islands (Orissa)

(Hosts: Damp wood of logs).

2. Beesoni Synder.

SYSTEMATICS.

Microcerotermes baeesoni Snyder, *Proc. U.S. nat. Mus.*,LXXXII. Art. 16 (2957): 12-14, fig. (1933)[1]; *Indian For. Rec. Ent.*, XX (11): 20-21 (1934)[2]. Margabandu, *J. Bombay nat. Hist. Soc.*, XXXVIII (1): 208-209 (1935). — [1, 2]Haldwani (U.P.).

3. bugnioni Holmgren.

SYSTEMATICS.

Microcerotermes bugnioni Holmgren, *in Escherich Termitenleben auf Ceylan*, Gustav Fisher, Publisher, Jena: 203-204, pl. 2, fig. 5. (1911); *Spolia zeylan*, VIII (32): 283-284 (1913). Hegh, *Lec Termites*, Bruxelles: 48 (1922), Kemner, *Bull. ent. Res.* XVI (4): 390-391 (1926). — Cylon.

BIOLOGICAL.

Green, *Spolia zeylan*, IX (33): 11 (1913). Bugnion, *Bull. Mus. Hist. nat. Paris*, XX (4): 185 (1914). Margabandu, *J. Bombay nat. Hist. Soc.* XXXVII (3): 709 (1934). Beeson, *Forest Insects*, Dehra Dun: 549-550 (1941).

(Host: Wood).

4. burmanicus Ahmad.

SYSTEMATICS.

Microcerotermes burmanicusm Ahmad, *Amer. Mus. Novit.*, (1342): 6 (1947). Ragnoon (Burma).

5. cameroni Snyder.

SYSTEMATICS.

Microcerotermes cameroni Snyder, *Indian For. Rec. Ent.*, XX (11): 21-32 (1934). North Vellore (Madras).

6. championi Snyder.

SYSTEMATICS.

Microcerotermes championi Snyder, *Proc. U. S. nat. Mus.*, LXXXII. Art. 16 (2957); 14-15., figs. (1933). Margabandu, *J. Bombay nat. Hist. Soc.*, XXXVIII (1): 208-209 (1935). Kumaon (U.P.)

7. crassus Snyder.

SYSTEMATICS.

Microcerotermes crassus Snyder, *Indian For. Rec. Ent.*, XX (11): 22-23 (1934). Meitkyina (Burma).

8. cylindriceps Wasman.

SYSTEMATICS.

Microcerotermes cylindriceps Wasmann, *Zool. Jb., Abt. I*, XVII (1): 121 (1902). Holmgren, *in Escherich Termitenleben auf. Ceylan.*, Gustav Fischer, Publsher, Jena: 204 (1911). Green, *Spolia zeylan.*, IX (33): 12 (1913). Hegh, *Les Termites*, Bruxellles. 48 (1922). Ceylon.

Eutermes cylindricpes Desneux, *Genera Insect.*, XXV: 41 (1904).

BIOLOGICAL.

Bugnion, *Bul. Mus. Hist. nat. Paris*, XX (4): 185 (1914). Margabandu, *J. Bombay nat. Hist. Soc.*, XXXVII (3): 709 (1934).

(Host: Wood in contact with soil).

9. fletcheri Holmgren.

SYSTEMATICS.

Microcerotermes fletcheri Holmgren, *Mem. Dep. Agric. India Ent.*, V (3); 170-171 (1917)[1]. Margabandu, *J. Bombay nat. Hist. J. Bombay nat. Hist. Soc.* XXXVII (3): 710 (1934).

[1]Bababudin Hills 4, ft. (Mysore).

(Host: Log of sandal wood).

10. greeni Holmgren.

SYSTEMATICS.

Microcerotermes greeni Holmgren, *Spolia zeylan.*, VIII (32): 284 (1913). Green, *Ibid.*, IX (33): 12 (1913). Hegh, *Les Termites*, Bruxelles: 48 (1922). Kemner, *Bull. ent. Res.*, XVI (4): 391 (1926).

Ceylon.

BIOLOGICAL.

Bugnion, *Bull. Mus. Hist. Nat. Paris*, XX(4): 185 (1914). Kemner, *Bull. ent. Res.*, XVI (4): 391 (1926). Jepson *Adm. Rep. Dir. Agric. Ceylon, 1933*: 118 (1934). Magabandu, *J. Bombay nat. Hist. Soc.*, XXXVII (3): 710 (1934).

Host: Dead and sometimes living *Camellia sinensis*, the tea plant).

11. heimi Wasmann..

SYSTEMATICS.

Microcerotermes heimi Wasmann, *Zool. Jb., Abt. I*, XVII (1): 120, pl. 4, fig. 3 (1902). Holmgren, J. *Bombay nat. Hist. Soc.*, XXI (3): 789-792 (1912)[1]. Assmuth, *Ibid.*, XXII (2): 374, 382 (1914)[2]. John, *Treubia*, VI (3-4): 418 (1925)[3]. Beeson, *Forest Insects*, Dehra Dun: 550 (1941).

[1, 2]Salsette Island (Bombay).
[2]Poona, Vadtal, Valan (Bombay).
[3]Ceylon.

Eutermes heimi Desneux, *Genera Insect.*, XXV: 42 (1904).

[4]Bababudin Hills 4200 ft. (Mysore).

BIOLOGICAL.

Holmgren, *J. Bombay nat. Hist. Soc.*, XI (3): 789-792 (1912). Assmuth, *Ibid.*, XXII (2): 374, 382 (1914). Holmgren, *Mem. Dep. Agric. Indian Ent.*, V (3): 171 (1917)[4]. John, *Treubia*, VI (3-4): 418 (1925). Margabandu, *J. Bombay nat. Hist. Soc.*, XXXVII (3): 710 (1934). Beeson, *Forest Insects*, Dehra Dun: 550 (1941).

(Hosts: Bamboo, log of wood and tree stumps)

12. tenuignathus Holmgren.

SYSTEMATICS.

Microcerotermes tenuignathus Holmgren, *J. Bombay nat. Hist. Soc.*, XXII (1): 116-117 (1913). Margabandu, *Ibid.*, XXXVII (3): 710 (1934). Vadtal (Bombay).

(Host: Dry wood).

Genus v. **Capritermes** Wasmann.

Capritermes Wasmann, *Abh. senckenb. naturf. Ges.*, XXI (1); 151 (1897). Silvestri,

Boll. Mus., Zool. Ant. Comp. Torino, XVI (389)[1]: 4 (1901), Wasmann, *Zool. Jb., Abt. I*, XVII (1): 125 (1902). Silvestri, *Redia*, 1: 62 (1903). Desneux, *Genera Insect.*, XXV: 29 (1904). Holmgren, *Zool. Jb., Abt.* I, XXIII: 37 (1906). Wasman, *Rev. Zool. Afr.*, I:150 (1911). Holmgren, *K. Svenska VetenskaAkad. Handl.*, XLVIII (4): 114, 116 (1912); *Ibid.*, L (2): 242 (1913); *Zool. Jb., Abt. I*, XXXVIL: 278 (1914). Sjöstedt, *K. Svenska VetenskaAkad. Handl.*, (3), III (1): 282 (1925). John, *Treubia*, VI (3-4): 414 (1925). Emerson, *Zoologica N. Y.*, VI (4): 445 (1925); *Bull. Amer. Mus. nat. Hist.*, LVII: 542 (1928). Kemner, *K. svensk VetenskAkad. Handl.*, XIII (4): 171 (1934). Hill, *Termites from the Australian Region*, Melbourne: 418-420 (1942).

1. ceylonicus Holmgren

SYSTEMATICS.

Copritermes ceylonicus Holmgren, *in Escherich Termitenleben auf Ceylan*, Gustav Fischer, Ceylon

Publisher, Jena: 204-205, pl. II, fig. Q (1911). Green *Spolia zeylan.*, IX Bruxelles: 48 (1922).

BIOLOGICAL.

Bugnion, *Bull. Mus. Hist. nat. Paris,* XX (4): 184 (1914). Margabandu, *J. Bombay nat. Hist. Soc.*, XXXVII (3): 711 (1934).

(Host: Wood in contact with soil).

2. distinctus Holmgren.

SYSTEMATICS.

Capritermes distinctus Holmgren, *K. svenska VetenskAkad. Handl,* L (2): 246, pl. VI, figs. 24, 30 (1913). Hegh, *Les Termites,* Bruxelles: 48 (1922). — Ceylon.

BIOLOGICAL.

Bugnion, *Bull. Mus. Hist. nat. Paris,* XX (4): 184 (1914). Margabandu, *J. Bombay nat. Hist. Soc.*, XXXVII (3): 711 (1934).

3. distortus Silvestri.

SYSTEMATICS.

Capritermes distortus Silvestri, *Rec. Indian Mus.*, XXIV: 541-543, fig. V, no. 1-15 (1922). Margabandu, *J. Bombay nat. Hist. Soc.* XXXVII (3): 711 (1934). — Kvalai 1,300-3,000 Cochin (S. India)

4. fletcheri Holmgren.

SYSTEMATICS.

Capritermes flectheri Holmgren, *Mem. Dep. Agric. India Ent.*, V (3): 169 (1917)[1]; Margabandu, *j. Bombay nat. Hist. Soc.*, XXXVII (3): 711 (1934). — [1]Puthupadi forest (Malabar), Anamalai Hills 4,000 ft. Madras, Bababuin Hills 4,400 ft. (Mysore).

(Hosts: Dead bamboo, log of wood, and roots of *Coffea robusta,* coffee).

5. garthwaitei Gardner.

SYSTEMATICS.

Capitermes garthwaitei Gardner, *Indian J. Ent.*, IV (1-2): 106-108, figs. 2-4 (1944). — Maymyo (Burma).

6. **gravelyi** Silverstri.

SYSTEMATICS.

Capritermes gravelyi Silverstri, *Rec. Indian Mus.*, XXIV (4): 535-537, fig. I no. 1-9 (1922). — Kas, 3,700 ft., Helvak & Vela 2,000-2,100 ft. (Satara).

7. **hustsoni** Kemner.

SYSTEMATICS.

Capritermes hutsoni Kemner, *Bull. ent.Res.*, XVI (4): 391-392, fig. 5 (1926)[1]. Margabandu *J. Bombay nat. Hist. Soc.*, XXXVII (3): 711 (1934). — Ceylon.

8. **incola** (Wasman).

SYSTEMATICS.

Eutermes incola Wasmann, *Wien ent. Ztg.*, XII (7): 242 (1893)[1]. Desneux, *Genera Insect.*, XXV: 43 (1904). — [1]Ceylon. [2]Khandala (Bombay)

Capritermes incola Wasmann, *Zool. Jb., Abt. I*, XVII (1): 128 (1902). Holmgren, *I, XVII* (1): 128 (1902). Holmgren, *in Escherich Termitenleben auf*

Ceylan, Gustav Fischer, Publisher, Jena: 204, pl. 2, fig. R (1911); *J. Bombay nat. Hist. Soc.*, XXI (3): 792—793 (1912)[2].

Green, *Spolia zeylan.*, IX (33): 13 (1913). Bugnion, *Bull. Mus. Hist. nat. Paris*, XX (4): 184-202 (1914). Hegh, *Les Termites*, Bruxelles: 48 (1922). John, *Treubia*, VI (3-4): 418 (1925).

Capritermes longicornis Wasmann, *Zool. Jb., Abt.* XVII (1): 126, pl. 4 fig. 7 (1902). Margabandu, *J. Bombay nat. Hist. Soc.*, XXXVII (3): 711 (1934).

Eutermes longicornis desneux, *Genera Insect.*, XXV: 43 (1904). Lefroy, *Indian Insect Life*, Calcutta, I: 120 (1909).*

---

* Lefroy is probably referrring to *Eutermes longicornis* Wasmann to which Desneux (1904) has also referred and which is *Capitermes longicornis;* the actual Eutermes longicornis unewly described by Holmgren in 1912.

BIOLOGICAL.

Assmuth, *J. Bombay nat. Hist. Soc.*, XXII (2): 374—375 (1913)[3]. Bugnion, *Bull. Mus. His. nat. Paris*, XX (4): 184—202 (1914). Holmgren, *Mem. Dep. Agric. India Ent.*, V (3): 170 (1917)[4]. John, *Treubia*, VI (3): 418 (1925). Margabandu, *J. Bombay nat. Hist. Soc.*, XXXVII (3): 711 (1934). — [3]Bombay. [4]Ketti 6,500 ft. (Nilgiri), Mercara 4,00 ft. (Coorg), Baba budin Hills 4,600 fit Mysore.

(Hosts: Roots of grass).

9. laetus Silverstri.

SYSTEMATICS.

Capritermes laetus Silvestri *Rec. Indian Mus.*, VIII (5): 434-435, figs. 9, 10 (1914). — Sukli E. Dawna Hills (Burma).

10. longirotris (Brauer).

SYSTEMATICS.

Termes longirostris Brauer, *Reise Novara, Neurpt., II:* 47. (1866). Wasmann, *Dtsch., ent. Z: 145 (1899).* Desneux, *Genera Insect.*, XXV: 37 (1904)[1]. — [1]Nicobar Islands.

Capritermes longirostris Silvestri, *Rec. Indian Mus.*, XXIV: 537 (1922).

var. **cornutela** Silvestri.

Capritermes longirostris var. *cornutella* Silvestri, *Rec. Indian Mus.*, XXIV: 537-538, fig. II. No. 1—8 (1922). Margabandu, *J. Bombay nat. hist. Soc.*, XXXVII (3): 711 (1934). — Ceylon.

11. obtusus Silverstri.

SYSTEMATICS.

Capritermes obtusus Silvestri, *Rec. Indian Mus.*, XXV: 299—231, fig. VI, no. 1-7 (1923)[1]. Annandale, *Ibid.*, XXV: 233, 245-250 (1923)[2]. Magabandu, *J. Bombay, nat. Hist. Soc.*, XXXVII (3): 711 (1934). — [1,2]Barkuda Islands (Orissa).

var. **abbrevitus** Silvestri.

Capritermes obtusus var. *abbreviatus* Silvestri, *Rec. Indian Mus.* XXV: 231-232, fig. VIII, no. 1—6 (1923)[1]. — [1,2]Barkuda Islands (Orissa).

Annandale, *Ibid.*, XXV; 233 (1923)[2]. Margabandu, *J. Bombay nat. Hist. Soc.*, XXXVII (3): 711 (1934).

12. punjabensis (Holmgren).

SYSTEMATICS.

Capritermes punjabensis Holmgren, Mem. *Dep. Agric., India Ent.*, V (3): 170 (1917). Margabandu, *J. Bombay nat. Hist. Soc.* XXXVII (3): 711 (1934). — Lyallpur (Punjab).

(Very close to *Capritermes nemorosus* but is distinguished by its smaller facetted eyes and 15 jointed antennae, vide Holmgren, 1917).

13. speciosus (Haviland).

SYSTEMATICS.

Termes speciosus Haviland, *J. linn. Soc. (Zool).* XXVI: 413 (1898)[1]. *Desneux, Genera Insect.*, XXV: 45, pl. 2, fig. 10 (1904). — [1, 3]Sarawak (Borneo).

Capritermes speciosus Green, *Spolia zeylan.*, IX (33): 13 (1913)[2]. John, *Treubia*, VI (3-4): 414 (1925)[3]. Margabandu, *J. Bombay nat. Hist. Soc.*, XXXVII (3): 711 (1934). — [2]Ceylon (Doughtful). [3]Sumatra.

14. tetraphilous Silverstri.

SYSTEMATICS.

Capritermes tetraphilous Silvestri, *Re. Indian Mus.*, XXIV: 543-544 (1922). Margabandu, *J. Bombay nat. Hist. Soc.*, XXXVII (3): 711 (1934). — Rangamati : Chitta gong-(Assam,) Bengal.

Genus vi. **Mirotermes** (Wasmann).

Termes (Mirotermes) Wasmann, *Abh. senckenb. natuf. Ges.*, XXI (1): 151 (1897): *Zool. Jb., Abt I,* XVII (1): 125 (1902).

Mirotermes Silvestri, *Redia,* I: 66 (1903). Desneux, *Genera Insect.*, XXV: 29 (1904). Wasman, *Rev. Zool. After,* I: 150 (1911). Holmgren, *K. svenska VetenskaAkad. Handl.*, XLVIII (4): 8, 10, 105 (1912); *Ibid.*, L (2): 234 (1913); *Zool. Jb., Abt. I,* XXXVI (1914).

Sjöostedt, *K. Svenska VetenskaAkad. Handl.*, (3), III (1): 275 (1925). Emerspon, *Zoologica N. Y.*, VI (4): 432 (1925). John, *Treubia*, VI (3-4): 409 (1925). Fuller, *S. Afr. J. Nat. Hist.*, V: 177 (1925). Emerson, *Bull. Amer. Mus.Nat. Hist.*, LVII: 502, 542 (1928). Hill, *Termitesfro the Australian Region*, Melbourne: 377-378 (1942).

Termes Kemner, *Ent. Tidskr.*, I: 19 (1934).*

Subgenus 1. **Mirotermes** (Wasmann) s. str.

Mirotermes (Mirotermes) Holmgren, *K. svenska VetenskaAkad. Handl.*, XLVIII (4): 107-108 (1912). Hill, *Termites from the Australian Region*, Melbourne: 377 (1942). (Head with well developed frontal process concealing fontanel from above, mandibles symetrical).

1. ceylonicus Holmgren.

SYSTEMATICS.

Mirotermes ceylonicus Holmgren, *Zool. Jb., Abt. I, XXXVI: 276* (1914).	Ceylon.

2. dehraensis Gardner.

SYSTEMATICS.

Mirotermes dehraensis Gardner, *Indian J. Ent.*, IV (1-2): 108-110, figs. 6-7 (1944).	Dehra Dun (U.P.).

3. fletcheri Holmgren.

SYSTEMATICS.

Mirotermes fletcheri Holmgren, *Mem. Dep. Agric. India Ent.*, V (3): 168-169 (1917). Margabandu, *J. Bombay nat. Hist. Soc.* XXXVII, (3): 710 (1934)	Handagalli and Humpi ruins 1,200 ft., Bellary (Madras).

---

* Note vide Hill, 1942:–

(With reference to Kemner' s papers following the discovery of Rolander's manuscript and the drawing of *Termes fatalis* Linn., Hill is not satisfied that the evidence justifies the change of name from *Mirotremes* to *Terms).*

4. **marjonae** Snyder.

SYSTEMATICS.

Mirotermes marjonae Snyder, *Indian For. Rec. Ent.*, XX (11): 18—22, fig. 10 (1934). — Selen (Burma).

5. **obtusus** Holmgren.

SYSTEMATICS.

Mirotermes obtusus Holmgren. *Mem. Dep. Agric. India Ent.*, V (3): 168 (1917). Margabandu, *J. Bombay nat. Hist. Soc.*, XXXVII (3): 710 (1934). — Hospet : Bellary (Madras).

Genus vii. **Anoplotermes** Muller.

Anoplotermes Muller, *Jen. Z. Naturw.*, VII: 347 (1873). Desneux, *Genera Insect.*, XXV: 39 (1904). Holmgren, *in Escherich Termitenleben auf Ceylan.*, Gustav Fischer, Publisher, Jena: 202-203 (1911). Silvestri, *Bull. dell Laboral di Zool. gener. et agr. dell R. sculoa sup., d' Agric.*, IX: 146 (1914). Hegh, *les Termites*, Bruxelles: 694 (1922). John, *Treubia*, VI (3-4): 406 (1925).

Speculitermes, Wasmann, *Zool. Jb., Abt.* I, XVII, (1): 161 (1902).

1. **cyclops** (Wasmann).

SYSTEMATICS.

Speculitermes cyclops Wasmann, *Zool. Jb., Abt. I,* XVII (1): (1902). Silvestri, *Rec. Indian Mus.* VIII (5): 431 (1914)[1]. — [1]Misty Hollow Dwana Hills (Lower Burma).

Eutermes cyclops Desneux, *Genera Insect.*, XXV: 41 (1904). Lefroy, *Indian Insect Life,* Calcutta, I: 120 (1909). Margabandu, *J. Bombay nat. Hist. Soc.*, XXXVII (3): 714 (1934). — [2, 5, 4]Ceylon. [5]Shevaroy Hills 3,000 ft.,Bellahunisi (Bellary), Suntikoppa Coorg), Bababudin Hills, 4,000 ft. (Mysore).

Anoplotermes (Speculitermes) cyclops Holmgren, *in Escherich Termitenleben auf Ceylan,* Gustav Fischer, Publisher, Jena: 202-203 (1911)[2]. John, *Treubia,* VI (3-4): 406 (1925)[3].

BIOLOGICAL.

Green, *Spolia zeylan,* IX (33): 13 (1913). John, *Ibid.,* (34): 103-104 (1913) . Bugnion, *Bull. Mus. Hist. Nat. Paris,* XX (4): 184 (1914)[4]. Holmgren, *Mem. Dep. Arric. india Ent.,* V (3): 164-165 (1917)[2]. Hegh, *Les Termites,* Bruxelles: 48 (1922). John, *Treubia,* VI (3-4): 406 (1925).

(Hosts: logs of wood and tree stumps).

Genus viii. **Eutermes** (Heer) Holmgren.

Termes (Eutermes) Heer, *Ins. Fauna, Oeningen U. Radoboj Croatien,* II: 32 (1849). hagen, *Linn. Ent.,* XII: 186 (1858). Wasmann, *Abh. senckenb. naturf. Ges.,* XXI (1): 151 (1897). Haviland, *J. Linn. Soc. (Zool),* XVI (169): 378 (1898). Desneux, *Genera Insect.,* XXV: 39 (1904).

Eutermes Hagen, *Monat. Ber. Akad. wiss. Berlin:* 480 (1853). Muller, *Jen. Z. Naturw.,* VII: 343 (1873). Froggatt, *Proc. Linn. Soc. N.S.W.,* XXII (4): (741) (1898). Sjöstedt, *K. Svenska VetenskAkad. Handl.,* XXXIV (4): 139 (1900). Wasmann, *Zool. Jb., Abt. I,* XVII (1): 128 (1902). Silvestri, *Redia,* I: 81 (1903). Holmgren, *Tran. Linn. Soc. Lond. Zool.,* (2), XIV: 146 (1910) ; *K. svenska VetenskAkad. Handl.,* XLVIII (4): 59 (1912); *Ibid.,* L (2): 153 (1913). Oshima, *Philipp. J. Sci.,* VIII (4): 279 (1913). Mjöberg, *Ark. Zool.,* XII (15): 33 (1920). Oshima, *Philipp. J. Sci.,* XVII (5): 550 (1920). Sjöstedt, *K. svenska VetenskAkad. Handl.,* (3), III (1): 290 (1925). John, Treubia, VI (3-4): 391 (1925). Kemner, *K. svenska VetenskAkad. Handl.,* (3), XIII (4): 114 (1934). Hill, Termites from the Australian Region, Melbourne: 202-204 (1942).

Nasutitermes Banks, *Bull. U.S. nat. Mus.,* 108: 69, 82 (1920). Snyder, *Ibid.,* 108: 189 (1920). fuller, *S. Afr. J. Nat. Hist.,* III (1): 19 (1921); *Ibid.,* (2): 107 (1922). Emerson, *Zoologica N.Y.,* VI (4): 375 (1925); *Bull. Amer. Mus. nat. Hist.* LVII: 476 (1928). Light, *Philipp. J. Sci.,* XLII (1): 16 (1930); *Univ. Calif. Publ. Zool.* VI (5): 114 (1933).

Vide Hill 1942:— Holmgren [1912, p. 59] divided the genus *Eutermes* into a number of subgenera, which have since been adopted by various writers as either subgenera or genera. This classification is considered by many authors to be unsatisfactory and has not been adopted by Hill [1942], for the following reasons: The species are more easily identified specifically than subgenerically; in some dimorphic species the large soldier could not be placed in the same subgenus as the small soldier, so it follows that if the winged adult is unkown the subgenus could not be determind; several with almost identical winged adults would be separated into different sub genera if the characters of the soldier only were considered. It may be mentioned that the type of one subgenus is now considered to be species of another genus *(Mirotermes),* and one subgenus *(Grallatotermes),* which is based on the soldier form ony, is characterized (twice) in different terms.

Characters of the genus vide Hill, 1942.

Winged adult. Head broadly oval, generally narrowed towards the front. Fontanel usually in the middle of the head. Ocelli variable in position. Eyes small to large. Post-clypeus variable in length from much less than half its width to more than half its width. Labrum widest in the middlees variable, Ist tooth usually about as long as 2nd, rarely much longer than 2nd. Antennae of 13—17 segments. Pronotum very slightly saddle-shapped, narrower than head. Meso and metanotum usually more or less widely indented posteriorly. Wing stumps almost equal size. Wings usually dark in colour, rarely hyaline, Radius usually absent, present only in *E. tumuli.* Median vein very close to the cubitus, all veins distinct, particularly the anterior most veins and the proximal branches of the cubitus. Cerci short, styli absent. Tibiae with two apical spurs.

Soldier. Nasute Mandibles rudimentary, with molar part more or less evident, and often with a point towards the apex. Antennae of 10—15 segments. Pronotum always saddle-shapped, Cerci present. Styli absent. Tibiae with two apical spurs.

Worker. Head wide and relatively flat. Fontanel in the middle of the head, more or less distinct. Postclypeus and mandibles as in winged adult. Antennae of 12—17 segments. Prototum saddle-shaped. Cerci short. Styli absent. Tibiae with two apical spurs.

1. **anamalaiensis** Snyder.

SYSTEMATICS.

Nasutitermes (Rotunditermes) anamalaiensis Snyder, Proc. U.S. *nat. Mus.*, LXXXII. Art. 16 (2957): 11-12, fig. 6 (1933). Magrabandu, *J. Bombay nat. Hist. Soc.*, XXXVIII (1): 208-209 (1935). — Anamalai Hills 2,400 ft.

2. **biformis** (Wasmann).

SYSTEMATICS.

Termes (Eutermes) biformis Wasmann, *Zool. Jb., Abt I,* XVII (1): 133, pl. 4, fig. 10, 10*a* (1902)[1]. Desneux, *Genera Insect.*, XXV: 41 (1904). — [1, 4, 6]Ceylon. [2]Khandala and Salsette Island (Bombay).

Eutermes biformis Lefroy, *Agric. J. India,* VII (3): 224-225, fig. (1912). Holmgren, *J. Bombay nat. Hist. Soc.*, XXI (3): 793 (1912)[2]. Assumth, *Ibid.*, XXII (2): 375, pl. I, fig. 10 (1913)[3]. Green, *Spolia zeylan,* IX (33): 14 (1913)[4], (possibly a variety of *rubidus*). Hegh, *Les Termites,* Bruxelles: 48 (1922). — [3, 7]Bombay.

Eutermes (Trinervitermes) biformis Holmgren, *Mem. Dep. Agric. India Ent.*, V (3-4): 162-163 (1917)[5]. John, *Treubia,* VI (3-4): 400 (1925)[6]. — [5]Shevaroy Hills 4,500 ft. (Madras), Saidapet, (Coorg).

Trinvervitermes biformes Beeson, *Forest Insect,* Dehra Dun: 552, fig. 151, no. 8 (1941). — [8]Commonest foraging species in India.

BIOLOGICAL.

Fletcher *Agric. J. India,* VII (3): 224-225 (1912). Assmuth, *J. Bombay nat. Hist. Soc.*, XXII (2): 375) (1913). Bugnion, *Bull. Mus. Hist. Nat.* Paris, XX (4): 180 (1914). Holmgren, *Mem. Dep. Agric. India Ent.*, V (3): 162-163 (1917). Hingston, *J. Bombay nat. Hist. Soc.*, XXXII (4): 717-725, Pl. 1 (1928)[7]. Margabandu, *Ibid.*, XXXVII (3): 712 (19340. Beeson, *Forest Insects,* Dehra Dun: 552 fig. 151, no. 8 (1941)[8].

(Host: Grass).

3. **biramanicus** Beeson.

SYSTEMATICS.

Hospitlitermes biramanicus Beeson, *Forest Insects,* Dehra Dun: 547 (1941). — Burma.

(Host: Teak saplings).

4. **brunneus** Snyder.

SYSTEMATICS.

Nasutitermes fletcheri Synder, *Proc. U.S. nat. Mus.,* LXXXII. Art. 16 (2957): 10 (1933). — India.

Nasutitermes brunneus Snyer *Indian For. REc. Ent.,* XX (11): 11 (1934).

(The specimens of *N. fletcheri* Holmpgen which he described in the previous paper in 1933 on re-examination found to be those of anew species.)

5. **ceylonicus** Holmgren.

SYSTEMATICS.

Eutermes ceylonicus Holmgren, *in Escherich Termitenleben auf Ceylan,* Jena: 197-198, pl. 3, fig. (1911). Greent, *Spolia zeylan.,* IX (33): 14 (1913). Hegh, *Les Termites,* Bruxelles: 48 (1922). John, Treubia, VI (3-4): 394 (1925). Kemner, *Bull. ent.* XVI (4): 388-389 (1926). — Ceylon.

BIOLOGICAL.

Bugnion, *Bull. Mus. Hist. nat. Paris,* XX (4): 172, 180, 183-184 (1914); *Ann. Soc. ent. Fr.,* LXXXIII: 361 (1914). John, *Treubia,* VI (3-4): 394 (1925). Hutson, *Adm. Rep. Dir. Agric. Ceylon, 1932:* 127 (1933). Margabandu, *J. Bombay nat. Hist. Soc.,* XXXVII (3): 712 (1934).

(Hosts: Bark of living stems of *Camellia sinensis,* tea plant and wood work of buildings).

6. **ceyloneillus** Kemner.

SYSTEMATICS.

Eutermes ceyloniellus Kemner, *Bull. ent. Res.,* XVI (4): 389 (1926). Margabandu, *J. Bombay nat. Hist. Soc.,* XXXVII (3): 712 (1934). — Ceylon.

(Host: Stems and roots of *Camellia sinensis,* tea plant.)

7. **crassicornis** Holmgren.

SYSTEMATICS.

Eutermes (Eutermes) crassicornis Holmgren, *Mem. Dep. Agric. India, Ent.*, V (3): 162 (1917). Margabandu, *J. Bombay nat Hist. Soc.*, XXXVII (3): 712 (1934). — Anamalai 4,000 ft. (Madras), Sidapur 3,500 ft. (Coorg) Bababudin 4,700 ft. (Mysore).

(Hosts: Trunk of a living tree and rotten logs).

8. **emersoni** Snyder.

SYSTEMATICS.

Nasutitermes (Nasutitermes) emersoni Snyder, *Indian For. Rec. Ent.*, CXX (11): 11-12 (1934) — Sundarbans (Bengal).

9. **escherichi** Holmgren.

SYSTEMATICS.

*Eutermes escherichi** Holmgren, *in Escherich Termitenleben auf Ceylan*, Gustav Fischer, Publisher Jena: 199-201, pl. 3, figs. LI, L2 (1911). Green, *Spolia zeylan.*, IX (33): 15 (1913). Bugnion, *Ibid.*, IX: 155 (1944). Hegh, *les Termites*, Bruxelles: 48 (1922). — Ceylon

Eutermes (Eutermes) escherichi John, Treubia, VI (3-4): 399 (1925).

BIOLOGICAL.

John, *Spolia zeylan.*, IX (34): 108 (1913). Bugnion, *Bull. Mus. Hit. nat. Paris*, XX (4): 180 (1914). John, Treubia, VI (3-4): 399 (1925). Margabandu, *J. Bombay nat. Hist. Soc.*, XXXVII (3): 712-713 (1934). (Found in the nest of *Termes obscuriceps*).

(Host: Wood in contact with soil).

---

*Note. *Eutermes escherichi* sp. n. described by Holmgren from soldier and worker. Holmgren described *Eutermes perparvus* sp. n. from winged adult. Later on Holmgren recorgnised the latter species an image of the former, whose image was hitherto unknown.

10. fletcheri Holmgren.

SYSTEMATICS.

Eutermes (Eutermes) fletcheri Holmgren, *Mem. Dep. Agric. India Ent.*, V (3): 161-162 (1917)[1]. — [1]Shevaroy 4,000 ft. (Madras).

Nasutitermes fletcheri Snyder, *Proc. U.S. nat. Mus.*, LXXXII. Art. 16 (2957): 10 (1933).

Nasutitermes (Nasutitermes) fletcheri Margabandu, *J. Bombay nat. Hist. Soc.*, XXXVII (3): 712 (1934); *Ibid.*, XXXVIII (1): 208-209 (1935).

(Host: Dead tree stumps)

11. gardneri Snyder.

SYSTEMATICS.

Nasutitermes (Subulitermes) gardneri Snyder, *proc. U. S. Mus.*, LXXXII. Art. 16 (2957): 12 (1933)[1]. Margabandu, *J. Bombay nat. Hist. Soc.*, XXXVIII (1): 208-209 (1935). — [1]Darjeeling and Rangirum 6,000 ft. (Bengal).

(Host: Wood).

12. grallatoriformis Holmgren.

SYSTEMATICS.

Eutermes (Grallatotermes) grallatoriformis Hollmgren, *Mem. Dep. Agric. India Ent.*, V (3): 163-164 (1917)[1]. Margabandu, *J. Bombay nat. Hist. Soc.*, XXXVII (3): 713 (1934). — [1]Tellkadi: Anamalai (Madras, Aryapur (S. Canara).

(Host: Bark of *Mangifera indica*, mango tree).

13. hantanae Holmgren.

SYSTEMATICS.

Eutermes hantanae Holmgren, *in Eascherich Termitenleben auf Ceylan*, Gustav Fischer, Publisher, Jena: 198, pl. 3, figs. Pl. P2 (1911). Green, *Spolia zeylan.*, IX (33): 14 (1913). Bugnion, *Ibid.*, IX: 149 (1914): *Ann. Soc. ent. Fr.*, LXXXIII: 362 (1914); *Spolia zeylan.*, IX (35): 155-162, pls. 24, 25 (1914). Hegh, *Les Termites*, Bruxelles: 48 (1922). — Ceylon.

BIOLOGICAL.

Bugnion, *Bull. Mus. Hist. nat. Paris.*, XX (4): 180-184 (1914), Margabandu. *J. Bombay nat. Hist. Soc.*, XXXVII (3): 712 (1934).

(Host: Wood in contact with soil.)

14. heimi Wasmann.

SYSTEMATICS.

Eutermes heimi Wasmann, *Zool. Jb., Abt. I*, XVII (1): 135 (1902). Desneux, *Genera Insect.*, XXV: 45 (1904).

[1,2]Coimbatore.
[1]Humpi ruins (Bellary), Maddur (Mysore).
[2]South India.

Eutermes (Trinervitermes) heimi Holmgren, *Mem. Dep. Agric. India Ent.*, V (3): 163 (1917).

BIOLOGICAL.

Lefroy, Indian Insect Life, Calcutta, I: 120 (1909). Holmgren, *Mem. Dep Agric. India Ent.*, V (3): 163 (1917)[1]. Ramakrishna Ayyar, *Madras Agric. J.*, XXI (10): 421-430 (1933)[2]. Margabandu, *J. Bombay. nat. Hist. Soc.*, XXXVII (3): 712 (1934). Beeson, *Forest Insects*, Dehra Dun: 551 (1941).

(Hosts: Sugarcane. The damageis caused by cutting off leaf blades. It is a foraging species sending out processions to collect grass).

15. inaniformis Holmgren.

SYSTEMATICS.

Eutermes (Subulitermes) inaniformis Holmgren, *K. Svenska VetenskAkad. Handl.*, L (2): 154, pl. 7, fig. I, (1913)[1]. John, Treubia, VI (3-4): 391 (1925)[2]. Margabandu, *J. Bombay nat. Hist. Soc.*, XXXVII (3): 713 (1934).

[1]India
[2]Singapore, Sumatra.

16. inanis (Haviland).

SYSTEMATICS.

Termes inanis Haviland, *J. linn. Soc. (Zool.)*, XXVI: 425 (1898)[1].

[1]Malaya.

Termes (Eutermes) inanis sub sp. *horni* Wasmann, *Zool. Jb., Abt I,* XVII (1): 137 (1902)[2]. — [2, 3, 4, 5]Ceylon.

Eutermes inanis Desneux, *Genera Insect.*, XXV: 42 (1904). Green, Spolia zeylan., IX (33): 14 (1913). Hegh, *Les Termites,* Bruxelles: 48 (1922)[3].

BIOLOGICAL

Bugnion, *Bull. Mus. Hist. nat. Paris,* XX (4): 180, 183 (1914)[4]; *Spolia zeylan.*, IX: 155 (1914); *Ann. Soc. ent. Fr.*, LXXXIII: 361 (1914). Green, *Rept. Trans. 3rd internat. Congr. trop. Agric. London:* 608-636 (1916)[5]. Margabandu, *J. Bombay nat. Hist. Soc.*, XXXVII (3): 712 (1934).

[Host: *Hevea* spp. (rubber), but not a serious pest].

17. Indicola Holmgren.

SYSTEMATICS.

Eutermes (Eutermes) indicola Holmgren, *Mem, Dep. Agric. India Ent.*, V (3): 161 (1917)[1]. Margabandu, *J. Bombaynat. Hist. Soc., XXXVII* (3): 712 (1934). Beeson, *Forest Insects,* Dehra Dun: 551 (1941). — [1]Mercara (Coorg), Anamalai 3,000-4,000 ft. (Madras).

(Hoss: Dead stumps and even live trees).

18. indicus Snyder.

SYSTEMATICS.

Nasutitermes (Trinervitermes) indicus Snyder, *Indian For. Rec. Ent.*, (11): 12-14, figs. A, B (1934). — Kalyanpur (U.P.)

19. jepsoni Snyder.

SYSTEMATICS.

Nasutitermes (Hospitalitermes) jepsoni ER, *Indian For. Rec. Ent.*, XX (11): 16 (1934). — Kaing (Burma).

20. kotuae Bugnion.

SYSTEMATICS.

Eutermes kotuea Bugnion, *Ann. Soc. Ent. Suisse,* XII: 192-200, 3 pls. (1914); *Ann. Soc. ent. Fr.*, LXXXIII: — Ceylon.

351-359, 362 (1914). Hegh, *Les Termites,* Bruxelles. 48 (1922).

BIOLOGICAL.

Bugnion, *Bull. Mus. Hist. nat. Paris,* XX (4): 180, 184 (1914). Margabandu, *J. Bombay nat. Hist. Soc.,* XXXVII (3): 712-713 (1934).

21. lacustris Bugnion.

SYSTEMATICS.

Eutermes lacustris Bugnion, *Rev. suisse Zool. Geneva,* XX: 487-505 pls. 2 (1912); *Compt. Rend. Sci. Biol. Fr.,* LXXII: 1091 (1912). Holmgren, *Spolia zeylan.,* VIII (32): 282-283 (1913). John, *Ibid.,* IX (34): 110-112 (1913). Bugnion, *Bull. Mus. Hist. nat. Paris,* XX (4): 180-183, pl. II, fig. 2 (1914); *Spolia zeylan.,* IX: 149-154, 157 (1914); *Ann. Soc. ent. Fr.,* LXXXIII; 360 (1914). Hegh, *Les Termites,* Bruxelles: 48 (1922). John, *Treubia,* VI (3-4): 397 (1925). — Ceylon.

Eutermes greeni Holmgren, *Spolia zeylan., XIII* (32): 282—28 (1913).

BIOLOGIAL.

Green, *Spolia zeylan.,* IX (33): 15 (1913). Bugnion, *Bull. Mus. Hist. nat. Paris,* XX (4): 180-182, pl. II, fig. 2 (1914). John, *Treubia,* VI (3-4): 397 (1925). Margabandu, *J. Bombay nat. Hist. Soc.,* XXXVII (3): 713 (1934). Beeson, Forest Insects, Dehra Dun: 5512 (1941).

[Hosts: *Anacardium* spp. (kaju nut), *Hevea* spp. (rubber), *Elaeocarpus* spp. and some unidentified trees].

22. longinotus Snyder.

SYSTEMATICS.

Nasutitermes (Trinervitermes) longinotus Snyder, *Indian For.Rec. Ent.,* XX (11): 14-15 (1934). — Poona (Bombay).

23. madrasi Snyder.

SYSTEMATICS.

Nasutitermes (Hospitalitermes) madrasi Snyder, *Indian For. Rec. Ent.,* XX, (11): 16-17 (1934). — Vellore (Madras).

24. monoceros (König).

SYSTEMATICS.

Temes monoceros König, *Schr. Berl. naturf.* IV (17): 25, pl. 1, figs. 10, 11 (1779). Hagen, *Wien. Zool. Bot. Ver.*: 471 (1858): *Linn. Ent.*, XII: 299 (1858); *Ibid.*, XIV: 113-117 (1860). Desneux, *Genera Insect.*, XXV: 43 (1904). Petch, *Ann. R. bot. Gdns. Peradeniya*, V: 395-420, pls. 9 (1913). Ceylon.

Termes monoceros. hospitalis, Haviland, J. *linn. Soc. (Zool*, XXVI: 437, pl. 25, figs. 83-86 (1898).

Eutermes monoceros subsp. hospitalis Wasmann, *Zool. Jb., Abt. I*, XVII: (131 (1902)[1]. [1]Singapore (Malaya), Sarawak (Borneo).

Eutermes monoceros Bugnion, *Ann. Soc. ent. Fr.*, LXXXVIII: 271-281, pls. 3 (1909). Holmgren, *in Escherich Termitenleben auf Ceylan*, Gustav Fischer, Publisher, Jena: 202, fig. 42, A, S, 99 (1911). Bugnion, *Bull. Soc. vaud. Sci. nat.*, XLVII: 417-437 (1911); *Ann. Soc., ent. Fr.*, LXXXIII: 359-360 (1914). Fletcher, *Proc. 4th ent. Mtgs. Pusa:* 332-33, pl. 52, figs. 10, 11 (1921), [Translation of König's (1779) paper on South Indian termites]. Hegh, *Les Termites*, Bruxelles: 48 (1922). John, *Treubia*, VI (3-4): 405 (1925). Besson, *forest Insects*, Dehra Dun: 548-549, fig. 151, no. 8 (1941).

BIOLOGICAL.

Bugnion, *Ann. Soc. Ent. Fr.*, LXXVIII; 271-281, 3 pls. (1909); *Bull. Soc. vaud Sci. nat.*, XLVII: 417-437 (1911). Fletcher, *Agric. J. India*, VII (3): 225-226 (1912). Bugnion *Bull. Soc. ent. suisse*, XII: 119-123, fig. 4 (1912); *Mitt. schweiz. ent. Ges.*, XII: 119-123 (1913). Green, *Spolia zeylan.*, IX (33): 14 (1913)2. John, *Ibid.*, (34): 112-115 (1913). Petch, *Ann. R. Bbot. Gdns. Peradeniya*, V: 395-420, pls. 9 (1913). Bugnion, *Bull. Mus. Hist. Nat. Paris*, XX (4): 180-182 (1914). John, *Treubia*, VI (3-4): 405 (1925). Bugnion, *Sauvenin de Sunigoda Terr et la Vic, Paris*, III: 192-206, 7 figs. (1933). Margabandu, *J. Bombay nat. Hist. Soc.*, XXXVII (3): 713 (1934). Beeson, Forest Insects, Dehra Dun: 548-549, fig. 151, no. 8 (1941). [2]All over Ceylon up to 2,000 ft.

(Hosts: Lichens, green algae and moulds. It does not eat wood).

25. **moratus** Sivestri.

SYSTEMATICS.

Eutermes moratus Silvestri, *Rec.Indian Mus.,* VIII (5): 431 (1914). Yambung : Abor 1100 ft. (Assam).

26. **oculatus** Holmgren.

SYSTEMATICS.

Eutermes oculatus Holmgren, *in Escherich Termitenleben auf Ceylan,* Gustav Fischer, Publisher, Jena: 200 (1911). Green, *Spolia zeylan,* IX (33): 15 (1913). John, *Ibid.,* (34): 109 (1913). Ceylon.

Eutermes (Eutermes) oculatus John, *Treubig,* VI (3-4): 391 (1925).

Eutermes longicornis Holmgren. *K. Svenska VetenskAkad. Handl.,* XLVII (4): 64 (1912); *Ibid.,* L (2): 172, pl. VII, fig. 5 (1913); *Spoloa zeylan.,* VIII (32): 283 (1913). Bugnion, *Ann. Soc. ent. Fr.,* LXXXIII: 361 (1914); *Spolia zeylan,* IX: 155 (1914). Hegh, *Les Termites,* Bruxelles: 48 (1922).

BIOLOGICAL.

Green, *Spolia zeylan.,* IX (33): 15 (1913). Bugnion, *Bull. Mus. Hist. nat. Paris,* XX (4): 180, 184 (1914). John, *Treubia,* VI (3-4): 391 (1925). Margabandu, *J. Bombay nat. Hist. Soc.* XXXVII (3): 713 (1934).

(Host: *Dendrocalanus giganteus,* giant bamboo).

27. **processionarius** Schmitz.

SYSTEMATICS.

Eutermes processionarius Schmitz., *Zool. Anz.,* LIX (11-12): 302-304, fig. II, 11A, 11B (1924). Upinangidi-Taluk : Canara (India).

28. **roboratus** Silvestri.

SYSTEMATICS.

Eutermes roboratus Silvestri, *Rec. Mus.,* VIII (5): 430-431 (1914). Moulmein (Lower Burma).

29. rubidus (Hagen).

SYSTEMATICS.

Termes rubidus Hagen, *Linn. Ent.*, XIV: 117-118 (1860). Desneux, *Genera Insect.*, XXV: 45 (1904). Ceylon.

Eutermesrubidus Holmgren, *in Escherich Termitenleben auf Ceylan.*, Gustav Fischer, Publisher, Jena: 201 (1911). Green *Spolia zeylan.*, IX (33): 14 (1913). Bugnion, *Ann. Soc., ent. Fr.*, LXXXIII: 362 (1914). Hegh, *Les Termites*, Bruxelles: 48 (1922).

Eutermes (Trinervistermes) rubidus John, *Treubia*, VI (3-4): 400 (1925). Kemner, *Bull. ent. Res.*, XVI (4): 389-390 (1926).

BIOLOGICAL.

Holmgren, *in Eschierich Termitenleben auf Ceylan*, Gustav Fischer, Publisher, Jena: 133-134 (1911). John, *Spolia zeylan.*, IX (34): 104-106 (1913). Bugnion, *Bull. Mus. Hist. nat. Paris*, XX (4): 180-182 (1914). John, *Treubia*, VI (3-4): 400 (1925). Magabandnu, *J. Bombay nat. Hist. Soc.* XXXVII (3): 713-714 (1934).

(Host: Grass).

30. singaporiensis (Haviland).

SYSTEMATICS.

Termes singaporiensis Haviland, *J. Linn. Soc. (Zool)*, XXVI: 429-430, pl. 25, figs. 76-79 (1898)[1]. [1]Singapore (Malaya) Ceylon

Eutermes isngaporensis Wasmann, *Zool. Jb., Abt. I*, XVII (1): 137 (1902). Desneux, Genera, Insect, XXV: 45 (1904).

Genus ix. **macrotermes** (Holmgren).

Termes (Macroterms) Holmgren, *Zool. Anz.*, XXXV: 286 (1910).

Macrotermes Fuller, *S. Afr. J. Nat. Hist.*, III (1): 17 (1921). John, *Treubia*, VI (3-4): 385 (1925).

1. annandalei Silvestri.

SYSTEMATICS.

Macroterms annandalei Silvestri, *Rec. Indian Mus.*, VIII (5): 427-428 (1914). Moulmein (Lower Burma).

2. *azarelii (Wasmann.)

SYSTEMATICS.

Termes azarelii, Wasmann, *Ann. Mus. Civ. Storia, Nat. Genoa,* XVI: 626-628 (1896); Ibid., Insect., XXV: 1904). Palon, Pegu (Burma)

3. convulsionarius (König).

SYSTEMATICS.

Termes Convulsionarius König, *Schr. Berl. naturf. Freunde,* IV (1): 24 (1779)[1]. BUGNION, *Bull. Soc. ent. suisse,* XII (4): 125-135, pl. 1 (1912)[2]; *Bull. Mus. Hist. nat. Paris,* XX (4): 185-188 (1914)[3]. FLETCHER, *Proc. 4th ent. Mtgs. Pusa:* 331 (1921)[4]. HEGH, *Les Termites,* Bruxelles: 48 (1922)[5].

Macrotermes convulsionarius John, *Treubia,* VI (3-4): 385 (1925)[6].

Hodotermes convulsionarius, HAGEN, *Linn. Ent.,* XII: 90 (1858). DESNEUX, *Genera Insect.,* XXV: 17 (1904).

Termes estherae DESNEUX, *Ann. Soc. ent. Belg.,* LI: 390-392 (1907)[7]. GREEN, *Spolia zeylan.,* IX (33): 11 (1913). HOLMGREN, *Ibid.,* VIII (32): 280 (1913)[8]; *Mem. Dep. Agric. India Ent.,* V (3): 145-146 (1917)[9]. FLETCHER, *Proc. 4th ent. Mtgs. Pusa*: 331 (1921).

[1, 4]Tanjore (S. India). [2, 3, 5, 6, 8]Ceylon. [7]Kaladgi, Huvin, Hippargi and Muddebihal, Bijapur (Bombay). [9]Hadagalli, Magala, Harpanahalli and Hospet (Madras.) [9]Coimbatore.

BIOLOGICAL

BUGNION, *Bull. Soc. ent. suisse,* XXI (4): 136-139 (1912); *Bull. Mus. Hist. nat. Paris,* XX (4): 185-188 (1914). JOHN, *Treubia,* VI (3-4): 385 (1925). MARGABANDU, *J. Bombay nat. Hist. Soc.,* XXXVII (3): 704-705 (1934).

(Hosts: Dead parts of plants in soil).

4. serrulatus Snyder.

SYSTEMATICS.

Macrotermes serrulatus SNYDER, *Indian For. Rec. Ent.,* XX (11): 6-8 (1934). BEESON, *Forest Insects,* Dehra Dun: 549 (1941). Magok and Hsipaw (Burma).

---

* The description of *Azarelii* Indicates that it belongs to the genus *Macrotermes.*

Genus X. **Microtermes** Wasmann.

Termes (Microtermes) WASMANN, *Abh. senckenb. naturf. Ges.*, XXI (1), (1897).

Microtermes WASMANN, *Zool. Jb., Abt. I,* XVII (1): 115 (1902). DESNEUX, *Genera Insect.*, XXV: 29 (1904). HOLMGREN, *Mem. Dep. Agric. India Ent.*, V (3): 159 (1917). JOHN, *Treubia,* VI (34): 390 (1925).

1. globicola Wasmann.

SYSTEMATICS.

Microtermes globicola WASMANN, *Zool. Jb., Abt. I,* XVII (1): 116 (1902). DESNEUX, *Genera Insect.*, XXV: 37 (1904). HEGH, *Les Termites,* Bruxelles: 48 (1922). JOHN, *Treubia,* VI (3-4): 390 (1925). — Ceylon.

BIOLOGICAL.

GREEN, *Spolia zeylan.*, IX (33): 11 (1913). BUGNION, *Bull. Mus. Hist. nat Paris,* XX (4): 185, 202 (1914). HOLMGREM, *Mem. Dep. Agric. India Ent.*, V (3): 159 (1917)[1]. JOHN, *Treubia,* VI (3-4): 390 (1925). MARGABANDU, *J. Bombay nat. Hist. Soc.*, XXXVII (3): 709 (1934). — [1]Maddur (Mysore), Coimbatore.

(Hosts: Logs of wood and grass).

2. incertus (Hagen)

SYSTEMATICS.

Termes incertus HAGEN, Monat. *Ber. Akad. wiss. Berlin*: 481 (1853). HAVLIAND, *J. linn. Soc. (Zool.),* XXVI: 388, pl. 23, figs. 15-18 (1898). SJÖSTED, *Mon. Term. Afr*: 135 (1900)[1]. — [1, 2, 3]Africa. [3]Bombay.

Termes (Eutermes) incertus HAGEN, *Linn. Ent.*, XII: 230 (1858); *Peters Reise Mossambique Neur.*, T.1, pl. 4, fig. 22 (1862)[2]. DESNEUX, *Genera Insect.*, XXV: 37 (1904).

Microtermes incertus HOLMGREN. *J. Bombay nat. Hist. Soc.*, XXI (3): 788-789 (1912)[3]; *Ibid.*, XXII (1): 113-114 (1913). MARGABANDU, *Ibid.*, XXXVII (3): 709 (1934). BEESON, *Forest Insects,* Dehra Dun: 550 (1941).

Microtermes incertoid HOLMGREN, *K. svenska VetenskAkad. Handl.*, L (2): 151 (1913); *J. Bombay nat. Hist. Soc.*, XXII (1): 113-114 (1913). MARGABANDU, *Ibid.*, XXXVII (3): 709 (1934).

(Host: Wood in contact with soil).

3. macronotus Holmgren.

SYSTEMATICS.

Microtermes macronotus HOLMGREN, *K. svenska VetenskAkad. Handl.*, XLVIII (4): 43 (1912). GREEN, *Spolia zeylan.*, IX (33): 11 (1911). HOLMGREN, *K. svenska VetenskAkad. Handl.*, L (2): 145, pl. VI, fig. 1, (1913). MARGABAN-DU, *J. Bombay nat. Hist. Soc.*, XXXVII (3): 709 (1934). — Ceylon.

4. mycophagus (Desneux).

SYSTEMATICS.

Termes mycophagus DESNEUX, *Ann. Soc. ent. Belg.*, XLIX: 348-352 (1905)[1]. LEFROY, *Indian Insect. Life*, Calcutta, I: 120 (1909)[2]. — [1, 2]Karachi (Sind).

Microtermes mycophagus HOLMGREN, *Mem. Dep. Agric. India Ent.*, V (3): 159-160 (1917)[3]. MARGABANDU, *J. Bombay nat. Hist. Soc.*, XXXVII (3): 705 (1934). — [3]Lyallpur (Punjab)

(Hosts: Wood and wooden structures in contact with soil).

5. obesi Holmgren.

SYSTEMATICS.

Microtermes obesi HOLMGREN, *J. Bombay nat. Hist. Soc.*, XXI (3): 787-788 (1912)[1]. — [1, 2, 3]Bombay.

K. svenska VetenskAkad. Handl., L (2): 150, pl. VI, fig. 4 (1913)[2]; *Mam. Dep. Agric. India Ent.*, V (3): 159-160 (1917)[3]. — [3]Kadur and Bangalore (Mysore), Pusa (Bihar), Coimbatore, Shevaroy 4200 ft. and Hadagalli (Bellary), Sidpur and Suntikoppa (Coorg),

Microtermes anandi HOLMGREN, *K. svenska VetenskAkad. Handl.*, L (2): 152 (1913); *J. Bombay nat. Hist. Soc.*, XXII (1): 114 (1913); *Men. Dep. Agric. India Ent.*, V (3): 159-160 (1917).

Hoshangabad and Nagpur (M.P.) Gurdaspur and Multan (Punjab), Mirganj (Bengal), Hyderabad.

BIOLOGICAL.

Microtermes obesi HOLM., FLETCHER, *Sci. Rep. Agric. Res. Inst. Pusa,* 1913-14: 69 (1914); *Ibid., 1917-18:* 101-106 (1918). MACKENNA, *Rep. Progr. Agric. India. 1917-18:* 92 (1918). FLETCHER, *Proc. 3rd ent. Mtgs. Pusa,* I: 313 (1919); *IBid.,* III: 888 (1919). Fletcher and Ghosh, *Ibid.,* II: 705-712, (1919). Fletcher, *Bull. Dep. Agric. India,* 100: 245 (1921). Dover and Mathur, *Indian For. Rec. Ent.,* XX (7): 3 (1934). Margabandu, *J. Bombay nat. Hist. Soc.,* XXXVII (3): 708-709 (1934). Husain, *proc. Wld's Grain Exhib & Conf., 1934,* II: 562 (1935). Beeson, *forest Insect,* Dehra Dun: 550 (1941).

Microtermes anandi Holm., Assmuth, *J. Bombay nat. Hist. soc.,* XIII (2): 374, 382 (1913). Annandale, *Rec. Indian Mus.,* XXV (2); 233, 245-251 (1923)[4]. Bose, *Ibid.,* XXV: 255-256 (1923). Silvestri, *Ibid.,* XXV: 225 (1923). Silvestri, *J. Bombay nat. Hist. Soc.,* XXXVII (3): 708-709 (1934). — [4]Barkuda Islands (Orissa).

(Hosts: Serious pest of wheat seedlings in India and even plants coming to ears are attacked; dead trees and stumps, logs and opther wooden articles in contact with ground; also recorded damaging growing cabbage and cauliflower).

var. **curvignathus** Holmgren.

Microtermes anandi var. curvignathus Holmgren, *K. Svenska VetenskAkad. Handl.,* L. (2): 153, pl. VI, fig. 9 (1913); *J. Bombay nat. Hist. Soc.,* XXII (1): 114 (1913). Margabandu, *Ibid.,* XXXVII (3): 709 (1934). — Bombay.

6. **pubescens** Snyder.

SYSTEMATICS.

Microtermes pubescens Snyder, *Proc. U.S. nat. Mus.,* LXXXII, Art. 16 (2957): 9-10 (1933)[1]. Margabandu, *J. Bombay nat. Hist. Soc.,* XXXVII (1): 208-209 (1935). — [1]Dehra Dun (U.P.)

(Host: Aerial roots of *Ficus* spp.).

7. **sindensis** (Desneux).

SYSTEMATICS.

Termes sindensis Desneux, *J. Ann. Soc. ent. Belg.* XLIX: 354-355 (1905)[1]. Lefroy, *Indian Insect Life,m* Calcutta, I: 120 (1909). Margabandu, *J. Bombay nat. Hist. Soc.*, XXXVII (3): 705 (1934)[2]. [1,2] Karachi (Sindi.)

Microtermes sindensis Homgren, *J. Bombay nat. Hist. Soc.*, XXI (3): 787-788 (1912).

8. **unicolor** Snyder.

SYSTEMATICS.

Microtermes unicolor Snyder, *Proc. biol. Soc. wash.*, XLVI: 92-93, figs 2 (1933)[1]. Beeson, *forest Insects,* Dehra Dun: 551 (1991). Host: Aerial roots of *Ficus benghalensis).* [1]U.P. (India).

Genus xi. **Termes** Linnaeus.

Termes Linnaues, Syst. Nat., ed. 10 (1758); *Ibid.*, ed 13, Gmelin, 1: 2911 (1788-93). Haviland, *J. linn. soc. (Zool).* XXVI: 377 (1898). Sjösted, *Mon. Term. Afr., 1900:* 67 (1900). Holmgren, *in Eshcerich Termitenleben auf Ceylan,* Gustav fischer, Publisher, Jena: 210 (1911).

Termitina (partim) Froggatt, *Proc. Linn. Soc. N.S.W.:* 518 (1896). Wasmann, *Abh. Senckenb. naturf. Ges.*, XXI (1): 150 (1897). Silvestri *Redia,* I: 37 (1903). Desneux, *Ann. Soc. ent. Belg.*, XLVIII: 285 (1904); *Genera Insect.*, XXV: 25, 35 (1904).

Termes (Termes) Hagen, *Linn. Ent.*, XII: 108-109 (1858). Desneux, *Genera Insect.*, XXV: 35 (1904). Bank and Snyder *Bull. U. S. nat. Mus.*, 108: 8 (1920). *Fuller, S. Afr. J. nat. Hist.*, III (1): 17 (1921). [*Termes capensis* considered as type of *Terms instead of* T. *fatalis*]. Light, *Philipp, J. Sci.*, XVIII: 245 (1921). Emerson, *Zoologica N. Y.*, VI, (4): 298 (1925), [*T. capensis* considered as type].

Odontotermes Holmgren, *K. svenska VetenskAkad. Handl.*, XLVIII (4). (1912); *Mem. Dep. Agirc. India Ent.*, V (3): 146 (1917).

Subgenus 1. **Euscaiotermes** Silvestri

Odontotermes (Euscaioterems) Silvestri, *Rec. Indian Mus.*, XXV (2): 221 (1923).

1. **primus** Silvestri.

SYSTEMATICS.

odontotermes (Euscaiotermes) primus Silvestri, *Rec.Indian Mus.*, XXV (2): 221-223 (1923)[1]. Annandale, *Ibid.*, XXV (2): 233 (1923)[2]. Margabandu, *J. Bombay rat Hist. Soc.*, XXXVII (3): 707 (1934). — [1, 2]Barkuda Islands (Orissa).

Subgenus 2. **Hypotermes 6** Holmgren.

Odontotermes (Hypotermes) Holmgren, *KI. svenska VetenskAkad. Handl.*, L (2): 141 (1913); *Mem. Dep. Agric. India Ent.*, V. (3): 158 (1917).

1. **marshalli** (Kemner).

SYSTEMATICS.

Odontermes marshalli Kemner, *Bull. ent. Res.* XVI (4): 387-388 (1926). — Ceylon.

Hypotermes marshalli Jepson, *Trop Agri-culturist* LXXVI (2): 679 (1931). Margabadu, *J. Bombay nat. Hist. Soc.*, XXXVII (3): 708 (1934). (Hosts: Bamboo fence and stumps of *Grevillea* spp.).

2. **obscuriceps** (Wasmann).

SYSTEMATICS.

Termes obscuriceps, Wasmann, *Zool. Jb., Abt I,* XVII (1): 113 (1902). Desneux, *Genera Insect.*, XXV: 38 (1904). Holmgren, *in Eshcerich Termitenleben auf Ceylan,* Gustav Fischer, Publisher Jena: 194, p. 3, fig. G (1911). Green, *Spolia zeylan.*, IX (33): 10 (1913). Hegh, Les *Termites* Bruxelles: 48 (1922). — Ceylon.

Odontote mes obsuriceps Holmgren, *Spolia zeylan.*, VIII (32): 281-282 (1913).

Odonotermes (Hypotermes) obscuriceps Holmgren, *K. Sevenska VetenskAkad. Handl.*, L (2); 141 (1913).

John, *Treubia,* VI (3-4): 390 (1925). Beeson, *Forest Insects,* Dehra Dun: 549 (1941).

BIOLOGOICAL.

Doflein, *Spolia Zeylan.,* III: 205 (1906). petch, *Ann. R. bot. Gdns. Peradeniya,* III: 189-262, pl. VII (1906). Desneux, *Ann. Soc. ent. Belg.,* LI: 394-400 (1907). Bugnion, *mem. Soc. zool. Fr.,* XXV: 210-232 (1912). Green, *Trop. Agriculturist,* XL: 50 (1913). John, *Spolia zeylan.,* IX (34): 116 (1913). Bugnion, *Bull. Mus. Hist. nat. Paris,* XX (4): 185-188, 193-196 (1914). Green, *Rep. Trans. 3rd internat. Congr. Agric. London: 608-636* (1916). Hutson, *Trop. Agriculturist,* LIX: 85-87 (1923); *Leaf. Dep. Agric. Ceylon,* 24: 3-4, pl. II (1923).Jepson, *Yerb. Dep. Agirc. Ceylon:* 8-10, pl. (1924). John, *Treubia,* VI (3-4): 390 (1925). Jepson, *Trop. Agriculturist,* LXXVII (2): 68 (1926. Rutherford, *Planter's Note Book,* 8th ed., Colombo, Ceylon; 234, 532 (1926). Bathellier, *Faune Colon. Franc.,* I (4): 125-364, pl. XIII. figs. (1927)[1]. Hutson, *Tech. Rep. Dep. Agric. Ceylon, 1928:* 7-8 (1929). Jepson, *Bull. Dep. Agric. Ceylon,* 85: 7-8, pl. II (1929); *Tech. Rep. Dep. Agric. Ceylon, 1929:* 12 (1930). Hutson, *Adm. Rep. Dir. Agric. Ceylon, 1930:* 74 (1931).

[1]Indo-China.

Jepson, *Trop. Agriculturist,* LXXVI (2): 67-69 (1931). Hutson, *Adm. Rep. Dir. Agric. Ceylon, 1931:* 114, 117-118 (1932); *Ibid.,* 1932: 128-129 (1933). Dover and Mathur, *Indian For. Rec. Ent.,* XX(7): 3 (1934). Jepson, *Adm. Rep. Dir. Agric. Ceylon, 1933:* 119-121 (1934). Margabandu, *J. Bombay nat. Hist. Soc.* XXXVII (3): 708 (1934). Jepson,*Adm. Rep. Dir. Agirc. Ceylon, 1934:* 119-120 (1935). King, *Tea Quart.,* X (3): 160-166, pl. II (1937). Joachim and Kandiah, *Trop. Agriculturist,* LXXXV (6): 333-338 (1940). Beeson, *Indian For. Rec. Ent.,* IV (2): 69 (1941); *Forest Insects,* Dehra Dun: 549 (1941).

[Hosts: Wood vegetable debris, leves and branches living and dead tea bushes, *Hevea* spp. (rubber). Burma timber such as *Xylia dolabriformis* and *Hopea odorata is* immune to its attack].

3. **xenoptermitis** (Wasmann).

SYSTEMATICS.

Termes xenoptermitis Wasmann, *Ann. Mus. Civ. Storia Nat. Genoa,* XVI: 628 (1896)[1]. *Ibid.,* XVII: 152 (1896)[2]. Desneux, *Genera Insect.,* XXV: 46 (1904). — [1, 2, 3,]Palon, Pegu (Burma).

Termes (Eutermes) xenotermitis Lefroy, *Indian Insect Life,* Calcutta, I: 120 (1909)[3].

Termes (Hypotermes) xenotermitis Holomgren, *mem. Dep. Agric. India Ent.,* V (3); 158-159 (1917)[4]. Margabandu, *J. Bombay nat. Hist. Soc.,* XXXVII (3): 708 (1934). — [4]Dacca (Bengal), Shillong and Gauhati (Assam).

Hypotermes xenotermitis Beeson, *Forest Insects,* Dehra Dun: 549 (1941).

[Hosts: Wood of *Morus indica* and *Tectona grandis).*

Subgenus 3. **Termes** (L.) Hagen s. str.

Termes Hagen, *Monat, Ber. Akad wiss. Berlin:* 480 (1853); *Peters Reise, Mossambique, Insekten:* 86 (1862).

Termes (Odontotermes) Holmgren, *Zool. Anz.,*XXXV: 286 (1910); *Mem. Dep. Agric. India Ent.,* V (3): 154 (1917). Ahmad, *Amer. Mus. Novit.,* (1392): 1-11 (1949).

Odontotermes (Odontotermes) Holmgren, *K. Svenska VetenskAkad. Handl.,* XLVIII (4) (1912).

Odontotermes (Cyclotermes) Holmgern, *Mem. Dep. Agric. India Ent.,* V (3): 146 (1917).

1. **almorensis** Snyder.

SYSTEMATICS.

Termes (Cycloterms) almorensis Snyder, *Proc. U. S. nat. Mus.,* LXXXII, Art. 16 (2957): 8-9 (1933)[1]. Margabandu, *J. Bombay nat. Hist. Soc.,* XXXVIII (1): 208-209 (1935). — [1]West Almora Forest (U.P.)

2. **anamallensis** Holmgren.

SYSTEMATICS.

Odontotermes (Odontotermes) anamallensis Holmgren, *Mem. Dep. Agric. India Ent.,* V (3): 157-158 (1917). Margabandu, *J. Bombay nat. Hist. Soc.,* XXXVII (3): 705 (1934). — Anamalai Hills 3000-4000 ft.

(Host: Tree stumps)

3. **assamensis** Holmgren.

SYSTEMATICS.

Odontotermes assamensis Holmgren, *K. svenska VetenskAkad. Handl.*, L (2); 114, pl. IV, fig. 16, pl. V, fig. 15 (1913). Silvestri, *Rec. Indian Mus.*, VIII (5): 428 (1914). Margbandu, *J. Bombay nat. Hist. Soc.*, XXXVII (3): 705 (1934).

Sadiya: Diburgarh and Yambung: Abor (Assam).

4. **assmuthi** Holmgren.

SYSTEMATICS.

Odontotermes assmuthi Holmgren, *K. svenska VetenskAkad. Handl.*, L (2): 140 (1913)[1]; *J. Bombay nat. Hist. Soc., Hist. nat. Paris,* XX (4): 185 (1914)[2]. Holmgren, *Mem. Dep. Agric, India Ent.*, V (3): 154-155* (1917)3. John, Treubia, VI (3-4): 389 (1925)[4].

[1]Bombay, Salsette Island (Bombay). [2]Ceylon.

BIOLOGICAL.

Fletcher, *Sci. Rep. agric. Res.* Inst.Pusa, 1917-18: 101 (1918); *Proc. 3rd ent. Mtgs. Pusa,* III: 887 (1919). John, *Treubia,* VI (3-4): 389 (1925). Margbandu, *J. Bombaynat. Hist. Soc.*, XXXVII (3): 705 (1934). Pruthi and Narayana, *Indian J. Agric. Sci.*, IX (1): 15-37, 1 fig. (1939).

(Hosts: Dead and rotten leaves and branches in soil, sugar cane and logs of *Dalbergia sissoo*).

[3]Shevaroy 4500 ft. mercara (Coorg), Pusa (Bihar), Gurdaspur & Lyallpur (Punjab). [4]Singapore (Malaya).

5. **bangalorensis** Holmgren.

SYSTEMATICS.

Odontotermes bangalorensis Holmgren. *K. svenska VetenskAkad. Handl.*, L (2): 112 pl. IV. ifg. 15 (1913)[1]; *J. Bombaynat. Hist. Soc.*, XXII (1): 105-109 (1913)[2].

Odontotermes (Cyclotermes) bangalorensis Holmgren, *J. Bombay nat. Hist. Soc.*, XXII (1): 105-109 (1913); *Mem. Dep. agric. India Ent.*, V (3): 151-152 (1917)[3].

[1, 2]Bangalore. [3]Coimbatore, Shevaroy 2000-4200 ft., Hospet & Hadagalli (Bellary), Mysore, Layallpur (Punjab).

---

* Doubtfull examples were determined by Bugnion as *Odontotermes ceylonicus* under which name this species was consequently referred to and figured by Fletcher, *Agric.* J. *India,* VII (3): 224, 233, fig. 6 (1912).

Fletcher, *Proc. 3rd ent. Mtgs. Puasa,* I: 314 (1919); *Bull. Dep. Agric. India,* 100: 246 (1921). Annandale, *Rec. Indian Mus.,* XXV (2): 247 (1923). Dover and Mathur, *indian For. Rec. Ent.,* XX (7): 3 (1934). Margabandu, *J. Bombay nat. Hist. Soc.,* XXXVII (3): 705 (1934). Beeson, *Indian For. Rec. Ent.,* IV (2): 68 (1941) *Forest Insects,* Dehra Dun: 545 (1941)[4].

[4]Dehra Dun (U.P.)

[Hosts: Wood in contact with soil, timbers such as *Acacia arabica, A. Catechu, Bombax malabaricum, Cedrus deodara, Dalbergia sissoo* and *Shorea robusta,* cattle dung, rope, bamboo chick, bark of *Acacia* spp. and even roots of *Andropogon* spp. (*Chotum*)].

6. **bellahunisensis** Holmgren.

SYSTEMATICS.

Odontotermes (Cyclotermes) bellahunisensis Holmgren, *Mem. Dep. Agric India Ent.,* V (3): 150-151 (1917)[1]. Margabandu, *J. Bombay nat. Hist. Soc.,* XXXVII (3): 705 (1934).

[1]Bellahunisi, Bellary (Madras) and Mysore.

7. **biunneus** (Hagen).

SYSTEMATICS.

Temes obesus Walker, *Cat. Neur. Brit. Mus.,* (3): 516 (1853). [name used earlier by Rambur (1842) for another species].

[1, 2]Bengal

Termes brunneus, Hagen, *Linn. Ent.,* XII: 133-136, pl. 3, fig. 13 (1858)[1]. Desneux, *Genera Insects.,* XXV: 36 (1904). Lefroy, *Indian Insect Life,* Calcutta, I: 119 (1909)[2]. Green, *Spol ia zeylan.,* IX (33): 11 (1913)[3]. Bugnion, *Bull. Mus. Hist. nat. Paris,* XX (4): 185 (1914)[4]. Margabandu, *J. Bombay nat. Hist. Soc.,* XXXVII (3): 704 (1934).

[3, 4]Ceylon.
[5]Khandala, Kirkee, Poona (Bombay).

Odontoteres brunneus Holmgren, *J. Bombay nat. Hist. Soc.,* XXI: 783-784 (1912)[5].

Odontotermes (Cycloterms) brunneus Holmgren, *Mem. Dep. Agric. India Ent.,* V (3): 153 (1917). Margabandu, *J. Bombay nat. Hist. Soc.,* XXXVII (3): 707 (1934).

Cyclotermes brunneus Beeson, *Forest Insects,* Dehra Dun: 545 (1941).

var. **wallonensis** (Wasmann).

SYSTEMATICS.

Termes obesus subsp. **wallonensis* Wasmann, *Zool. Jb., Abt. I,* XVII (1): 106 (1902). Desnex, *Genera Insect.,* XXV;: 38 (1904). [1]Ceylon. [2]Bangalore Gujrat (Bombay).

Termes wallonensis Bugnion, *Bull. Mus. Hist. nat. Paris,* XX (4): 185 (1914)[1].

Odontotermes brunneus subsp. *wallonensis* Holmgren, *J. Bombay nat. Hist. Soc.,* XXII: 109-110 (1914)[2]. [3]Coimbatore,Maddur and Bababudin 4400 ft. and Kollegal (Mysore), Bellary.

Odontotermes (Cyclotermes) wallonensis, Holmgren, *Mem. Dep. Agric. India Ent.,* V (3): 152-153 (1971)[3].

Cyclotermes wallonensis, Beeson, *Forest insects,* Dehra Dun: 547 (1941).

BIOLOGICAL.

Holmgren, *j. Bombay nat. Hist. Soc.,* XXII: 109-110 (1914); *Mem. Dep. Agric. India Ent.,* V (3): 152-153 (1917). Margabandu, *J. Bombay nat. Hist. Soc.,* XXXVII (3): 706-707 (1934). Beeson, *forest Insects,* Dehra Dun: 547 (1941).

(Hosts: *Casuarina equisetifolia* and wood work of buildings).

8. ceylonicus (Wasmann).

SYSTEMATICS.

Termes ceylonicus Wasmann, *Zool. Jb, Abt I,* XVII (1): 113 (1902). Desneux, *Gener Insect.,* XXV: 36 (1904). Bugnion, *Rev. Suisse Zool./,* XIX: 383-395, pls. 2 (1911). Holmgren, *in Escherich Termitenleben auf Ceylan,* Gustav Fischer, Publisher, Jena: 194, pl. 3, fig. F (191). Green, *Spolia zeylan.,* IX (33): 9 (1913). Hegh, *Les Termites,* Bruxelles: 48 (1922). Ceylon.

Odontotermes ceylonicus Holmgren, *j. Bombay nat. Hist.Soc.,* XXI (3): 784 (1912); *Spolia zeylan.,* VIII

---

* *wallonensis* Wasmann first treated as synonym of *Terms obesus* Rambur by Desneux (1904) but Holmgren (1914) grouped it with *brunneus* as sub-species.

(32): 280-281 (1913). John, *Treubia,* VI (3-4): 387 (1925). Kemner, *Bull. Ent. Res.,* XVI (4): 386 (1926).

BIOLOGICAL

Bugnion, *Bull. Mus. Hist. nat. Paris,* XX (4): 185, 187, 202 (1914). John, ***Treubia,*** VI (3-4): 387 (1925). Hutson, *Tech. Rep. Dep. Agric. Ceylon, 1928:* 7-8 (1929). Jepson, *Bull. Dep. Agric, Ceylon,* 85: 7-8 (1929). Margabandu, *J. Bombay nat. Hist. Soc., Adm. Rep Dir. Agric Ceylon, 1934:* 117-120 (1935). Beeson, *forest Insects,* Dehra Dun: 551 (1941).

[Hosts: Wood or timber in contact with ground, wooden articles of building, dead bushes of *Camellia Sinensis* (tea) and bark of *Hevea* spp. (rubber) tree].

9. dehraduni Snyder..

SYSTEMATICS.

Termes (Termes) dehradui Snyder, *Proc. U.S. nat. Mus.,* LXXXII, Art. 16 (2957): 6-7 (1933). Margabandu, *J. Bombay nat. Hist. Soc.,* XXXVIII (1): 208-209 (1935). — Dehra Dun (UP).

10. distans Holmgren.

SYSTEMATICS.

Odontotermes (Cyclotermes) distans Holmgren, *Mem. Dep. Agric. India Ent.,* V (3): 153 (1917). Margabandu, *J. Bombay nat. Hist. Soc.,* XXXVII (3): 706 (1934). — Shevaroy 4000 ft. (Madras).

11. escherichi Holmgren.

SYSTEMATICS.

Termes escherichi Holmgren, *in Escherich Termitenleben auf Ceylan,* Gustav Fischer, Publisher, Jena: 195-196, pl. 3, fig. 1 (1911). Green, *Spolia zeylan.,* IX (33): 10 (1913). Hegh, *Les Termites,* Bruxelles: 48 (1922). — Ceylon.

Odontotermes (Cyclotermes) escherichi John, *Truebia,* VI (3-4): 386 (1925). Kemner, *Bull. ent. Res.,* XVI (4): 386-387 (1926).

BIOLOGICAL.

Bugnion, *Bull Mus. Hist. nat. Paris,* XX (4): 185, 187 (1914). John, *Treubia,* VI (3-4): 386 (1925). Kemner, *Bull. ent. Res.,* XVI (4): 386-387 (1926).

(Hosts: Dead and decaying trees, logs of wood, stems of *Grevillea* spp. and *Camellia sinensis).*

12. fatalies König.

SYSTEMATICS.

Termes fatale König, Schr. Berl. naturf. Forade, IV (1): 1, pl. I, figs. 1-9 (1779)[1]. Fabricius *Ent. Syst.,* II: 87 (1793).

Termes fatalis hagen, *Linn. Ent.,* XII: 143-145 (1858); *Ibid.,* XIV: 105 (1860). Haltermann, *Botanische Untersuchungen, S. Schwendener:* 411-420 (1899)[2]. Desneux, *Ann. Soc. ent. belg.,* XLIX: 359-360 (1905). *Petch, *Ann. R. Bot.Gdns. Peradeniya,* III: 191 (1906). Lefroy, *Indian Insect Life,* Calcutta, I: 120 (1909)[3]. † Green *Spolia zeylan.,* IX (33): 11 (1913). Bugnion, *Bull. Mus. Hist. nat. Paris,* XX (4): 185 (1914)[4]. Emerson, *Zoologica N. Y., VI* (4): 298 (1925). Margabandu, *J. Bombay nat. Hist. Soc.,* XXXVII (3): 705 (1934)[5].

Temes maracianus Walker *Cat. Neur. Brit. Mus.,* 3: 517 (1853).

Termes königi Desneux, *Ann. Soc. ent. Belg.,* LXIV: 336 (1905).

[1, 3, 4]Ceylon and India.
[2]Borneo and Java.
[5]Tanjore (India).

13. feae Wasmann.

SYSTEMATICS.

Termes feae Wasmann, *Ann, Mus. Civ Storia Nat. Genoa,* XVI: 625-626 (1896)[1]. Desneux, *Genera Insect.,* XXV: 37 (1904). Lefroy, *Indian Insect Life, Calcutta,* I: 120 (1909)[2].

[1, 2]Burma.
[3, 6]Khandala (Bombay).
[4]Abor (Assam).
[5, 7]Poona and Bombay.

---

* Petch quotes Haltermann [1899] who has stated that it is found in Borneo and Java.

† Green states that Holmgren gives Ceylon as the locality of *fatalis* on the authority of Hagen; it is common mound building species in Tanjore, India, and its occurrence in Ceylon is doubtful.

Odontotermes feae Holmgren, *J. Bombay nat. Hist. Soc.*, XXI (3): 786-787 (1912)[3]. Silvestri, *Rec. Indian Mus.*, VIII (5): 428 (1914)[4]. Beeson, *Forest Insects, Dehra Dun: 551,* fig 151 (1941). [5]Salsette Island (Bombay).

BIOLOGICAL.

Assmuth, *J. Bombay nat. Hist. Soc.*, XXII (2): 374, 383 (1913)[5]. Holmgren, *Ibid.*, XXII (1): 110-111 (1913)[6]. Assmuth, *Ibid.*, (4): 690-692 (1915)[7]. Holmgren, *Mem. Dep. Agric. India Ent.*, V (3): 156 (1917)[8]. Annandale, *Rec. Indian Mus.*, XXV (2): 233-251 (1923). Silvestri, *Ibid.*, XXV (2): 223 (1923). Annandale, *J. Bombay nat. Hist. Soc.*, XXX (1): 27, 33-34 (1924)[9]. Dover and Mathur, *Indian For. Rec. Ent.*, XX (7): 3 (1934)[10]. Margabandu, *J. Bombay nat. Hist. Soc.*, XXXVII (3): 708 (1934). Manning, *Pamphl. Tea Sci. Sect. U.P. A.S.I. Madras,* 2: 16-18 (1940)[11]. Beeson, *forest Insects,* Deha Dun: 551, fig. 153 (1941).

[Hosts: Wooden articles, dead trees, vegetable debris, books and even seedling of *Eucalyptus rostrata* (red gum)].

[7]Ranchi (Bihar), South Canara. [8]Coimbatore, Margalli and Mercara (Coorg), Shevaroy 4200 ft., Bangalore and Bababudin 4,000-4,700 ft. (Mysore), Dharwar and Belgaum (Bombay), Harra (Bengal), Maymyo (Burma). [9]Barkuda Islands (Orissa)., [10]Calcutta. [11]Coonor (South India).

14. feaeoides Holmgren.

SYSTEMATICS.

Odontotermes (Odontotermes) feaeoides Holmgren, *Mem. Dep. Agric. India Ent.*, V (3): 156-157 (1917)1. Margabandu, *J. Bombay nat. Hist. Soc.*, XXXVII (3): 706 (1934). [1]Coorg (South. India).

(Hosts: Log of wood lying on ground.)

15. ferruginosus (Latreille).

SYSTEMATICS.

*Termes ferruginosum*Latreille, *Hist. Nat. Crust. Ins.*, XIII: 70 (1804)[1]; *Dict. Hist. Nat.* XXII: 63. [1,2]India.

Termes ferruginosus Desneux, *Genera Insect.*, XXV: 39 (1904). Lefroy, *Indian Insect Life,* Calcutta. I: 120 (1909)[2].

16. flavomaculatus Holmgren..

SYSTEMATICS.

Odontotermes (Cyclotermes) flavomaculatus Holmgren, *mem. Dep. Agric. India Ent.*, V (3): 150 (1917)[1]. Margabandu, *J. Bombay nat. Hist. Soc.*, XXXVII (3): 706 (1934). [1]Assam.

(Note: Soldiers and workers indistinguishable from those of *obsesus* Rambur.)

17. gravelyi Silvestri.

SYSTEMATICS.

Odontotermes gravelyi Silvestri, *Rec. Indian Mus.*, VIII (8): 428-430, fig. 4 (1914). Sukli 2,100 ft. E. Dawna Hills (Lower Burma).

18. horni Wasmann.

SYSTEMATICS.

Termes horni Wasmann, *Zool. Jb., Abt I,* XVII (1): 111 (1902). Desneux, *Genera Insect.*, XXV: 37 (1904). Lefroy, *indian Insect Life,* Calcutta,I: 120 (1909). Holmgren, *in Escherich Termitenleben auf Ceylan,* Gustav Fischer, Publisher, Jena: 194, Pl. 3, fig. E (1911). Bugnion, *Mem. Soc. zool. Fr.*, XXV: 210-232 (1912); *Rev. suisse Zool.*, XXI: 299-330, 3 pls. (1913). Green, *Spolia zeylan.*, IX (33): 9 (1913). Ceylon.

Odontotermes horni Holmgren, *J. Bombay nat. Hist. Soc.*, XXI (3): 783 (1912); *Spolia zeylan.*, VIII (32): 280 (1913).*

Termes (Odontotermes) horni John, *Treubia,* VI (3-4): 387 (1935).

Termes Peradeniyae Holmgren, *in Escherich Termitenleben auf Ceylan,* Gustav Fischer Publisher, Jena: 197 (1911); *Spolia zeylan.*, VIII (32): 280 (1913).

BIOLOGICAL.

John, *Spolia zeylan.*, IX (34): 103 (1913). Bugnion, *Bull. Mus. Hist. nat. Paris,* XX (4): 185-186, 193, pl. [1]Bangalore (India).

---

* *Termes horni* may later on be proved to be synonym of *T. taprobanes.*

VIII, figs. 1-4 (1914). Green, *Rep. Trans. 3rd internat. Congr. trop. Agric. London:* 608-636 (1916). Hegh, *Les Termites,* Bruxelles: 48 (1922). Hutson, *Leafl. Dep. Agric. Ceylon,* 24: 3, pl. II (1923). Jepson, *Yearb. Dep. Agric. Ceylon:* 8-10, pl. II (1924). John, *Treubia,* VI (3-4): 387 (1925). Jepson, *Trop. Agriculturist,* LXVII (2): 68 (1926). Rutherford, *Planter's Note Book,* 8th ed., Colombo, Ceylon: 234, 532 (1926). Hutson, *Tech. Rep. Dep. Agric. Ceylon, 1928:* 7-8 (1929). Jepson, *Bull.Dep. Agric. Ceylon,* 85: 7-8 (1929);*Tech. Rep. Agric. Ceylon, 1929: 7, 12* (1930). Hutson, *Adm. Rep.Dir. Agric. Ceylon, 1930:* 74 (1931); *Ibid., 1931:* 116-118 (1932); *Ibid., 1932:* 127-128 (1933). Dover and Mathur *Indian For. Rec. Ent.,* XX (7): 3 (1934)[1]. Jepson, *Adm. Rep. Dir. Agric. Ceylon, 1933:* 119-121 (1934). Margabandu, *J. Bombay nat. Hist. Soc.,* XXXVII (3): 706 (1934). Jepson, *Adm. Rep. Dir. Agric. Ceylon, 1934:* 117-120 (1935). Beeson, *Indian For. Rec. Ent.,* IV (2): 69 (1941); *Forest Insects,* Dehra Dun: 552 (1941).

(Hosts: Wood, dead branches of trees in contact with the ground; bark of living trees and tea bushes but rarely destroying living tissues; rubber trees but not as a serious pest ; half of the five months' old crop of *Coix lacrumajobi* was once destroyed. Burma timber *Hopea odo rata* resistant to this species).

var. **hutsoni** Kemner.

Odontotermes horni var. *hutsoni* Kemner, *Bull. ent. Res.,* XVI (4): 384-385 (1926)[1]. Margabandu, *J. Bombay nat. Hist. Soc.,*XXXVII (3): 706 (1934). [1]Ceylon.

(Hosts: Coconut log, living plant of *Camellia sinensis,* tea).

var. **minor** Kemner.

Odontotermes horni var. *minor* Kemner, *Bull. ent. Res.,* XVI (4): 385-386 (1926)[1]. Margabandu, *J. Bombay nat. Hist. Soc.,* XXXVII (3): 706 (1934). [1]Ceylon.

(Hosts: Bark of *Hevea* spp. and dead wood of living plant of *Camellia sinensis,* tea).

19. latigula Snyder.

SYSTEMATICS.

Termes latigula Snyder, *Indian For. Rec. Ent.*, XX (11): 9-10 (1934), [species distinct) but belongs to *Termes* (*Cylotermes*) *obsesus* Rambur group]. — Andaman Islands.

20. malabaricus Holmgren..

SYSTEMATICS.

Odontotermes (Odonototermes) malabaricus Holmgren, *Mem. Dep. Agic. India Ent.*, V (3): 155 (1917).[1] Margabandu *J. Bombay nat. Hist. Soc.*, XXXVII (3): 706 (1934).

Termes (Termes) malabaricus Snyder, *Proc. U.S. nat. Mus.*, XLI. 16 (2957): 7-8 (1933)[2]. Magrabandu, *J. Bombay nat. Hist. Soc.*, XXXVIII (1): 208-209 (1935). Beeson, *Forest Insects*, Dehra Dun: 552 (1941).

[1]Malabar, Yercaud, Shevaroy 4500 ft., Bababudin 4000 ft. (Mysore). [2]Dehra Dun (U.P.).

(Hosts: Coconut stems and logs of *Grevillea* spp. lying on ground).

21. microdens Holmgren..

SYSTEMATICS.

Odontotermes microdens Silvestri, *Rec. Indian Mus.*, VIII (5): 428 (1914) — Chaumuhani (Bengal), Gurdaspur (Pubjab).

22. mirganjensis Holmgren..

SYSTEMATICS.

Odontotermes (Odontotermes) mirganjensis Holmgren, *Mem. Dep. Agric. India Ent.*, V (3): 159 (1917)[1]. Margabandu, *J. Bombay nat. Hist. Soc.*, XXXVII (3): 706 (1934). — [1]Mirganj (Bengal).

(Host: Dead Branch of a tree).

23. obesus (Rambur).

SYSTEMATICS.

Termes obesus Rambur, *Histoire Naturalle Neveroptires:* 304 (1842). Hagen, *Linn. Ent.*, XII: 145-147, pl. 3, fig. 36 (1958). Wasmann, *Dtsch. ent. Z: 145* (1899). Desneux, *Genera Insect.*, XXV: 38 (1904). Lefroy, — [1]India., [2]Ceylon. [3, 5, 6]Bombay province.

Indian Insect, Life, Calcutta, I: 120-121, figs. 47, 49 (1909)[1]. Bugnion, *Bull. Mus. Hist. nat. Paris,* XX (4): 185 (1914)[2].

Odontoteres obesus Holmgren, *J. Bombay nat. Hist. Soc.,* XXI (3): 778-785 (1912)[3], [forms a, b, and c.]

Odontotermes (Cyclotermes) obesus Silvestri *Rec. Indian Mus.,* XXV: 223-224, fig. II (1923)[4].

Cyclotermes obesus Beeson, *Forest Insects,* Dehra Dun: 535-536, 546-547, fig. 151, no. 5 (1941).

[4,13]Barkuda Islands (Orissa). [7,8]Luskerpore Valley. [9, 10, 12]Malabar Coimbatore. [9]Bellary; Bangalore Maddur and Bababudin 4,000-4,600 ft., (Mysore); Kathiawar ; Gurdaspur and Kasur (Punjab); Hoshangabad and Nagpur (M.P.), Malda, Harra and Bogra (Bengal). [9, 12, 12]Lyallpur (Punjab). [10, 12]Hadagalli. [9, 11, 15, 16]Dehra Dun (U.P.), [11]Pusa (Bihar). [14]Andaman Islands.

BIOLOGICAL.

Lefroy, *Indian Insect Life,* Calcutta, I: 120-121, text figs. 47-49, pl. VIII (1909). Assumth *J. Bombay nat Hist. Soc.,* XXII, (2): 372-375 (1913)[5]. Holmgren, *Ibid.,* XXII, (1): 104-105 (1914)[6]. Andrew, *J. Indian, Tea Ass.,* 1: 24-25 (1916)[7]. *Ibid.,* 2: 54-72 (1916)[8]. Holmgren, *Mem. Dep. Agric, India Ent.,* V. (3): 146 -149 (1917)[9]. Fletcher, *Proc. 3rd ent. Mtgs. Pusa,* I: 314 (1919)[10]. Fletcher and Ghosh, *Ibid.,* II: 705-712 (1919)[11]. Fletcher *Bull. Dep. Agric. India,* 100: 246 (1921)[12]. Annandale, *Rec., Indian Mus.,* XXV, (2): 233-251 (1923)[13]. Bose, *Ibid.,* XXV (2): 253-258 (1923). Silvesri, *Ibid.,* XXV (2): 223-224, fig. II (1923). Annandale, *J. Bombay nat. Hist. Soc.,* XXX (1): 26-34, fig. 1 (1924)[14]. Dover, *Indian For.,* LVII (7): 341-351 (1931)[15]. Ramakrishna Ayyar, *Madras Agric.* J., XXI (10): 421-430 (1933). Margabandu., *J. Bombay nat. Hist. Soc.,* XXXVII (3): 707 (1934). Beeson, *Indian For. Rec. Ent.,* IV, (2): 68 (1941)[16], *Forest Insects,* Dehra Dun: 535-536, 546-547, fig. 151, no. 5 (1941).

(Hosts: Dead trees, stumps, all types of wood in contact with soil, bark of lving plants and trees, sugarcane setts, fallen leaves and branches and dry cow dung, fungi such as *Xylaria-nigripes* and *Collybia albuminosa*).

var. **gurdaspurensis** Holmgren

Odontotermes obesus var. *gurdaspurensis* Holmgren, *Mem. Dep. Agric. India Ent.,* V (3): 149-150 (1917). Margabandu, *J Bombay nat. Hist. Soc.,* XXXVII (3): 707 (1934).

Gurdaspur (Punjab).

var. **oculatus** Silvestri.

Odontotermes (Cyclotemes) obesus var. *oculatus* Silvestri, *Rec. Indian Mus.*, XXV: 224, fig. II, no. 3 (1923)[1]. Beeson, *Forest Insects*, Dehra Dun: 546, fig. 152, 154 (1941). [1, 2, 3]Barkuda Islands (Orissa). [4]Andaman Islands.

BIOLOGICAL.

Annandale, *Rec. Indian Mus.*, XXV (2): 233 (1923)[2]. Bose, *Ibid.*,XXV: 253 (1923)[3]. Annandale, *J. Bombay nat. Hist. Soc.*, XX (1): 25-34, figs. 2, 4 (1924)[4]. Margabandu, *Ibid.*, XXXII (3) ; 707 (1934). Beeson, *Forest Insects*, Dehra Dun: 546, figs. 152, 154 (1941).

(Typically large mound building termite with fungus *Xylaria nigripes* in combs).

24. oblongatus Holmgren.

SYSTEMATICS.

Odontotermes oblongatus Holmgren, *K. svenska VetenskAkad. Handl.*, L. (2): 139, pl. V, fig. 5 (1913)[1]. Annandale, *Rec., Indian Mus.*, XXV (2): 233 (1923)[2]. Silvestri, *Ibid.*, XXV: 223 (1923)[3]. John, *Treubia*, VI (3-4): 389 (1925)[4]. Margabanndu, *J. Bombay nat. Hist. Soc.*, XXXVII (3): 708 (1934). [1]India. [2,3]Barkuda Islands (Orissa). [4]Malaya.

25. orissae Snyder.

SYSTEMATICS.

Termes orissae Snyder, *Indian For. Rec. Ent.*, XX (11) ; 10-11 (1934). Goilkera (Bihar),and Orissa.

(Soldier as in *Termes obesus* Rambur).

26. parvidens Holmgren.

SYSTEMATICS.

Odontotermes (Odontotermes) parvidens Holmgren, *Mem. Dep. Agric. India Ent.*, V. (3): 154 (1917)[1]. [1]Mysore, Gurdaspur (Punjab), Chaumuhani (Bengal), Kamakhya 1800 ft. (Assam).

BIOLOGICAL.

Fletcher, *Proc. 3rd ent. Mtgs. Pusa*, I: 313 (1919)[2]; Bull Dep. Agric, India, 100: 245 (1921). Margabandu, *J. Bombaynat. Hist. Soc.*, XXXVII (3): 708 (1934). Beeson, *Forest Insects*, Dehra Dun: 552 (1941). [2]Gauhati (Assam).

(Hosts: A serious pest of *Cajanus cajan (indicus)*, eating roots; bark of living trees, and timbers such as *Duabanga sonneratioides, Melanorrhoea glaba, Pinus longifolia, Shorea robusta* and *Sterculia villosa)*.

27. preliminarius Holmgren.

SYSTEMATICS.

Termes preliminarius Holmgresn, *in Escherich Termitenleben auf Ceylan.*, gustav Fischer, Publisher, Jena: 196 (1911)1. *Green, Spolia Zeylan.*, IX (33): 10 (1913)[2]. Margabandu, *J. Bombay nat. Hist. Soc.*, XXXVII, (3): 705 (1934). [1,2]Ceylon.

28. Redemanni Wasmann.

SYSTEMATICS.

Termes redemanni Wasmann, *Wien ent., Ztg.*, XII (7): 239-242 (1893). Desneux, *Genera Insect.*, XXV: 38 (1904). Holmgren, *in Escherich Termitenleben auf Ceylan*, Gustav Fischer, Publisher, Jena: 193, Pl. 3, fig. H (1911). Green, *Spolia zeylan.*, IX (33): 10 (1913). Hegh, *Les Termites*, Bruxelle: 48 (1922). Ceylon.

Odontoterms redemanni Beeson, *Forest Inestcs*, Dehra Dun: 547, fig. 155 (1941).

BIOLOGICAL.

Termes redemanni Petch, *Ann. R. Bot. Gdns. Peradeniya*, III : 189-262, Pl. VIII (1906). Desnuex. *Ann. Soc. ent. Belg.*, LI : 394-400 (1907). Bugnion, *mem. Soc. zool. Fr.*, XXV: 210-232 (1912). Green, *Trop. Agriculturist*, XL: 5 (1913). John, *Spolia Zeylan.*, IX (34): 116 (1913). Bugnion, *Bull. Mus. Hist. nat. Paris*, XX (4) : 185-188, 191-202, pl. II, fig. 1, pl. IX (1914). Green, *Rept. Trans. 3rd internat. Congr. trop Agric. London* : 608-636 (1916). Hutson, *Leafl. Dep. Agric. Ceylon*, 24: 3-4, pl. II (1923). Jepson, *Yearb. Dep. Agric. Ceylon* : 8-10, Pl. II (1924). Dover and Mathur, *Indian For. Rec. Ent.*, XX (7) : 3 (1934).

Odontotermes redemanni Holmgren, *J. Bombay nat. Hist. Soc.*, XXI (3); 783 (1912). Annandale, *Rec. Indian Mus.*, XXV : (2) : 284 (1923). Hutson, *Trop.*

Agriculturist, LIX : 85-87 (1923) ; *Ibid.,* LIX : 294 298 (1923). Annandale, *J. Bombay nat. Hist. Soc.,* XXX (1) : 27, 30, fig. 3 (1924). Jepson, *Trop. Agriculturist,* LXVII (2) : 68 (1926). Kemner, *Bull. ent. Res.,* XVi (4): 386 (1926). Rutherford, *Plantes' Note Book,* 8th ed., Colombo Cylon : 234, 534 (1926). Hutson, *Tech. Rep. Dep. Agric. Ceylon, 1928* 7-8 (1929). Jepson, *Bull. Dep. Agric. Ceylon,* 85 : 7-9, pl. I (1929); Tech. Dep. Agric. Ceylon, *1929* : 12 (1930). Margabandu, *J. Bombay nat. Hist. Soc.* XXXVII (3): 706 (1934). Mukerjee and Mitra *Proc. zool. Soc. Bengal,* 2 (1): 9-28 (1949).

Odonotermes (Cyclotermes) redemanni John, *Treubia,* VI (3-4) : 386 (1925).

Termes (Cyclotermes) redemanni Hutson, *Adm. Rep. Dir. Agric. Ceylon, 1931* : 117-118 (1932) ; *Ibid., 1932 : 128-129* (1933). Jepson, *Ibid., 1933* : 119-122 (1934).

Cylcotermes redemanni Hutson, *Adm. Rep. Dir. Agric. Ceylon, 1930 : 74* (1931). Jespon, *Trop. Agriculture,* LXXVI (2): 67-69 (1931); *Adm. Rep. Dir. Agric. Ceylon, 1934 : 119-10* (1935); *Ibid.,* 1935 : 51 (91936). Hutson, *Ibid., 1936:* 25-26 91937. Joachimand Kandiah, *Trop. Agriculturist,* LXXXXV (6) : 333-338, pl. I, figs. 1-3, pl. II (1940). Beeson,Forest *Insects,* Dehra Dun : 547, fig. 155 (1941).

[Hosts : Wood and wooden structures of buildings in contact with soil, dead and living plants of *Caemllia sinensis* (tea) and *Hevea* spp. (rubber), seedlings of *Cocos nucifera* (coconut palm) and *Crotalari brownei.* Burma timbers such as *Hompea odorata* and *Xylia dolabriformis* have been found immune to the attack of this species.]

29. teprobans Walker.

SYSTEMATICS.

Termes taprobanes Walker, *Catalogue, Neur. Insects Brit. Mus.,* C 3 : 522 (1853)[1]. Hagen, *Linn. Ent.,* XII : 136-138, pl., fig. 16 (1858) ; *Ibid.,* XIV : 103-104 (1860). Haltermann, *Botanische Untersuchungen, S.*

[1, 8]india and Ceylon.
[1, 2]Boreno.
[2]Java.

Schurendener : 411-420 (1899)[2]. Anonymous, *Bull. Dep. Ld. Rec. Agric. C.P.*, 6 : 3 (1902). Desneux, *Genera Insect.*, XXV : 38 (1904). Petch,* *Ann. R. bot. Gdns. Peradeniya*, III : 191 (1906). Lefroy, *Indian Insect Life*, Calcutta, I: 120 (1909)[3]. Green, *Spolia zeylan.*, IX (33) : 13 (1913†) Holmgren, *Ibid.*, VIII (32): 280 (1913[+]). Bugnion, *Bull. Mus. Hist. nat. Paris.*, XX (4): 185 (1914) Margabandu, *J. Bombay nat. Hist. Soc.*, XXXVII (3): 705 (1934).

[Hosts : Wood in contact with soil)

Note : A thorough study of the type sp[ecimens fo *Termes Taprobanes* and *Odontotemes horni* is definitely required so as to pass judgment wheather these are on enad the same species or different species.

Notes.– A through study of the type specimens of *Terms taprobanes* and *Odantotermes horni* is definitely required so as pass judgement whether these are one and the same species or different species.

*Petch quotes Haltermann [1899] who has mentioned that this species is also found in Borneo and Java.

† Holmgren [1913] points out that *Odontotermes horni* may be synonym of *Termes taprobanes.*

PART XXVIII–ANTHRIBIDAE (COLEOPTERA)

Sharp, *Ent. Mon. Mag.* 10, pp. 29-32, (1873), Redtenbacher, *Fauna Austr.* ed. 3, II, pp. 384-388, (1874), Seidlitz, *Fn. Balt.*, pp. 129-130, 486-490, (1875); *Fn. Transsylv.*, p. 170,-749, (1891), Leconte & Horn, *Proc. Amer. Phil. Soc.* 15, pp. 391-409, (1876). Sharp, *Ann. Mag. Nat. Hist.* (4) 17, pp. 422-439, (1876). Wollaston, *Col. St. Helena*, pp. 174-207, (1877), Ritsema in snelleman. *Midden-Sumatra* IV, Col., p. 126, (1887). gundlach. *Contrib. Ent. Cubana*, 5, pp. 268-271, (1891). Broun, *Man. New Zeal. col.* 5, pp. 1254-1271, (1893). Stierlin, *Col Helvet.* 8, pp. 420-424, (1894). Jordan *Now. Zool.* 1, pp. 591-651, (1894); 4, pp. 173-177, (1897); 5, pp. 359-373, (1898): 10, pp. 127- 130, 415-434, (1903); 11, pp. 230-309, (1904); 13, pp. 408-410, (1906); 16, pp. 307-308, (1909); 18, pp. 90-116, 601-604, (1911); 19, pp. 63-67, 87-90, 137-145, (1912); 20, pp. 257-277, 582-585, (1913); 21, pp. 224-234, (1914); 23, pp. 342-349, 359-363, (1916); 27, pp. 260-264, (1920); 30, pp. 167-185, 216-221, (1923); 31, pp. 227-260, (1924); 32, pp. 238-257, 290-291, (1925); 33, pp. 46-170, (1926); 34, pp. 77-128, 151-158, (1927-28); 35, pp. 244-249, (1929-30); 36, pp. 143-145, 281-310, (1930); 38, pp. 295-313, 362-383, (1932-33); 39, pp. 86-89, 275-291, 311-329, (1933-36); 40, pp. 199-261, 333-343, (1936-37); 41, pp. 140-147, 423-432, 437-442, (1938-39); *Stett. Ent. Zeit.* 56, pp. 122-204, 245-265, 369-401, (1895); *Ann. Mus. Geneva,* (2) 38, pp., 624-643, (1897); (3) 41, pp. 80-91, (1904); (3) 49 pp. 46-56, (1920); *Deutsche Ent. Zeitschr*, 2, pp. 369-373, (1901); 1, pp. 76-78, (1902); *Biol. Cenr. Amer. Col.* 4, pp. 299-383, (1906); *Nova Guinea V. Zollogie,* pp. 351-352, (1908); *Tijdschr. Ent.* 55, pp. 127-142, (1912); 58, pp. 44-47, 48-51, (1915); 59, pp. 160-162, (1916); *Rec. Ind. Mus.* 8, pp. 197-198, (1913); 9. pp. 203-216, (1913); *Trans. Linn. Soc. Lond.* 16, pp. 247-267, t. 15, (1914); *Voy. All. et Jeann, Afr. or Col.* 10, pp. 329-346, (1914); *Opusc. Inst. Scient Indochine, fn. Ent.* 1, pp. 71-109, (1923); *Ann. Mag. Nat. Hist.* (9) 13, pp. 606-620 ; (1924); (9) 16, pp. 257-264, (1925); *Ins. Samoa,* 4, Col. 2, pp. 161-172, (1928). Schaeffer, *Trans. Am. Ent. Soc.* 32, pp. 267-278, (1906). Schaufuss in *calwer Käferb.* ed. 6. pp. 1024-1029, (1913). Kuhnt, *Illustr. Best.-Tab. Käf Deuschl.*, pp. 899-902, (1913). Stebbing. *Indian Forest Insects,* pp. 386-388, (1914). Reitter, *Fauna Germ.*, 5, pp. 2-10, (1916). Henriksen, *Danm. Fauna,* 22, pp. 8-15, (1918). Heller, *Tijdschr. Ent. 61,* pp. 242-265, (1918); *Stett. Ent. Zeit.* 84, pp. 14-17, (1923); *Ent. Blätt.* 21, pp. 97-108, (1925). Wolfrum, *Ent. Mitt.* 11, pp. 56-63, (1922); 13, pp. 35-40, (1924). Everts, *Col. Neerland* 3, pp. 500-502, (1922). Munster. *Norsk Ent. Tidskr.* 1, P. 86, (1921). Jordan, *Nat. Hist. of Juan Fernandez and Easter Island,* 3, pp.

479-480, (1926); *Treubia,* 7, pp. 331-340, (1929); *Boll. Soc. Ent. Ital.* 61, pp. 165-166, (1929); *Proc. R. Ent. Soc. Lond.,* 4, pp. 50-52, (1929); *mem. Mus. Nat. paris,* 9, pp. 171-173, (1939); *Proc. R. Ent. Soc. Lond.* (A) 18, pp. 99-105, (1943); (B) 14, pp. 10-20, 111-112, (1945); *Ent. Mon. Mag.* 78, pp. 182-191, (1942), *Entomologist,* 77, pp. 141-142, (1944); *Proc. Haw. Ent. Soc.* 12(3), pp. 517-524, (1946); *Ark. För Zool.* 41 A (6), pp. 1-9, (1949). Pierce, *Proc. U. S. Nat, Mus.* 77, (Art. 17), (2840); 34, pp. (1930), Wolfum, *Ent. Blätt.* 26(2), pp. 89-91, (1930); 30 (1), pp. 2-19, (1934); 34, pp. 67-76, (1938); 41-44, pp. 133-148, (1945-48). Gardner, *Indian For Rec.* 16 (11), pp. 327-334 (larvae), (1932); 2 (2), (N.S.), pp. 99-112, (larvae), (1936); 3 (6), pp. 127-128, (larvae), (1937). Zimmerman, *Proc., Haw. Ent. Soc.* 10(1), p. 152, (1938); *B. P. Bishop Mus.,* Occ. Papers, 14(13), pp. 219-250, (1938); *Bull. Bishop Mus.* No. 172, pp. 65-72, (1942). Beeson, *Forest Insects,* pp. 43-50, (1941). Andreson, *Proc. Ent. Soc. Wash.* 46 (7), pp. 171-175, (larvae), (1943); *Ann. Ent., Soc. Amer.* 40 (3), pp. 489-517, (larvae), (1947). Hoffmann, *faune de France,* 44, pp. 1-184, (1945). Bradley, *Bull. Brooklyn Ent. Soc,* 41, pp. 96-99, (1946). Mathur, *Indian For Rec.* 8(3), pp. 13-16, (larvae) (1950).

See Wolfum, *Coleopterorum Catalogus,* 102, (1929), and Supplement (1953), for detailed synonymy.

1. **Acorynus** Schönherr.

Acorynus, Schönherr, *Gen Curc.* 1, p. 123, (1833). Imhoff, *Gen. Curc.* 1, p. 45, (1842). Lacordaire, *Gen. Col.* 7, p. 513, (1866). Jordan, Nov. Zool. 1, p. 608, (1894). Bovie, *Ann. Soc. Ent. Belg.* 49, p. 235, (1905). Jordan, *Rec. Ind. Mus.* 9. pp. 205 206, (1913). Wolfrum. *Col. Cat.* 102, p. 23, (1929).

Homoeotropis Kolbe, *Stett. Ent. Zeit.* 55, p. 381, (1895). Jordan, *Vay. All. et Jean Afr. Or. Col.* 1 (10), p. 335, (19140.

anacis Jordan

Acorynus anacis Jordan, *Nov. Zool.* 33, p 162, (1926)[1]. Wolfrum, *Col. Cat.* 102, p. 23, (1929).

[1]Khasis Hills: Assam.

angulieer (Walker)

Xenocerus angulifer Walker, *Ann. Mag. Nat. Hist.* (3) 3, p. 262, (1859)[1].

[1]Ceylon.

Sympaector angulifer Bovie. *Ann. Soc. Ent. Belg.* 49, p. 266, (1905). wolfrum, *Col. Cat.* 102, p. 52, (1929).

Acorynus angulifer (Walker).

apicalis Jordan

Acorynus apicalis Jordan, *Nov. Zool.* 1, pp. 623-624, (1894)[1]; *Stett. Ent. Zeit.* 56, pp. 154—157, (1895); *Nov. Zool.* 10, p. 423, (1903). Bovie, *Ann. Soc. Ent. Belg.* 49, p. 236, (1905). wolfrum, *Col. Cat. 102,* p. 23, (1929).

[1]Sikkim.

carinifrons Jordan

Acorynus carinifrons Jordan, *stett. Ent. Zeit.* 56, pp. 155-156, (1895)[1]; *Nov. Zool.* 11, p. 232, (1904). **Bovie,** *Ann. Soc. Ent. Belg.* 49, p. 236, (1905). Jordan, *Rec. Ind. Mus.* S, p. 197, (1913)[2]. Wolfrom, *Col. Cat.* 102, p. 24, (1929). gardner, *Indian For. Rec.* (N. S.)[2], p. 107, pl. 2, figs. 24, 25, (larval characters) (1936). Beeson, *Forest Insects,* p. 45, (biology) (1941). [Food-plants ; *Anthoecphalus cadamba,* "Phendua", *Quercus spicata*].

[1]Khasis Hills: Assam.

[2]Dejoo: Assam.

[Distribution: Samsing 1800 ft. (Kalimpong Division); Tista Valley 700 ft. Darjeeling Division: Bengal].

ceylonicus Jordan

Acorynus ceylonicus Jordan, *Nov. Zool.* 1, p. 622, (1894)[1]; *Stett. Ent. Zeit.* 56, p. 154, (1895). Bovie, *Ann. Soc. En.t Bleg.* 49, p. 236, (1905). Jordan, *Nov. Zool.* 33, p. 158, (1926). Wolfrum, *Col. Cat.* 102, p. 24, (1929).

[1]Ceylon.

cingalsu Jordan

Acorynus cingalus Jordan, *Nov. Zool.* 33, p. 158, (1926)[1]. Wolfrum *Col. Cat.* 102, p. 24, (1929).

[1]Kandy: Ceylon.

dohertyi Jordan

Acorynus dohertyi Jordan, *Nov. Zool* 33, p. 150, (1926)[1]. wolfrum *Col. Cat.* 102, p. 24, (1929)[2]. **Jordan**, *Ark. för Zool.* 41A(6), p. 3, (1949)[3].

[1]Manipur: Assam.
[2]Burma
[3]Kambaiti: N. E. Burma.

dohrni Jordan

Acorynus dohrni Jordon, *Sett. Ent. Zeit.* 56, p. 151, (1895), Bovic, *Ann. Soc. Ent. Belg.* 49, p. 237, (1905). Wolfrum, *Col. Cat.* 102, p. 24, (1929).

[1]Ceylon.

labidus Jordan

Acorynus Labidus Jordan, *Nov. Zool.* 31, p. 231, (1924)[1]; *Nov. Zool.* 33, p. 162, (1926). Wolfrum, *Col. Cat.* 102, p. 24, (1929). Beeson, *Forest Insects,* p. 49, (biology) (1941). [Food-plants: *heritiera fomes, Tectona grandis, Terminalia tomentosa,* unknown wood.]

[1]Nalanda, Weligama: Ceylon.

[Distribution: Sunderbans: Bengal. Amarapalam, Nilambur, Wynaad: Madras, Dehra dun, Jhajra: U.P. Saingy-yam (N. Toungoo): Burma].

laenatus Jordan

Acorynus laenatus Jordan, *Nov. Zool.* 41, p. 140, (1938)[1]. Beeson, *Forest Insects,* p. 45, (biology) (1941). [Food-plant: *Albizzia lebbek].*

[1]Yanaungmyin, Pyinmana: Burma.

malaisei Jordan

Acornyus malaisei Jordan, *Ark. för. Zool.* 41A (6), p. 3, (1949)[1].

[1]Kambaiti: Burma.

Passerinus (Pascoe)

Litocerus passerinus Pascoe, *Ann. Mag. Nat. Hist.* (3) 5, p. 47, (1860)[1]. Bovie, *Ann. Soc. Ent. Belg.* 49, p. 242, (1905).

[1]Borneo.

Acorynus passerinus Jordan, *Rec. Ind. Mus.,* 9, p. 206, (1913)[2]; *Nov. Zool.* 31, p. 231, (1924); *Nov. Zool.* 33, p. 162, (1926). wolfrum *Col. Cat.* 102, p. 25, (1929).

[1]Peradeniya: Ceylon.

similis Jordan

Acorynus similis Jordan, *Nov. Zool. 1, pp* 620-621 (1894)[1]. Bovie, *Ann. Soc. Ent. Belg. 49,* p. 238, (1905). Wolfrum, *Col. Cat.* 102, p. 26, (1929).

[1]Sikkim.

solutus Jordan

Acorynus solutus Jordan, *Ark. för Zool.* 41A (6), p. 4, (1949)[1].

[1]Kambaiti: Burma

2. **Androceras** Jordan

Androceras, Jordan, *Nov. Zool.* 34, p. 83, (1928). Wolfrum, *Col. Cat.* 102, p. 31, (1929).

gerrhus Jordan

Mucronianus gerrhus Jordan, *Nov. Zool.* 30, p. 217, (1923)[1]. *Androceras gerrhus,* Jordan, *Nov. Zool.* 34, p. 84, (1928). Wolfrum, *Col. Cat.* 102, p. 31, (1929).

[1]Nengpoh, Khasis Hills: Assam.

khasianus Jordan

Micronianus khasianus Jordan, *Nov. Zool.* 10, pp. 427-428, (1903)[1]. Bovie, *Ann. Soc. Enty. Belg.* 49, p. 245, (1905).

[1]Khasis Hills: Assam.

Androceras Khasianus Jordan, *Nov. Zool.* 34, p. 83, (1928)[2]. Wolfrum, *Col. Cat.* 102, p. 31, (1929) Gardner, *Indian For.* Rec. (N. S.) 2 (2) p. 103, Pl. 1, figs. 13, 14, (larval characters) (1936).

[2]Tonkin. Indochina

Beeson, *Forest Insects,* 45, (biology) (1941). [Food plants: *Terminalia myriocarpa,* Unknown wood]

[Distribution: Sadiya Division: Assam. Gajalduba Jalpaiguri Division); Samsing 1800 ft. (Kalimpong Division): Bengal.

pulcherrimus Jordan

Androceras pulcherrimus Jordan. *Nov. Zool.* 38, pp. 310-311, (1932)[1]. Beesn, *Forest Insects,* pp. 45, (biology) (1941). [Food-plant: *Alnusnepalensis*].

[1]Lopchu 5000 ft. Darjeeling Division: Bengal.

3. **Anthribus** Fabricius*

Anthribus, Fabricius, *Ent. Syst.* 1(2), p. 375, (1792); *Syst. El.* 2, p. 404, (1801) (partim). Imhoff, *Gen. Curc.* 1, p. 46, (1842). Redtenbacher, *Fauna Austr.* p. 472, (1849), ed. 2, p. 675, (1858), ed 3, II, p. 386, (1874). Jacq. du Val. *Gen. Col. d'Eur.* 4, p. 5, (1854—68). Bose-Gutfl. *Käf. Deutschl.*p. 458, (1859). C. G. Thoms. *Skand Col.* 1, p. 17, (1865). Lacordaire, *Gen. Col.* 7, p. 574, (1866). Seidl, *Fn. Balt.* p. 469, (1872—1875). C. Schenkl. *Deutsche Käferw,* p. 223, (1885). Siedil, *Fn. Transsylvan.* p. 748, (1888—91). Sharp, *Trans. Ent. Soc. Lond.* p. 318—319, (1891). Stierlin, *Col. Helvet.* p. 421, (1892). Broun, *Man. New Zeal. Col. 5,* p. 1254, (1893). Acloque, *Fn. France Col.* p. 355, (1896). Lameere, *Fn., Belg.* 2, p. 473,

* (nec Geoffroy 1762 nec Föester 1771); according to Dr. Jordan.

(1900). Bovie, *Ann. Soc. Ent. Belg.* 49, p. 292, (1905). Jordan, *Rec. Ind. Mus.* 9, p. 212, (1913). Wolf rum, *Col. Cat.* 102, p. 77, (1929).

Macrocephalus, Oliveir, *Ent.* 4 (80), p. 5, (1795). Bedel, Fn. *Col. Bassin Seine,* 6, pp. 7, 13, (1882-88). Everts, *Col. Neerland,* 2, p. 533, (1903); 3, p. 501, (1922).

Platystomus, Schaufuss, in *Calwer, Käferb.* ed. 6, p. 1028, (1913). Kuhnt, *Illustr. Best.-Tab. Käf. Deutschl.* p. 901, (1913). Reitter, *Fn. Germ.* 5, p. 7, (1916). Henriksen, *Denm. fauna,* 22, p. 12, (1918).

macrocerus macrocerus Jordan

Anthribus macrocerus macrocerus Jordan, *Nov. Zool.* 1, p. 235, (1904)[1]. Bovie *Ann. Soc. Ent. Belg.* 49, p. 293, (1905). Jordan, *Rec. Ind. Mus.* 8, p. 198, (1913)[2]; 9, p. 213, (1913)[3]. Wolfrum, *Col. Cat.* 102, p. 78, (1929). [Food-plant: *Amoora rohituka*].

[1, 3]Sikkim.
[2]Dejoo: Assam. Darjeeling: Bengal. Distribution: Okkyi (Shwegu Division): Burma].

macrocerus andamanensis Jordan

Anthribus macrocerus andamanensis Jordan, *Rec. Ind. Mus.* 9, p. 213 (1913)[1]. wolfrum, *Col. Cat.* 102, p. 78, (1929).

[1]Andamans.

Wallacei malaicus Jordan

Anthribus wallacei malaicus Jordan, *Nov. Zool.* 11, p. 235, (1904)[1]. Bovie, *Ann. Soc. ent. Belg.* 49, p. 293, (1905). Jordan. *Rec. Ind. Mus.* 9. p. 213, (1913)[2]; *Tijdschr. Ent.* 59 p. 161, (1916)[3]. Wolfrum *Col. Cat.* 102, p. 78, (1929)[4].

[1, 2]Borneo.
[1]Sumatra. Malacca.
[3]Tjigembong:Java.
[4][Distribution: Pyonchaung (N. Toungoo Division): Burma].

4. **Apolecta** Pascoe

Apolecta, Pascoe, *Ann. Mag. Hist.* (3) 4, 431, (1859). Lacordaire, *Gen. Col.* 7, p. 555, (1866). Bovie, *Ann. Soc. Ent. Belg. 49,* p. 317, (1905), Jordan, *Nov. Zool,* 23, pp. 342—349, (Key to species), (1916). Wolfrum, *Col. Cat.* 102, p. 102, (1929).

andrewesi Jordan

Apolecta anadrewesi Jordan, *Nov. Zool.* 23, pp. 342-343, 345, 346, (1916)[1]. Wolfrum, *Col. Cat.* 102, p. 102, (1929).

[1]Nilgiri Hills; South India.
[Distribution: Coorg]

neitneri Lacordaire

Apolecta neitneri Lacordaire *Gen. Col.* 7, p. Atlas tab 80, Figs., 3, 30 (1866) Jordan stett 28, *Ent. Zeit.* 56, pp. 179, 180, (1895); *Nov. Zool.* 11, p. 237, (1904). Bovie, *Ann. Soc. Ent. Belg.* 49, p. 317, (1905). Jordan, *Nov. Zool.* 23, pp. 342, 343, 345, 346, (1916). wolfrum, *Col. Cat.* 102, p. 102, (1929).

[1]Ceylon.

5. **Araecerus** Schönherr

Araecerus, Schönherr, *Disp. Meth.* p. 40, (1826); *Gen. Curc.* 1, p. 173, (1833). *Bovie, Ann. soc. Ent. Belg.* 49, p. 320, (19050. Jordan, *Biol. Centr—Am. Col.* 4. (6), p. 381, (1906); *Rec. Ind. Mus.* 9, p. 215. (1913); *Trans. Linn. Soc. Lond.* 16, p. 267, (1914); *Nov. Zool.* 34, p. 94, (1928). wolfrum, *Col. Cat.* 102, p. 105, (1929). Pierce, *Proc. U.S. Nat. Mus.* 77 (17), pp. 31-32, (biology), (1930). Auturoi, *Rev. Ent. s. Paulo,* 1, pp. 52—61, (description, biology), (1931). Gardner, *Indian For. Rec.* 2 (2) p. 100 (Biology) p. 110 (larval characters) (1936). Jordan, *Proc. How. Ent. Soc.* 12 (3), pp. 517—524, (1946). Zimmerman, B. P. *Bishop Mus. Occ. Papers,* 14, (13), pp. 241-242, (1938); *Bull. Bishop Mus.* 172, p. 71, (1942).

Araeocerus, Schonherr, *Gen. Curc.* 5, p. 273, (1839). Imhoff. *Gen. Curc.* 1, p. 55, (1842) Walker, *Ann. Mag. Nat. Hist.* 3 (3), p. 262, (1859). Lacordaire, *Gen. Col.* 7, p. 590, (1866). Broun, *Man. New Zeal. Col.* 1, p. 558, (1880). Bedel, *Fn. Col. Bassin Seince.* 6, pp. 11—16, (1882-83). Leconte & Horn, *Classif. Col. N. Amer.,* p. 530, (1883). Seidl. *En. Transsylv.* p. 749, (1888—91), Stierlin, *Col. Helvet,* p. 423, (1894). Jordan, *Stett. Ent. Zeit.* 56, p. 203, (1895). Blackburn, *Trans. R. Soc. S. Austr.* 24, p. 162, (Table), (1900), Everts, *Col. Neerland,* 2, p. 535, (1903). Schaufuss, *Kâferb.* ed. 6, p. 1029, (1913). Kuhnt, *Illustr. Best-Tab. Käf. Deutschl.* p. 902, (1913). Reitter. *Fauna Germ.* 5, p. 10, (1916). Hernriksen, *Danm. Fauna,* 22, p. 14, (1918).

areolatus Pascoc

Araecerus areolatus Pascoe, *Ann. Mag. Nat. Hist.* (3) 4, pp. 438-439, (1859).[1] Bovie, *Ann. Soc. Ent. Belg.* 49, p. 320, (1905). Wolfrum, *Col. Cat.* 102, p. 106, (1929). [Food-plant: Unknown seeds].

[1]Borneo.

[Distribution: — Middle Andaman].

bradytus Jordan

Aracerus bradytus Jordan, *Nov. Zool.* 31, pp. 247, 248, (1924)[1]. Wolfrum, *Col. Cat.* 102, p. 106, (1929). — [1]Nilgiris: Madras,

candicans Jordan

Araecerus candicans, Jordan, *Nov. Zool.* 38, pp. 312-313, (1932)[1]. Beeson, *forest Insects,* p. 45, (biology) (1941). [Food-plant: *Termindalia procera*]. — [1]North Andaman.

fasciculatus DeGeer

Araecerus fasciculatus DeGeer, *Mem. Ins.* 5, p. 276, tab. 16 fig. 2, (1775) *(Curuculio fasciculatus)*[1] Wollaston., *Ann. Mag. Nat. Hist.* (3) 5, p. 18, (1870); *Col. St. Helena,* p. 174, (1877). Leconte, *Pro. Amer. Phil. Soc.* 15, p. 407, (1876). Bedel. *Fn. Col. Bassin Seine,* 6, p. 11, 16, (1882—1888). Letzner, *Jahresber, Schles. Ges.* 60, p. 308, (1883). Sharp, *Trans. Ent. Soc. Lond.* p. 322, 323, (1891). Seidl, *Fauna Transsylv.* p. 749, (1888—91), Stierlin, *Col. Helvet,* p. 424, (1894). Chittenden, *Bull. U.S. Dep. Agric. Ent.* 8, p. 36, (1897). Barlow, *Ind. Mus. Notes,* 4 (3), p. 125, (1899). Zehntner, *Bull. Proefstat, Salatiga,* 3, pp. 1—17, (1902). Everts, *Col. Neerland,* 2, p. 536, (1903); 3, p. 502 (1922). Jordan, *Biol. Centr. Amer.* Col. 4 (6), p. 381, (1906); *Rec. Ind. Mus.* 9, p. 216, (1913)[2]; *Voy. All et Jeann. Afr. or. Col.* 1 (10)p. 344, (1914); *Trans. Linn. Soc. Lond.* 13, p. 267, (1914)[3]; *Tijdschr. Ent. 58,* p. 51, (1915); 59, p. 162, (1916); *Ann. Mus. Genova,* (3), 9, p. 56, (1920); *Ent. Mitt.* 11, p. 157, (1922); *Opusc. Inst. Scient. Indochine, Fn. Ent.* 1, p. 109, (1923)[4]; *Ins. samoa, 4, Col. 2,* p. 164, (1928). C. Schaufuss, in *Calwer, Käferb.* ed. 6, 1907—1916 p. 1029, (1913). Kuhnt, *Illustr. Best. Tab. Käf. Deutschl.* p. 902, (1913). Reitter *Fauna Germ.* 5, p. 10, 5. 153, f, 11, (1916)[5]. Henriksen. *Danm. Fauna,* 22, p. 14, (1918). Bovie, *Ann. Soc. Ent. Belg.* 49, p. 320, (1905). Lefroy, *Ind. Ins. Life,* 1, p. 380 (1909). Beeson, *Ind. forester,* 45 (2), p. 50, (1919). Cotton, *Jour. Agr. Res.* 20, p. 606, (larval characters; (1921). Jordan, *Nov. Zool.* 31, p. 246, (1924); 34, p. 104, (1928)[6]; *Proc. Ent. survey Publ.* 6 (12), p. 69, (1933); *Nov. Zool,* 39, p. 275, 291, (1936)[7], *Proc. Haw. Ent. Soc.* 12 (3), p. 517, (1946); *Ark. för Zool.* 41 A (6), p. 9, (1949)[8]. Wolfrum, *Col. Cat.* 102, p. 106-107, (1929). Petch, *Ta Quarterly,* 2

[1]Surinam: Guiana

[2]Paradeniya ; Ceylon. Calcutta: Bengal.

[3]Throughout the tropics.

[4]Tonkin. Laos.

[5]Germany.

[6]Kuala Lumpur, Selangar.

[7,9]Mauritius.

[8]Sedan: N.E. Burma.

(1), p. 22, (1920). Garnder, *Ind. For. Rec.* 16 (11), p. 328, (1932); 2(2), p. 100, 110, (larval characters and biology) (1936). Vinson, *Roy. Soc. Arts & Sci. Mauritius,* (c) 3, p. 188, (1935)[9]. zimmerman, *B. P. Bishop Mus., Occ.* Papers, 14 (13), p. 242, (1938); *Bull. Bishop Mus.* 172, p. 71, (1942). Bianchi, *Proc. Haw. Ent. Soc.* 13 (1), p. 17, (1947). Anderson, *Ann. Ent. Soc. Amer.* 40 (3), p. 514, (1947). Masayoshi Uno, *ent. Rev. Japan,* 5 (1), p. 49, (1950)[10].

[9]Mauritius.

[10,11]Japan.

Bruchus cacao Fabricius, *Syst. Ent.* p. 64, (1775); *Syst. El.* 2, p. 397, (1801). Olivier *Ent.* 4, (80), p. 15 t 2, f, 21 a-b, (1775).

Anthribus coffeae Fabricius Syst. *EI. 2,* p. 411, (1801). Gyllenhal, in *Scch. Gen. curc.* 1, p. 175, (1883). Imhoff. *Gen. Curc.* 1, p. 55, (1842).

Araccerus griesus Stephens, *Ill. Brit. Ent. Mandib.* 4, p. 211, t. 21, f. 2, (1834).

Amblycerus japonicus Thunberg, *Nova Acta Ups.* 7, p. 122, (1815).

Brachytarsus niveovariegatus Roelofs, *Ann. Soc. Ent. Belg.* 22, (Bull), p. 55, (1878)[11].

Araecerus seminarius Chevrolat, *Ann. Soc. Ent. Belg.* 14, p. 7. pl. 1, fig. 5, (1871).

BIOLOGY

Lucas, *Bull. Soc. Ent. France,* p. 67, (1877). Glover, Monthly Rep. U.S. Dept. Agr. p. 367, (1872); *Second Rep. Commiss. Agr.* for 1872, 114f, (1873). Muller, *Proc. Ent. Soc. Lond.* p. 9. (1873); *Canad. Ent. 5,* p. 156, (1873). Schwarz, *Proc. Amer. Phil.* Soc. 17, p. 469, (1878). Blackburn & Sharp, *Trans. Roy. Dublin Soc.* (3)3, p. 360, (1885). Everts. *Tijdsehr. Ent.* 28, p. 107, (18850. Chittenden. *Bull. U.S. Dept. Agr. Div. Ent.* 4, p. 129 (1896); 8, p. 36, (1897). Howard, *Bull. 33 Office Ept. Stat. U.S. Dept. Agr.,* p. 348, (1896). Barber, *Ind. Mus. Notes,* 4, p. 125, t. 11, f. 3a-d, (1899). smith, *Ann. Rep. Agr. New Jersey* (1899), 27, p. 367, (1900).

Perkins, *Fauna Howaiensis,* 2 (3), p. 182, (1900). Delacroix, *Les maladies des Coffeiers,* p. 154, (1900). Seurat, *Bull. Mus. Paris,* 6, p. 410, (1900). Koltze, *Verh. Ver. Nat. Unterh. Hamburg,* 11, p. 50, (1901). Zehntner, *Bull. Proefstate. Salatiga,* 3, pp. 1-17, (1902). Theobald, *First Rep. Econ. Zool.,* p. 137, (1903). Verrill,

Trans. Connecticut Acad. 11, p. 786, f, (1903). Broun, *Jour. N. York Ent. Soc.* 14, p. 116, (1906). Reh, *Zeischr. wiss. Ins. Bio.* 3, pp. 23—25, (1907). Van dine, *Ann. Rep. Haw. Agr. Expt. State* p. 48, (1907); p. 31, (1908). Giffard, *Proc. Hawaiian Ent. Soc.* 1, p. 181, (1908). Day. Ent. Mon. Mag. 44, pp. 265—267, (1908). Tucker, *Bull. U.S. Dept. Agr.* 64 (7), p. 61, t. 3, (1909); *Jour. Econ. Ent.* 2, pp. 373-381, (1909). Lefroy, *Ind. Ins. Life,* 1, p. 380, (1909). Jordan, *Rec. Ind. Mus.* 9, p. 216, (1913), Van der Goot, *Meded. V. h-Proefst. Miden Java,* BUll. 26, pp. 1—36, (1917). Swezey, *Proc. Haw. Ent. Soc.* 4 (3), p. 452, (1921). Rama Krishna Ayyar, *Rept. Proc. Fourth Ent. Meet. Pusa,* p. 33, (1921). Jordan, *Opusc. Inst. Scient. Indochine, Fn. Ent.* 1, p. 109, (1923). Autuori, *Rev. Ent. S. Paulo.* 1 (1), pp. 52—61, (1931). Hutson, *Trop. Agric.* 78, p. 140, (1932). Kempers, *Tijdschr. Ent.* 75 (suppl.), pp. 60—70, 18 figs, (1932). Taher el Sayed, *Ann. Appl. Biol.* 22, pp. 557—577, (1935). Buck & Cotton, *U.S. Dept. Agr. Farmer's Bull.* 1260, p. 5, (1940). Beeson, *Forest Insects,* pp. 45-46, (1941). Usman, *Indian Coffee Board, Monthly Bull.* pp. 1-8, (1949). Downes & Williams, *Colonial Plant and Animal Products,* 1(1), p. 40, (1950). Downes, *Eighth Intern. Congr. Ent.* 1949. Stockholm, p. 2, (1950).

[Food-plants: *Anona squamosa, Areca catechu, Artocarpus hirsuta, Capsicum frutescens* (fruit), *Carica papaya* (dry stems), *Cassia occidentalis, Citrus sp.* (dry fruits), *Coffea arabica* (seeds), *Carchorus capsularis* (capsules), *Crotolaria striata, Gardenia* (flower buds). *Indigofera* sp., *Leucaena glauca, Melanorhoea usitala* (seeds), *melia* sp. (fruits), Parwal (stems), *Paplionaceae, Pentacme suavis* (seeds), *Pinus longifolia, Tephrosia, candida, Terminalia catappa* (seeds)].

[Distribution: Andamans, Calcutta: Bengal. Coimbatore: Kas/aragod, S. Mangalore: Madras, Dehra Dun : Uttar Pradesh. West Katha Division: Burma. Ceylon. Cosmopolitan].

fragalis (Walker)

Tropideres fragalis Walker, *Ann. Mag. Nat. Hist.* (3) 4, p. 220, (1859)[1]. Pascoe, *Ann. Mag. Nat. Hist.* (3) 4, p. 439, (1859). Bovie, *Ann. Soc. Ent. Belg.* 49, p. 269, (1905).

Araecerus fragalis Wolfrum, *Col. Cat.* 102, p. 107, (1929).

[1]Ceylon.

intangens Walker

Araecerus intangens Walker, *Ann. Mag. Nat. Hist.* (3). 3, p. 262, (1859).[1] Bovie, *Ann. Soc. Ent. Belg.* 49, p. 321, (1905). wolfrum *Col. Cat.* 102, p. 107, (1929). Gardner, *Indian For. Rec.* (N. S.) 2 (2), p. 110, (larval characters) (1936). — [1]Ceylon.

[Food plants: *Jatropha curcas* (Fruits), *Punica granatum* (decaying fruits)]. — [Distribution: — Dehra Dun: U.P.].

irrsolutus (Walker)

Hylesinus irresolutus Walker, *Ann. Mag. Nat. Hist.* (3) 3, p. 261, (1859)[1]. — [1]Ceylon.

Aracerus irresolutus Jordan, *Stett. Ent. Zeit.* 56, p. 204, (1895). Bovie, *Ann. Soc. Ent. Belg.* 49, p. 321, (1905). Wolfrum, *Col. Cat.* 102, p. 107, (1929).

***pardalis** Pascoe

Araecerus pardalis Pascoe, *Ann. Mag. Nat. Hist.* (4) 17, pp. 58-59, (1876)[1]. Broun, *Man. New Zealand Col.* 1, p. 559, (1880).[2] Bolvie, *Ann. Soc. Ent. Belg.* 49, p. 321, (1905). Wolfrum, *Col. Cat.* 102, p. 108, (1929). — [1]Auckland. Ceylon. [2]New Zealand.

pumilus Jordan

Aracecerus pumilus Jordan, *Nov. Zool.* 31, p. 247, (1924).[1] Wolfrum, *Col. Cat.* 102, p. 108, (1929). — [1]Weligama: Ceylon.

Simulatus Gyllenhal

Araecerus simulatus Gyllenhal, in *Schh. Gen. Curc.* 1, p. 174, (1833).[1] Bovie, *Ann. Soc. Ent. Belg.* 49, p. 322, (1905). Jordan, *Nov. Zool.* 31, p. 248, (1924). Wolfrum, *Col. Cat.* 102, p. 108, (1929)[2]. Jordan, *Nov. Zool.* 36, p. 303, (1931). Beeson, *Forest Insects*, p. 46, (biology), (1941). [Food-plant: *Dalbergia fusca* (seeds)]. — [1]Java., [2]Sumatra. Philippines. [Distribution: Cochin: S. India. North Toungoo Division: Burma. Peradeniya: Ceylon].

suturalis Boheman

Araecerus suturalis Boheman, in *schh. Gen. Curc.* 5, p. 273, (1839)[1]. Bovie, *Ann. Soc. Ent. Belg.* 49, p. 322, (1905). Jordan, *Nov. Zool.* 31, pp. 227, 250, — [1]Calcutta: Bengal.

---

* Not Indian species. Occurs only in New Zealand, "Ceylon", error, according to Dr. Jordan.

(1924)[2]. Welfrum, *Col. Cat.* 102, p. 108, (1929). Gardner, *Indian For. Rec.* 16 (11), pp. 327-328, pl. 1, figs. 1-5, (larval character), (1932). Jordan, *Indian For. Rec.* 18(3), p. 7, (1933)[3]. Gardner *Indian For. Rec.* (N. S.) 2 (2), p. 110, (larva), (1936). Beeson, *forest Insects.* p. 46, (biology), (1941).

[2]Rodriguez Island.

[3]Fraserpet: Coorg. Aiyur, Jawalagiri: Madras. Pusa: Bihar Dehra Dun: Uttar Pradesh. Travancore. Ceylon

[Food-plants: *Acacia arabica, Aleurites fordii* (fruits), *A. moluccana* (fruits), *Anona squamosa* (dry fruits), *Carica papaya* (dry stems), *Ficus glomerata* (fallen ripe fruits), *Jatropha curcas* (fruits), *Mangifera, indica* (dead buds), *Melia azedorach* (fruits), *M. indica* (seeds), *Trachycarpus excelsa* (seeds)].

[Distribution: Matala, Saduganga: Ceylon].

6. **Atinellia** Jordan

Atinnellia, Jordan, *Nov. Zool.* 32, p. 252, (1925). Wolfrun, *Col. Cat.* 102, p. 60, (1929).

senex senex Jordan

Atinellia senex senex Jordan, *Nov. Zool.* 32, p. 252, (1925)[1]. Wolfrum. *Col. Cat.* p. 60, (1929) Jordan, *Indian For. Rec.* 18 (3), p. 3, (1933).

[1]Kandy: Ceylon.

senex grisea Jordan

Atinellia senex grisca Jordan, *Indian For. Rec.* 18 (3), p. 3, (1933)[1].

[1]Fraserpet: Coorg. Denkanikota, Jawalagiri, Noga-noor : Madras.

7. **Autotropis** Jordan

Autotropis, Jordan, *Nov. Zool.* 31, p. 241, (1924). Wolfrum, *Col. Cat.* 102, p. 59, (1929). Gardner, *Indian For. Rec.* 2 (2), p. 100, (biology), (1936).

downingi Jordan

Autotropis downingi, Jordan, *Nov. Zool.* 31, p. 242, (1924)[1]. Wolfrum, *Col. Cat.* 102, p. 59, (1929).

[1]Nilgiri Hills: S. India

modesta modesta Jordan

Autotropis modesta modesta Jordan, *Nov. Zool.* 31, pp. 241, 242, (1924)[1]. Wolfrum, *Col. Cat.* 102, p. 59, (1929). Jordan, *Nov. Zool.* 40, p. 203, (1937). [Food-plant: Tamarindus indica (branches)].

[1]Philippine.

[Distribution: Port Blair Andaman].

moedsta conspersa Jordan

Autotropis modesta conspersa Jordan, *Nov. Zool.* 34, p. 93, (1928)[1]. Wolfrum, *Col. Cat.* 102, p. 59. (1929). Gardner, *Indian For. Rec.* (N.S.) 2 (2), pp. 102-103, figs. 26—28, (larval characters), (1936). Beeson, *Forest Insects,* p. 46, (biology), (1941). [Food-plants: *Acrocarpus fraxinifolius, albizzia* sp. (dead sticks), *Tamarindus, indica, Tephrosia candida* (dry sticks), unknown wood and creeper].

[1]Hoa Binh, Tonkin.

[Distribution: Samsing 1800 ft. (Kalimpong Division): Bengal. Dehra Dun, Jhajra Lachiwala:U.P.].

notalis Jordan

Autotropis notalis Jordan, *Nov. Zoo.* 40, p. 203, (1937)[1]. Beeson, *Forest Insects,* p. 46, (biology), (1941). [Food-plant: *Acrocarpus fraxinifolius*].

[1]Samsing Bengal

[Distribution: Gazal-duba (Jalpaiguri Division): Bengal].

8. Baseocolpus Jordan

Basecolopus, Jordan, *Ark. för. Zool.* 41A (6), p. 6, (1949).

clivosus Jordan

Baseocolpus clivosus, Jordan, *Ark. För. Zool.* 41A (6), p. 6, fig. 1, (1949)[1].

[1]Kambaiti: Burma.

8. Basitropis, Jekel

Basitropis, Jekel, *Ins. sand.* 1, pp. 90-92, (1855). Lacordaire, *Gen. Col.* 7, p. 567, (1866), Black burn, *Trans. R. Soc. S. Austr* 24, p. 153, (1900). Bovie, *Ann. Soc. Ent. Belg.* 49, p. 294, (1905). Jordan, *Nova Guinea,* 5, p. 352, (1908); *Rec. Ind.* Mus. 9, p. 214, (1913). Wolfrum, *Col. Cat.* 102, p. 80, (1929).

affinis Jordan

Basitropis affinis Jordan, *Nov. Zool.* 10, pp. 432, 434, (1903)[1]. Bovie, *Ann. Soc. Ent. Belg.* 49, p. 295, (1905). Jordan, *Rec. Ind. Mus.* 9, p. 215, (1913)[2]; *Nov. Zool. 23,* p. 263, (1916)3; *Opusc. Inst. Scient. Indochine, fn. Ent.* 1, p. 107, (1923)[4]; *Nov.* Zool. 31, p. 243, (1924). Wolfrum, *Col. Cat* 102, p. 80,

[1]Andamans, Celebes, Sumatra.
[2]Andaman Island. Johore: Malay Peninsula.
[3]Cambodia, Vientiane.

(1929). Beeson, *Forest Insects,* p. 46, (biology), (1941).
[Food-plant: *Chloroxylon swietenia*].

[4]Tonkin. Laos. Cambodia.
[Distribution: Nilambur: Madras. Ceylon.]

hamata Jordan

Bastiropis hamata Jordan, *Nov. Zool.* 10, p. 432, (1903)[1]. Bovie, *Ann. Soc. Ent. Belg.* 49, p. 295, (1905). Jordan, *Rec. Ind. Mus.* 8, p. 198 (1913)[2]; 9, p. 215, (1913)[3]; *Opusc. Inst. Scient. Indochine, Fn. ent.* 1, p. 107, (1923)[4]; *Nov Zool.* 33, p. 168, (1926); *Nov. Zool.* 34, p. 93, (1928)[5]. Wolfrum, *Col.Cat.* 102, p. 80, (1929). Jordan, *Indian For. Rec.* 18 (3), p. 4, (1933)[6]; *Nov. Zool.* 40, pp 202, 203, (1937). Gardner, *Indian For. Rec.* (N. S.) (6), pp. 127-128, figs. 33, 34, (larval characters), (1937). Beeson *Forest Insects.* p. 46, (biology), (1941).

[1]Calcutta: Bengal.
[2]Dejoo, N. Lakhimpur: Assam. Nilgiri Hills Madras.
[3]Andaman Island.
[4]Tonkin. Laos.
[5]Tonkin.
[6]Jawalaigiri: Madras.
Widely distributed in India and Indochina.

[Food-plants: *Bombax malabaricum, Butea frondosa, ficus rumphii, Garuga pinnata, Largerstroemia parviflora, Lannea grandis, Pongamia glabra, Spondias mangifera, Zizyphus xylopyra*].

[Distribution: Haflong (Cachar District): Assam. Bhaihar (Balaghat Division); S. Chanda Division): Madhya Pradesh Kashmir. Balukhand (Puri Division): Orrisa. Dehra Dun; Ranipur (Saharanpur Division): U. P.].

modica Jordan

Basitropis Modica Jordan, *Nova. Zool.* 40, pp. 202-203, (1937)[1]. Wolfrum. *Col. Cat.* 102 (Suppl.), p. 36, (1953).

[1]Baihar, Balaghat; Chota Nagpur; Motinala M. P.

nitidicustis Jekel

Basitropis nitidicutis Jekel, *Ins. Saund.* 1, pp. 92—94, t. 2, f, 2, 2a, (1855)[1]. Motschulsky, *Bull. Mosc.* 48 (2), p. 241, (*? oreintalis*), (1874). Sharp, *Trans. Ent. Soc. Lond.*, p. 320, (*dispar*), (1891)[2]. Jordan, *Ann. Mus. Genova* ,38, pp. 642, 643, (1897); *Nov. Zool.* 10, pp. 432, 434, (1903). Bovie, *Ann. Soc. Ent. Belg.* 49, p. 295, (1905). Jordan, *Nov. Zool.* 19, p. 144, (1912)[3]; *Rec. Ind. Mus.* 9, p. 215, (1913)[4]; *Nov. Zool.* 23, p. 362, (1916)[5]; *Opusc. Inst. Scient.*

[1]Java and India.
[2]Japan.,
[3]Formosa.
[4]Chatrapur, Ganjan

Indochine, Fn. Ent. 1, p. 106, (1923)[6]; *Nov. Zool.* 1, p. 243, (1924); *Nov. Zool.* 33, p. 168, (1926). Wolfrum, *Col. Cat.* 102, p. 81, (1929). Jordan, *Nov. Zool.* 36, p. 299. (1931)[7]. Beeson, *forest Insects.* p. 46, (biology), (1941). Jordan, *Ark. Zool.* 41A (6), p. 9, (1949)[8].

District: Orissa. Calcutta: Bengal. Andamans. Ceylon. [5]Tonkin., [6]Tonkin. Laos. Cochin-china. [7]Java. [8]Myitkyna: Burma.

[Food-plants: *Artocarpus lakoocha, Dalbergia, paniculata, Eugenia jambolana, Ficus asperrima, Ficus hispida, ficus, infectoria, Heriteira fomes, Mangifera indica, Pongamia glabra, Shorea robusta, Spondias Mangifera,* unknown wood].

[Distribution: Raghunandan Plantation (sylhet Division): Assam. Dacca; Bagdogra (Kurseong Division); Gazalduba (Jalpaiguri Division); Tista Valley (Darjeeling Division); Sunderbans: Bengal. Dandeli, Kanara, Sambrani (N. Kanara Division); Tansa (E. Thana Division): Bombay. Mukki (Balaghat Division): M. P. Amarampalam (Nilambur Division); Bangalore; Kannoth (North Malabar Division); Nigiri Hills: Madras. Barkote, Nagisdh, Phanduwala. (Dehra Dun Division; Tharali, (Garhawal): U.P.].

rotundata Jordan

Basitrapis rotundata, Jordan, *Nov. Zool.* 10, P. 434, (1903)[1]. bovie. *ann. Soc. Ent. Belg.* 49, p. 295, (1905). Jordan, *Opusc. Inst. cie nt. Indochine, Fn. en.* 1, p. 107, (1923)[2]; *Nov. Zool.* 33, pp. 167, 169, (1926); *Nov. Zool.* 34, pp. 93, 103, (1928)[3]. Wolfrum, *Col. Cat.* 102, p. 81 (1929)[4]. Beeson, *Forest Insects,* p. 47, (biology), (1941).

[1]Sumatra Borneo. [2]Tonkin., [3]Tamarat, Khao Luang: Malay Peninsula. [4]Philippine.

[Food-plant: *Lannea grandis*]. [Distribution: Yanaungmyin Reserve, Pyinmana: Burma].

10. **Botriesa** Jordan

Botriessa Jordan *Nov. Zool.* 34, p. 127, (1928). Wolfrum. *Col. Cat.* 102, p. 75, (1929).

sepidiopsis Jordan.

Botriessa sepidiopsis Jordan, *Nov. Zool.* 34, pp. 127, 128 (1928)[1]. Wolfrum, *Col. Cat.* 102, p. 75, (1929). [1]Ruby Mines: Burma.

11. **Caccorhinus** Sharp

Caccorhinus, Sharp, Trans. *Ent. Soc. London,* p. 321, (1891). Jordan, *Stett. Ent. Zeit.* 56, p. 202, (1895). Bovie, *Ann. Soc., Ent. Belg.* 49, p. 296, (1905). Wolfrum, *Col. Cat.* 102 p. 91, (1929).

***brunnipennis** (Motschulsky)

Euparius brunnipennis Motschulsky, *Bull. Moscoa,* 36 (2), p. 499, (1863)[1]. [1]Ceylon

Coccorhinus brunnipennis Jordan, *Stett. Ent. Zeit.* 56, p. 202, (1895). Bovie, *Ann. Soc. Ent. Belg.* 49, p. 305, (1905). Wolfrum, *Col. Cat.* 102, p. 91, (1929).

disconotatus Jordan

Caccorhinus disconotatus. Jordan, *Stett, Ent. Zeit.* 56, p. 203, (1895)[1]. Bovie, *Ann. Soc. Ent. Belg.* 49, p. 296, (1905). Wolfrum, *Col. Cat.* 102, 92, (1929). [1]Ceylon

lateralis Jordan

Caccorhinus lateralis Jordan, *Nov. Zool.* 31, pp. 243-244, (1924)[1]. Wolfrum, *Col. Cat.* 102, p. 92, (1929). [1]Kanara: Bombay.

[Food-plant: Rotton unknown wood]. Distribution: Gwaldam Garhwal: U.P.].

tessellatus Motschulsky

Caccorinus tessellatus Motschulsky, *Bull. Mosc.* 48(2), p. 236, (1874)[1]. Bovie, *Ann. Soc. Ent. Belg.* 49, P. 296, (1905). Wolfrum, *Col. Cat.* 102, p. 92, (1929). [1]India.*

---

* Identification not definite. Perhaps *Mauia satelles* Blackburn, according to Dr. Jordan.

* "India" error. Recorded from East Indies. suggested by Dr. Jordan.

12. **Cedus** Pascoe

Cedus, Pascoe, *Ann. Mag. Nat. Hist.* (3) 5, p. 37, (1860). Lacordaire, *Gen. Col.* 7, p. 515, (1866). Bovie, *Ann. Soc.. Ent. Belg.* 49, p. 243, (1905). Wolfrum, *Col. Cat.* 102, p. 29, (1929).

Pachygenia, Motschuslky, *Bull. Mosc.* 48, p. 233, (1874).

diversus Jordan

Cedus diversus Jordan, *Nov. Zool.* 18, p. 111, (1911)[1]; *Nov. Zool.* 19, p. 140, (1912)[2]; *Nov. Zool.* 34, p. 83, (1928)[3]. Wolfrum *Col. Cat.* 102, p. 30, (1929). Jordan, *Nov. Zool.* 38, p. 301, (1932)[4].

[1]Madura: Madras.
[2]Formosa.
[3]Tonkin. Indochina.
[4]Java.
[Distribution: Jiri forests, Cachar district: Assam].

valens Jordan

Cedus valens Jordan, *Nov. Zool.* 32, pp. 239-240, (♂), 1925)[1]; 33, p. 163 , (1926). Wolfrum, *Col. Cat.* 102, p. 30, (1929).

[1]Jiri Forest, Haflong, Cachar: Assam.

Acorynus silvanus Jordan, *Nov. Zool.* 32, pp. 290-291 (♀) (1925)[2].

[2]Haflong, Cachar Assam.

13. **Dendrotrogus** Jekel

Dendrotrogus, Jekel, *Ins.* saund. 1, pp. 80-82, (1855). Lacordaire, *Gen. Col.* 7, p. 564, (1866). Jordan, *Stett. Ent. Zeit.* 56, pp. 189-191, (1895). Bovie, *Ann. Soc. Ent. Belg.* 49, p. 291, (1905). Jordan, *Rec. Ind. Mus.* 9, p. 211, (1913). Wolfrum, *Col. Cat.* 102, p. 75, (1929). Gardner, *Indian For. Rec.* 2(2), p. 104, (larval characters), (1936).

angustripennis Jordan,

Dendrotrogus angustipennis Jordan, *Stett. Ent. Zeit.* 56, pp. 190, 191, 192, (1895)[1]. Bovie, *Ann. Soc. Ent. Belg.* 49, p. 291, (1905). Jordan, *Nov. Zool.* 19. p. 144 (1912)[2]; *Rec. Ind. Mus.* 9, p. 212, (1913)[3]; *Nov. Zool.* 23. p. 362, (1916)[4]; *Opusc. Inst. Scient. Indochine, Fn. Ent.* 1, p. 93, (1923)[5]. Wolfrum, *Col. Cat.* 102, p. 75, (1929). Jordan, *Nov. Zool.* 38, p. 303, (1932)[6]. Beeson, *Forest Insects,* p. 47,

[1]Carin Cheba: Burma.
[2]Formosa.
[3]Kurseong: Bengal. Sinkip Island; Johore: Malay Peninsula.
[4]Cambodia Vientiane.

(biology), (1941). [Food-plants: *dischopsis polyantha, Eugenia jambolana, Lannea grandis,, Machilus odoratissima, Shorea assamica, S. robusta, Sonneratia apetala, Stephegyne parvifolia*].

Indochina. [5]Laos. Annam. Cambodia. Cochinchina., [6]Java. [Distribution: Margherita (Lakhimpur Division): Assam. Chittgong; Sunderbans: Bengal. Dandeli (N. Kanara Division): Bombay. Sappal 1700 ft. Madras Jhajra, Nagsidh, Lachi. wala, Thano (Dehra Dun Division); Lakhanmandi Block (Haldwani Division): Uttar Pradesh. Teinzo, Pyinmana. Burma].

colligens (Walker)

Eucorynus collignes Walker, *Ann. Mag. Nat. Hist.* (3), 3, p. 261, (1859)[1]. *Sharp Trans. Ent. Soc. London*, p. 318, (1891)[2].

[1]Ceylone

[2]Higo: Japan.

Dendrotrogus colligens Jordan, *Stett, Ent. Zeit.* 56, pp. 189—192, (1895). Bovie, *Ann. Soc. Ent. Belg.* 49, p. 291, (1905). Wolfrum, *Ann. Soc. ent. Belg.* 49, p. 291, (1905). Wolfrum, *Col. Cat.* 102, p. 75, (1929). Gardner, *Indian For. Rec.* (N. S.) 2 (2) p. 104. figs. 7—9, (larval characters), (1936). Beeson, *Forest Insects*, p.-47, (biology). (1941). [Food-Plants: *Chloroxylon swietenia, Shorea rubusta, S. tumbuggaia, Terminalia chebula, Vitex pinnaa,* and unknown wood].

[Distribution: — Kanar: Bombay. Supkhar, Raigarh (Balaghat Division): Madhya Pradesh. Cuddapah (Palghat Division); Nilambur: Madras. Dehra Dun, Thanop (Dehra Dun Division); Banki (Gorakhpur Division): Uttar Pradesh Bondaung (N. Toungoo Division): Burma.]

feae Jordan

Denderotrogus feae Jordan, *Stett. Ent. Zeit.* 56, p. 190, 192, (1895)[1]. Bovie *Ann. Soc. Ent. Belg.* 49, p. 291, (1905). Jordan. *Rec.* Ind. Mus. 9. p. 212, (1913)[2]; *Opusc. Inst. Scient. Indochine. Fn. Ent.* 1, p.93, (1923)[3]. Wolfrum, *Col. Cat.* 102, p. 76, (1929).

[1]Teinzo: Burma.
[2]Sibsagar: Assam.
[3]Laos.

hypocrita Jekel

Dendrotrogus hypocrita Jekel, *Ins. Saund.* 1, pp. 82—84, t. 2 f. 1, la (*fallax*), (indescript), (1855)[1]. Jordan, *Stett. Ent. Zeit.* 56, p. 189-192, (1895); *Ann. Mus. Genova,* 38, p. 642, (1897)[2]. Bovie, *Ann. Soc. Ent. Belg.* 49, p. 291, (1905). Jordan, *Nov. Zool.* 20, p. 584, (1913); *Rec. Ind. Mus.* 9, p. 212, (1913)[3]; *Tijdschr. Ent.* 59, p. 161, (1916)[4]; *Nov. Zool.* 30, p. 178, (1923); *Nov. Zool.* 34, p. 103, (1928)[5]. wolfrum, *Col. Cat.* 102, p. 76, (1929). Jordan, *Nov. Zool.* 36, p. 297, (1931)[6]; *Nov. Zool.* 38, p. 303 (1932).

[1]India?
[2]Siboga: Sumatra.
[3]Johore: Malay Peninsula. Sinkip Island.
[4]Bekassa: Sumatra.
[5]Malay Penishula.
[6]Java.
[Distribution: Lushai Hills: Assam].

perfolicornis Fabricius

Dendrotrogus perfolicornis Fabricius, *Syst. El* 2, p. 407, (*Anthribus*) (1801)[1]. Fahraeus, in *Sch.. Gen. Curc.* 5, p. 162,* (*Dendropemon,* nec. Dendropaemon Perty 1830), (1839). Jordan, *Stett. Ent. Zeit.* 56, pp. 189, 190, (1895); *Ann. Mus. Genova,* 38, p. 642, (1897)[2]. Bovie, *Ann. Soc. Ent. Belg. 49,* p. 291, (1905). Jordan, *Rec. Ind. Mus.* 9, p. 211 (1913)[3]; *Opusc. Inst. Scient. Indochine, fn. Ent.* 1, p. 94, (1923)[4]; *Nov. Zool,* 34, p. 102, (1928)[5]. Wolfrum, *Col. Cat.* 102, p. 76, (1929). Gardner, *Indian For. Rec.* 2(2), p. 104, (larval characters) (1936). Beeson, *forest Insects,* p. 47, (biology), (1914). [Food-plants: *Acrocarpus fraxinifolius,*

[1]Sumatra.
[2]Siboga: Sumatra.
[3]Andamans.
[4]Tonkin. Laos. Cochin-china.
[5]Pahang ; Peninsular Siam.
[Distribution: — Haflong (Cachar Division): Assam. Samsing 1800 ft.

*Suggested by Dr. Jordan.

Chikrassia tabularis, Terminalia pyrifolia unknown wood]. (Kalimpong Division): Bengal. Zibyaung (West Salween Division): Burma].

14. Deropygus Sharp.

Deropygus, Sharp, *Trans. Ent. Soc. London,* p. 326, (1891). Jordan, *Stett. Ent. Zeit.* 56, pp. 396-397, (1895). Bovie, *Ann. soc. Ent. Belg.* 49. 323, (1905). Wolfrum, *Col. Cat.* 102, p. 109, (1929); *Ent. Blatt.* 30. p. 12, (key to species), (1934).

chaerilus Jordan

Deropygus chaerilus Jordan, *Nov. Zool.* 40, p. 335, (1937)[1]. Beeson, Forest *Insect,* p. 47, (biology), (1941). [Food-plant: *Kydia calycina*]. — [1]Sampagi: Coorg.

curvatus Jordan

Deropygus curvatus Jordan, *Nov. Zool.* 40, pp. 204-205, (1937)[1]. Gardner, *Indian For. Rec.* 2 (2), pp. 106-107, (*? Tropideres paviei* Lesne), (larval characters), (1936); *Indian For. Rec.* (N. S.) 3 (6), p. 127, (1937). [Food-plant: Unknown wood]. — [1]Nakronda, Dehra Dun: Uttar Pradesh.

15. Directarius Jordan.

Directarius, Jordan, *Nov. Zool.* 1, p. 634, (1894). Bovie, *Ann. Soc. Ent. Belg.* 49, 261, (1905). Wolfrum, *Col. Cat.* 102, p. 50, (1929).

basiplaga Jordan.

Directarius basiplaga Jordan, *Nov. Zool.* 16, p. 308, (1909)[1]. Wolfrum, *Col. Cat.* 102, p. 50, (1929). — [1]Nilgiri Hills: Madras.

bifoveatus (Walker)

Araeocerus Bifoveatus Walker., *Ann. Mag. Nat.* Hist. (3), 3, p. 262, (1859)[1]. — [1]Ceylon.

Directarius bifoveatus Jordan, *Stett Ent. zeit.* 56, pp. 203, 204, (1895). Bovie, *Ann. Soc. Ent. Belg.* 49. p. 261, (1905). Wolfrum *Col. Cat.* 102, p. 50, (1929).

signatus Jordan

Directarius signatus, Jordan, *Nov. Zool.* 1, p. 635, (1894)[1]; *Stett. Ent. Zeit.* 56, p. 166, (1895). Bovie, *Ann. Soc. Belg.* 49, p. 261, (1905). Jordan, *Nov. Zool.* 32, p. 238, (1925). Wolfrum, *Col. cat.* 102, p. 50, (1929). Jordan, *Nov. Zool. 36,* p. 291, (1931)[2]. — [1]Carin Cheba: Burma. [2]Java.

15. **Disphareon,** Jordan.

Disphareon, Jordan, Deutsche, Ent. zeitchr. p. 77, (1902). Bovie, *Ann. Soc. Ent. Belg.* 49, p. 290, (1905). Wolfrum, *Col. Ca.t* 102, p. 75, (1929).

andrewesi Jordan

Disphaerona andrewesi Jordan, *Nov. Zool.* 16, p. 308, (1909)[1]. Wolfrum, *Col. Cat.* 102, p. 75, (1929).

[1]Nilgiri Hills: Madras.

cyrta Jordan

Disphaerona cyrta Jordan, *Nov. Zool.* 34, p. 125, (1928)[1]. Wolfrum, *Col. Cat.* 102, p. 75, (1929).

[1]Madura: S. India.

picta Jordan

Disphaerona picta Jordan, *Nov. Zool.* 34, p. 125, (1928)[1]. Wolfrum, Col. Cat. 102, p. 75, (1929).

[1]Dikoya: Ceylon.

punctata Jordan

Disphaerona punctata Jordan, *Deutsche Ent. zietschr.* p. 77, (1902)[1]. Bovie, *Ann. Soc. Ent. Belg.* 49, p. 290, (1905). Wolfrum, *Col. Cat.* 102, p. 75, (1929).

[1]Anuradhapura: Ceylon.

verrucella Jordan

Disphaerona verrucella Jordan, *Nov. Zool.* 19, p. 89, (1912)[1]; *Nov. Zool.* 34, p. 125, (1928). Wolfrum, *Col. Cat.* 102, p. 75, (1929).

[1]Nilgiri Hills: Madras.

verrucosus (Karsch)

Tropideres verrucosus Karsch, *Berl. Ent. Zeitschr.* 26, p. 388, (1882)[1]. Bovie, *Ann. Soc. Ent. Belg.* 49, p. 272, (1905). Jordan, *Opusc. Inst. Scient. Indochine, Fn. Ent.* 1, p. 82, (1923)[2].

[1]Ceylon.

[2]Tonkin

Disphaerona verrucosus Jordan, *Nov. Zool. 34,* pp. 125, 128, (1928). Wolfrum, *Col. Cat.* 102, p. 57, (1929), Beeson, *Forest Insects,* p. 47, (biology), (1941). [Food-plaint: Acacia *melanoxylon*].

17. **Epiplaterus** Jordan

Epiplaterus, Jordan, *Ark. för Zool.* 41A (6), p. 7, (1949).

modicus Jordan

Epiplaterus modicus Jordan, *Ark. För Zool.* 41A (6), p. 8, fig. 2, (1949)[1].

[1]Kambaiti: Burma.

18. **Eucorynus** Schönherr

Eucorynus, Schöherr, *Disp. Meth.* p. 41, (1826). Imhoff, *Gen. Curc.* 1, p. 46, (1842). Lacordaire, *Gen. Col.* 7,

p. 564, (1866). Blackburn, *Trans. R. Soc. South Austral.* 24, p. 144, (1900). Jordan, *Deutsche Ent. Zeitchr.* p. 77, (1902). Bovie, *Ann. Soc. Ent. Belg.* 49, p. 290, (1905). Jordan, *Rec. Ind. Mus.* 9, p. 211, (1913). Wolfrum, *-Col. Cat.* 102, p. 75, (1929).

crassicornis Fabricius

Eucorynus crassicornis Fabricius, Syst. *El.* 2, p. 407, (*Anthribus*), (801)[1]. Fahraeus, in *Schh. Gen. Curc.* 5, p. 170, (1839). Imhoff, *Gen. Curc.* 1, p. 46, (1842). Jekel, *Ins. Saund.* 1, p. 83, (1855). Jordan, *Ann. Mus. Genova,* 38, p. 642, (1897)[2]. Bovie, *Ann. Soc. Ent. Belg.* 49, p. 290, (1905). Lefroy, *Indian Insect Life.* 1, P. 380 (1909). Jordan, *Rec. Ind. Mus.* 8, p. 198, (1913)[3]; 9, p. 211, (1913)[4]; *Tijdschr. Ent.* 59, p. 162, (1916)[5]; *Nov. Zool.* 23, p. 362, (1916)[6]. Beeson, *Indian Forester,* 45(2), p. 50, (1919)[7]. Jordan, *Opusc. Inst. Scient. Indochine, Fn. Ent.* 1, p. 95, (1923)[8]. Bihar *Trans. Ent. soc. Lond.* p. 150, (biology), (1924). Jordan, *Nov. Zool.* 34, P. 103, (1928)[9]. Wolfrum, *Col. Cat.* 102, p. 75, (1929). Jordan, *Treubia,* 7 (4), p. 337, (1929)[10]; *Nov. Zool.* 39, pp. 275, 276, 277, (1936)[11]. Gardner, *Indian For. Rec.* 2(2), p. 105, figs. 2, 23, (larval characters), (1936); *Indian For. Rec.* 3(6), p. 128, (1937). Jordan, *Nov. Zool.* 40, p. 336, (1937)[12]. Beeson *Forest Insects,* p. 47, (biology), (1941). Jordan, *Ent. Mon. Mag.* 78, p. 189, (1942)[13].

Eucornynuis clavator Faimaire, *Rev. d'Ent.* 22, p. 43, (1903)[14].

Eucorynus setosulus Pascoe, *Ann. Mag. Nat. Hist.* (3) 4, p. 434, (1859)[15]

[1]Sumatra.
[2]Siboga, Kaju Tanam: sumatra Mentawei.
[3]Rotung: Abor.
[4]Sikkim. Sibsagar: Assam. Andamans. Sarawak.
[5]Tjigembong: java.
[6]Annam., Cambodia. Vientiane, Indochina.
[7]India. Burma. Malaya.
[8]Tonkin. Laos. Annam.
[9]Indo-Malayan countries., [10]Buru.
[11, 12]Mauritius.
[13]Sarawak: Borneo.
[14]Mauritius.
[15]Philippines.

[Food-plants: *Albizzia lebbok, Alianthus Malabrica, Artocarpus lakoocha, Boswellia serrata, Buchannania latifolia, Euphoribia pucherrima, Ficus carica, Ficus bengalensis, Garuga pinnata, Lannea grandis, Mallotus philippinensis, Mangifera indica, Manihot glaziovii, Millettia auriculata, Pavetta indica, Quercus, serrata, Ricinus communis, Sarcocephalus cordatus, Semecorpus anacardium, Shorea robusta, Spondias mangifera.*

[Distribution: Apalchand (Jalapaiguri Div.); Kuntimari; Bengal. Pusa: Bihar. Kirwati, E. Kanara; Manor N. Thana: Bombay. Amanata 1400 ft. (Bilaspur Div.); Baihar (Balaghat Division); Betul; Kanha, Banjar (S. Mandla Div.): M. P.

[Food-plants ; *Euphorbia pulcherrima, Girardinia heterophylla, Lanea grandis, Michelia champaca,* unknown wood].

[Distribution: — Bagdogra (Kurseong Division): Bengal. Assam. Dehra Dun, Lachiwala (Dehra Dun Division): Uttar Pradesh]

horni Jordan

Exillis horni Jordan, *Deutsche Ent. Zeitschr.* p. 77, (1902)[1]. Bovie *Ann. Soc. Ent. Belg.* 49, p. 309, (1905). Wolfrum, *Col. Cat.* 102, p. 93, (1929). Jordan, *Indian For. Rec.* 18, (3), p. 4, (1933).

[1]Bentota, Kandy, Matale, Puttalam Weligama: Ceylon.

21. Gibber Jordan

Gibber, Jordan, *Stett. Ent. Zeit.* 56, p. 176, (1895). Bovie, *Ann. Soc. Ent. Belg.* 49. p. 259, (1905). Wolfrum, *Col.* 102, p. 44, (1929).

callistus Jordan

Gibber callistus Jordan, *Nov. Zool.* 31, pp. 238-239, (1924)[1]. Wolfrum, *Col. Cat.* 102, p. 44, (1929). Jordan, *Indian For. Rec.* 18 (3), p. 2, (1933)[2]. Gardner, *Indian For. Rec.* (N. S.) 2(2), p. 108, (larval characters), (1936). Beeson, *Forest Insects,* p. 47, (biology), (1941).

[1]Dihing Reserve, Lakhimpur: Assam. [2]Fraserpet: Coorg Madras. Assam.

[Food-plants: *ficus religiosa, Mezenga xanthoxylum, Michelia champaca,* unknown wood].

[Distribution: Bagdogra (Kurseong Division); Samsing, 1800 ft. (Kalimpong Division): Bengal. Amarampallam, Nilambur (Nilambur Division): Madras].

frenatus Jordan

Gibber frenatus, Nov. Zool. 31, pp. 238, 239, (1924).[1] Wolfrum, *Col. Cat.* 102, p. 44, (1929). Beeson, *Forest Insects,* p. 47, (biology), (1941). [Food-plants: *Albizzia* sp. *Dalbergia sissoo, Girardinia heterophylla, Mezenga xanthorylum*].

[1]Dihing Reserve, Lakhimpur: Assam. [Distribution: Anagmalai Hills, 2400 ft. Madras. Dehra Dun. U.P.].

22. **Habrissu** Pascoe

Habrissus, Pascoe, *Ann. Mag. Nat. Hist.* (3) 4, p. 432, (1859). Bovie, *Ann. Soc. Ent. Belg.* 49, p. 275, (1905). Jordan, *Rec. Ind. Mus.* 9, p. 208 (1913). Wolfrum, *Col. Cat.* p. 63, (1929).

Habrocerastes, Lacordaire, *Gen. Col.* 7, p. 547, 548, (1866).

Tripoderes, Sharp, *Trans. Ent. Soc. London,* p. 312, (1891), (partim).

andrewesi Jordan

Harbissus andrewesi Jordan, *Nov. Zool.* 19, p. 90, (1912). Bovie. *Ann. Soc. Ent.Belg.* 49, p. 275, (1905). Wolfrum, *Col. Cat.* 102, p. 63, (1929). Beeson, *Forest Insects,* p. 48, (biology), (1941).

Habrissus sellifer Jordan (nec Lacordaire, 1866), *Nov. Zool.* 13, p. 409, (1906)[1]. — [1]Nilgiri Hills: Madras.

[Food-plant: *Erythrina indica*]. — [Distribution: — Manor (N. Thana Division): Bombay].

asellus Jordan

Habrissus asellus Jordan, *Stett. Ent. Zeit.* 56, pp. 177-178, (1895)[1]. Bovie, *Ann. Soc. Ent. Belg.* 49, p. 275, (1905). Wolfrum, *Col. Cat.* 102, p. 63, (1929). — [1]Ceylon.

indicus Jordan

Habrissus indicus Jordan, *Nov. Zool.* 10, p. 420, (1903)[1]; *Nov. Zool.* 11, pp. 234, 235, (1904). Bovie, *Ann. Soc. Ent. Belg.* 49, p. 275, (1905). Jordan, *Nov. Zool.* 13, p. 409, (1906). Wolfrum, *Col. Cat.* 102, p. 63, (1929). — [1]Khasis Hills: Assam.

molitor Jordan

Habrissus molitor Jordan, *Nov. Zool. 19,* p. 90, (1912).[1] Wolfrum, *Col. Cat.* 102, p. 63, (1929). — [1]Karkur Ghat, Nilgiri Hills: Madras.

sellifer Lacordaire

Habrissus sellifer Lacordaire, (*Gen. Col.* 7, p. 541, note 1, Atl. t. 80, f2, 2a, (1866)[1]. Jordan, *Stett. Ent. Zeit.* 56, p. 178, (*sellatus*) (1895). Wolfrum *Col. Cat.* 102, p. 63, (1929). — [1]Malacca, but probably Ceylon.

tibialis Jordan

Habrissus tibialis Jordan, *Deutsche Ent. Zeitschr.* p. 76, (1902)[1]; *Nov. Zool.* 10, p. 420, (1903). Bovie, *Ann. Soc. Ent. Belg.* 49 p. 275, (1905). Wolfrum, *Col. Cat.* 102, p. 63, (1929). — [1]Anuradhapura: Ceylon.

23. **Hucus** Pascoe

Hurcus, Pascoe, *Ann. Mag. Nat. Hist.* (3) 4, pp. 436-437, (1859). Lecordaire, *Gen. Col.* 7, p. 539, (1866). Bovie, *Ann. Soc. Ent. Belg.* 49, p.260, (1905). Wolfrum, *Col. Cat.* 102, p. 44, (1929); *Ent. Blätt.* 34 (2), p. 68, (1938).

bicolor bicolor Jordan

Hucus bicolor bicolor Jordan, *Nov. Zool,* 40, p. 334, (1937)[1]; *Nov Zool.* 41, pp. 140, 141, (1938). — [1]Makum Lakhimpur: Assam.

bicolor numatus Jordan

Hucus bicolor numatus, Jordan *Nov. Zool.* 41, pp. 140-141, (1938)'. Beeson, *forest Insects,* p. 48, (biology), (1941). — [1]Salween: Burma.

[Food-plants: *Croton oblongifolium, Erythrina* sp.].

insulanus Jordan

Hucus insulanus Jordan, *Nov. Zool.* 38, p. 308, (1932).[1] — [1]North Andaman.

loratus Jordan

Hucus loratus Jordan, *Nov. Zool.* 39, pp. 314-315, (1936)[1]. Beeson *forest Insects,* p. 48, (biology), (1941). [Food-plant: *Polyalthia simiarum*]. — [1]Samsing, Kalimpong: Bengal.

24. **Hybosternus** Jordan

Hybosternus, Jordan *Proc. R. Ent. Soc. Lond.* B, 14 (1-2), p. 19, (1945).

mesosternalis Jordan

Xenocerus mesosternalis Jorden, *Nov. Zool.* 1, p. 638, 645, (1894)[1]. Bovie, *Ann. Soc. Ent. Belg.* 49, p. 285, (1905). Jordan, *Rec. Ind. Mus.* 9. p. 209, (1913)[2]. Wolfrumm, *Col. Cat.* 102, p. 69, (1929). — [1]Java (error). [2]Ceylon.

Hybosternus mesosternalis Jordan, *Proc. R. Ent. Soc. Lond.,* B, 14 (1-2), p. 19, (1945).

25. **Hypseus** Pascoe

Hypseus, Pascoe, *Ann. Mag. Nat. Hist.* (3) 5, pp. 39-40, (1860). Lacordaire, *Gen. Col.* 7, p. 539, (1866).

Bovie, *Ann. Soc. Ent. Belg.* 49, p. 259, (1905). Wolfrum, *Col. Cat.* 102, p. 43, (1929).

dilectus Jordan

Hypseus dilectus Jordan, *Nov. Zool.* 38, pp. 377-378, (1933)[1].

[1]Bojuwantalura, Dikoya: Ceylon.

picticollis Jordan

Hypseus picticollis Jordan, *Nov. Zool.* 16, pp. 307-308 (1909)[1]. Wolfrum, *Col. Cat.* 102, p. 43, (1929).

[1]Nilgiri Hills: Madras.

26. **Litocerus** Schönherr

Litocerus, Schönherr, *Gen. Curc.* 1, p. 125, (1833). Lacordaire, *Gen. Col.* 7, p. 514, (1866). Jordan *Nov. Zool.* 1, p. 608. Bovie *Ann. Soc. Ent. Belg.* 49, p. 239, (1905). Jodran, *Rec. Ind. Mus.* 9, p. 206, (1913). Wolfrum, *Col. Cat.* 102, p. 27, (1929).

annulipes Jordan

Litocerus annulipes Jordan, *Stet. Ent.* Zeit. 56, p. 147, (1895)[1]. Bovie, *Ann. Soc. Ent. Belg.* 49, p. 239, (1905). Wolfrum, *Col. Cat.* 102, p. 27, (1929).

[1]Ceylon.

fuliginosus Oliveir

Litocerus fuliginosus Oliveir, *Ent.* 4, (80), p. 11, t. 2, f. 13 a-b, (1795)[1]. Schönherr, *Gen. Curc.* 1, p. 127, (1833). Bovie, *Ann. Soc. Ent. Belg.* 49, p. 240, (1905), Jordan, *Nov. Zool,* 18, p. 106, (1911). Wolfrum, *Col. Cat.* 102, p. 27, (1929).

[1]India.

hamearis Jordan

Litocerus hanearis Jordan, *Nov. Zool,* 19, p. 88, (1912)[1]. Wolfrum, *Col. Cat.* 102, p. 27 (1929).

[1]Karkur Ghat, Nilgiri Hills: Madras.

khasianus Jordan

Litocerus khasianus Jordan, *Nov. Zool.* 10, pp. 424-425, (1903)[1]. Bovie, *Ann. Soc. Ent. Belg.* 49., p. 241, (1905). Jordan, *Rec. Ind. Mus.* 8, p. 197, (1913)[2]; 9, p. 207, (1913); *Nov. Zool.* 30, p. 170, (1933); *Nov. Zool.* 32, p.239, (1925); *Nov. Zool.* 34, p. 81, (1928). Wolfrum, *Col. Cat.* 102, p. 28, (1929). Beeson, *Forest Inscets,* p. 48, (biology), (1941).

[1]Khasis Hills: Assam.

[2]Dejoo: Assam.

[Food-plants. *Juglans regia,* unknown wood.] [Distribution: — Samsing 1800 ft. (Kalimpong Division): Bengal. Hpenmaw, Myitkyina Burma].

leucospilus Jordan

Litocerus, leucospilus Jordan, *Nov. Zool.* 10, p. 425, (1903)[1]. Bovie, *Ann. Soc. Ent. Belg.* 49, p. 241, (1905). Wolfrum, *Col. Cat.* 102, p. 28, (1929). [1]Khasis Hills: Assam.

macrophthalmus luteus Jordan

Litocerus macrophathalmus luteus Jordan, *Re.c Ind. Mus.* 9, p. 206, (1913)[1]. Wolfrum, *Col. Cat.* 102, p. 28, (1929). [1]Andamans.

maculatus Olivier

Litocerus maculatus Olivier, *Ent.* 4, (80), p. 11, t. 2, f. 14 a-b, (1795)[1]. Pascoe, *Ann. Mag. Nat. Hist.* (3)4, p. 330, (1859). Jordan, *Nov. Zool. 1,* p. 615, (1894)[2]; *Nov. Zool,* 5, p. 365, (1898). Bovie, *Ann. Soc. Ent. Belg,* 49, p. 241, (1905). Wolfrum *Col. Cat.* 102, p. 28, (1929). [1]India. [2]Andaman Islands. Celebes.

moestus andamanicus Jordan

Litocerus moestus andamanicus Jordan, *Rec. Ind. Mus.* 9. p. 207, (1913)[1]. Wolfrum, *Col. Cat.* 102, p. 28, (1929). [1]Andamans.

nilgriensis Jordan

Litocerus nilgiriensis Jordan, *Nov. Zool.* 16, p. 207, (1909)[1]. Wolfrum, *Col. Cat.* 102, p. 28, (1929). [1]Nilgiri Hills: Madras.

phelus Jordan

Litocerus phelus Jordan, *Nov. Zool.* 32, p. 239, (1925)[1]; *Nov. Zool.* 34, p. 98, (*Acortynus*), (1928)[2]. Wolfrum, *Col. Cat.* 102, p. 28, (1929). Gardner, *Indian For Rec.* 16, (11), pp. 330-331, pl. 1, figs. 13—15, (larval characters), (1932); *Indian For. Rec.* (N. S.) 2 (2), p. 107, (larval characters), (1936). [1]Naga Hills: Assam. [2]Pahang; Malaya Peninsula.

philippinensis crucicollis Jordan

Litocerus philippinensis crucicollis Jordan, *Sett. Ent. Zeit.* 56, p. 145, (1895)[1]. Bovie, *Ann. Soc. Ent. Belg.* 49, p. 242, (1905). Jordan, *Nov. Zool.* 18, p. [1]Ceylon.

106 (1911); *Nov. Zool.* 19, p. 139, (1912). Wolfrum, *Col. Cat.*102, p. 29, (1929). Beeson *Forest Insects,* p. 48, (biology). (1941). [Food-plants: *Pterocarpus marsupium,* unknown wood]. [Distribution: Kanara Bombay. Mercara: Coorg. Nilgiri Hills: Madras].

taeniatus Jordan

Litocerus taeniatusf Jordan, *Nov. Zool.* 19, p. 88, (1912)[1]. Wolfrum, *Col. Cat.* 102, p. 29, (1929) [1]Nilgiri Hills: Madras.

verrucosus Jordan

Litocerus verrucosus Jordan, *Stett. Ent. Zeit.* 56, p.382, (1895)[1]. Bovie, *Ann. Soc. Ent. Belg.* 49, p. 243, (1905). Wolfrum, *Col. Cat.* 102, p. 29, (1929). [1]Khasis Hills: Assam.

27. **Litotropis** Fairmaire

Litotropsis, Fairmaire, *Ann. Soc. Ent. France,* (5) 10, p. 337, (1880); Le Natur. 2, p. 316, (1880). Bovie. *Ann. Soc. Ent. Belg.* 49, p. 294, (1905). Wolfrum, *Col. Cat.* 102, p. 80, (1929).

semiustus Jordan

Litotropis semiustus Jordan, *Stett. Ent. Zeit.* 56, pp. 193-194, (1895)[1]. Bovie, *Ann. Soc. Ent. Belg.* 49, p. 294, (1905). Wolfrum, *Col. Cat.* 102, p. 80, (1929). [1]Ceylon.

28. **Mauia** Blackburn

Mauia, Blackburn, *Trans. R. Dublin Soc.* (2)3, p. 194, (1885). Bovie, *Ann. Soc. Ent. Belg.* 49, p. 300, (1905). Wolfrum, *Col. Cat.* 102, p. 84, (1929). Zimmerman, B.P. Bishop *Mus.,* Occ. Papers, 14 (13), pp. 239-240, (1938).

Contexta, Jordan, *Deutsche Ent. Zeits.* p. 78, (1902).

satelles Blackburn

Mauia satelles Blackburn, *Trans. R. Dublin Soc.* 2 (3), p. 195, (1885)[1]. Perkins, *Fauna Hawaiensis,* 2 (3), p. 183, (1900)[2]. Wolfrum, *Col. Cat.* 102, p. 84, (1929). [1,2]Hawaii.

Contexta murina Jordan,*Deutsche Ent. Zeits,* p. 78, (1902).[3] [3]Puttalam: Ceylon.

29. **Mecocerus** Schönherr

Mecocerus, Schönherr, *Gen. Curc.* 1, p. 115, (1833). Lacordaire, *Gen. Col.* 7, p. 496, (1866). Jordan, *Nov. Zool.* 1, p. 595, (1894). Bovie, *Ann. Soc. Ent. Belg.* 49, p. 225, (1905). Jordan, *Nov. Zool.* 20, p. 257, (1913); *Rec. Ind. Mus.* 9, p. 204, (1913). Wolfrum, *Col. Cat.* 102, p. 12, (1929).

Acanthothorax, Gaede, *Mag. Zool.* 2. t. 15, (1833). Jordan, *Nov. Zool.* 20, p. 257, (1913).

allectus elegans Jordan

Mecocerus allectus elegans Jordan. *Nov. Zool.* 13, p. 408, (1906)[1]. Jordan, *Rec., Ind. Mus.* 9, p. 204, (1913)[2]. [Food-plant: *Terminalia belerica*].

[1]Anamalai, Nilgiri Hills: Madras. [2]Malabar: Madras. [Distribution: Wynaad: Madras. Travancore].

allectus maculatus Jordan

Mecocerus allectus maculatus Jordan, *Nov. Zool.* 1, p. 599, (1894)[1]. Bovie, *Ann. Soc. Ent. Belg.* 49, p. 225, (1905). Jordan, *Nov. Zool.* 13, p. 408, (1906); *Nov. Zool.* 20, p. 259, (1913)[2]; *Rec. Ind. Mus.* 8, p. 197, (1913)[3]; 9, p. 204, (1913). Wolfrum, *Col. Cat.* 102, p. 42, (1929). Jordan, *Ark. For Zool.* 41A (6), p. 2, (1949)[4].

[1]Ruby Mines, Carin Cheba: Burma. [2,4]Tenasserim: Burma. [3]Dawna Hills: Burma.

[Distribution: Tavoy: Burma].

asmenus asmensu Jordan

Mecocerus asmenus asmenus, Jordan, *Rec. Ind., Mus.* 9, pp. 204-205, (1913)[1]. Wolfrum, *Col. Cat.* 102, p. 12, (1929).

[1]Cachar, Sadiya: Assam.

30. **Mecotropis** Lacordaire

Mecotropis, Lacordaire, *Gen. Col.* 7, p. 495, (1866). Jordan, *Stett. Ent. Zeit.* 56, p. 372, (1895). Bovie. *Ann. Soc, Ent., Belg.* 49, p. 223, (1905). Jordan, *Rec. Ind. Mus.* 9, p. 208, (1913); *Nov. Zool.* 20, p. 265, (1913). Wolfrum, *Col. Cat.* 102, p. 10, (1929).

bipunctatus Lacordaire

Mecotropis bipunctatus Lacordaire, *Gen. Col.* 7, p. 496, Atl. 8, t. 78, f. 4a, (1866)[1]. Bovie, *Ann. Soc.*

[1]Ceylon.

Ent. Belg. 49, p. 223, (1905). Jordan, *Rec. Ind. Mus.*, 9, p. 208, (1913)[2]. Wolfrum, *Col. Cat.* 102, p. 11, (1929). — [2]S. India. Ceylon.

ephippim Jordan

Mecotropis ephippium Jordan, *Rec. Ind. Mus.* 9, p. 208, (1913)[1]. Wolfrum, *Col. Cat.* 102, p. 11, (1929). — [1]Andamans.

gardneri Jordan

Mecotropis gardneri Jordan, *Nov. Zool.* 38, p. 305, (1932)[1]. [Food-plant: Unknown wood.] — [1]Anamalai Hills, 2, 400 ft. Madras.

guttifer Jordan

Mecotropis guttifer Jordan, *Stett. Ent. Zeit.* 56, p. 370, (1895) (Java error)[1]. Wolfrum, *Col. Cat.* 102, p. 11, (1929). — [1]Ceylon.

nieroglyphicus Hope

Mecotropis hieroglyphicus Hope in Gray, *Zool. Miscellany,* p.26, (*Macrocephalus*) (1831)[1]. — [1]Nepal.

vitticollis Jordan.

Mecotropis vitticollis Jordan, *Stett. Ent. Zeit.* 56, pp. 371-372, (1895)[1]. Bovie, *Ann. Soc. Ent. Belg.* 49, p. 224, (1905). Jordan, *Rec. Ind. Mus.* 9, p. 208, (1913)[2]; *Nov. Zool.* 23, p. 359, 360, fig. 2 (1916); *Opusc. Inst. Scient Indochine, Faune Ent.* 1, p. 75, (1923). Wolfrum, *Col. Cat.* 102, p. 12, (1929). Jordan, *Nov. Zool.* 38, p. 305, (1932). — [1]Khasis Hills. Assam. [2]Assam.

xanthomelas Jordan

Mecotropis xanthomelas Jordan, *Rec. Ind. Mus.* 9. pp. 208-209, (1913)[1]. Wolfrum, *Col. Cat.* 102, p. 12, (1929). — [1]Andamans.

31. **Meganthribus** Jordan

Dolichocera, Gray, in *Griff. Anim. Kingd. Ins.* 2, p. 65, (1832).

Eugigas, Thomson, *Arch. Ent.* 1, p. 435, (1857) (partim). Jordan, *Stett. Ent. Zeit.* 56, p. 132, (1895) (partim). Bovie, *Ann. Soc. Ent. Belg.* 49, p. 223, (1905).

Meganthribus, Jordan, *Nov. Zool.* 20, p. 263, (1913); *Rec. Ind. Mus.* 9, p. 204, (1913). Wolfrum, *Col. Cat.* 102, p. 10 (1929).

harmandi harmandi (Lesne)

Eugigas harmandi Lesne, *Bull. Soc. Ent. Franc,* 60, pp. 91-92, (1891)[1]; *Nouvo. Arch. Mus. Paris,* 3 (3), p. 227, (partim), (1892). Jordan, *Stett. Ent. Ziet.* 56, p. 132, (1895)[2]. Lesne, *Mission Pavie,* 3, p. 129, (t. 8, f. 15-16, (1904). Jordan, *Nov Zool.* 11, p. 230 (1904). Bovie, *Ann. Soc. Ent. Belg.* 49, P. 223, (1905).

[1]Indochina. Cambodia.

[2]Cochinchina.

Meganthribus harmandi harmandi Jordan, *Nov. Zool.* 23, p. 359, (1916)[3]. *Opusc. Inst. Scient. Indochine, Fn. Ent.* 1, p. 75, (1923); *Nov. Zool.* 34, pp. 105, 106, (1928). Wolfrum, *Col. Cat.* 102, p. 10, (1929).

[3]Sambor. Cambodia.

harmanid schanus Jordan

Meganthribus harmandi schanus Jordan, *Nov. Zool.* 34, p. 106, (1928)[1]. Wolfrum, *Col. Cat.* 102, p. 10, (1929). Jordan, *Ark. för Zool.* 41A (6), p. 1, (1949)[2].

[1]Tenasserim: Burma.

[2]Moulmein: Burma.

harmandi tessellatus Jordan

Eugigas harmandi tessellatus Jordan, *Stett, Ent. Zeit.* 56, pp. 369-370. (1895)[1]. Bovie, *Ann. Soc. Ent. Belg.* 49, p. 223, (1905).

[1]Khasis Hills: Assam. Calcutta; Darjeeling: Bengal.

Meganthribus harmandi tessellatus Jordan, *Rec. Ind. Mus.* 9, p. 204, (1913)[2]; *Opusc. Inst. Scient. Indochine, Fn. Ent.* 1, p. 75, (1923). Wolfrum, *Col. Cat.* 102, p. 10, (1929), Gardner, *Indian For. Rec.* 2 (2), p. 109, figs. 1—6, (larva characters), (1936). Beeson, *Forest Insects,* p. 48, (biology), (1941).

[2]Sikkim. Assam. Calcutta; Darjeeling: Bengal.

[Food-plants: *Macaranga denticulata* (rotten wood), unknown wood].

Distribution: Shillong: Assam. Bagdogra (Kurseong Division); Samsing (Kalimpong Div.): Bengal. Dehra Dun: U.P.].

sulphureus (Waterhouse)

Mecocerus sulphureus Waterhouse, *Trans. Ent. Soc. Lond.* pp. 24-25, (1876) (partim, nec Cambodia)[1]; *Aid Ident, Ins.* t. 171, f. 3, (1886).

[1]Andaman Islands. Nicobar Islands.*

* Suggested by Dr. Jordan.

Eugigas sulphureus, Jordan, *Nov. Zool,* I, p. 595, (1894); *Stett. Ent. Zeit.* 56, p. 132, (1895). Bovie, *Ann. Soc. Ent. Belg* 49, p. 223, (1905).

Meganthribus sulphureus Jordan, *Nov. Zool,* 20, pp. 263, 264 (1913)[2]; *Rec. Ind. Mus.* 9, p. 204, (1913)[3]. Wolfrum, *Col. Cat.* 102, p. 10, (1929). — [2, 3]Andaman Islands.

32. **Melanosacus** Jordan

Dysnos, Jordan *stett. Ent. Zeit.* 56, pp. 400-401, (1895) (partim).

Melanopsacus, Jordan, *Ann. Mag. Nat. Hist.* (9) 13, p. 608, (1924). Wolfrum, *Col. Cat.* 102, p. 110, (1929). Zimmerman, *B. P. Bishop Mus.,* Occ. Papers, 14, (13), p. 245, (1938).

anthracinus Jordan

Melanopsacus anthracinus Jordan, *Ann. Mag. Nat. Hist.* (9) 13, pp. 612—613, (1924)[1], Wolfrum, *Col. Cat* 102, p. 110, (1929). Beeson *Forest Insects,* p. 48, (biology) (1941). — [1]Gopaldhara: Sikkim.

[Food-plant: *Shorea robusta*]. — [Distribution: — Horai (Haldwani Division): U.P.].

ceylonicus Jordan

Dynos ceylonicus Jordan, *Stett. Ent. Zeit.* 56, pp. 400-401, (1895)[1]. Bovie, *Ann. Soc. Ent. Belg.* 49, p. 324, (1905). — [1]Ceylon.

Melanopsacus ceylonicus Jordan, *Ann. Mag. Nat. Hist.* (9) 13, p. 616, (1924). Wolfrum, *Col. Cat.* 102, p. 110, (1929). Jordan, *Nov. Zool.* 40, p. 204, (1937).

gagates Jordan

Melanopsacus gagates Jordan, *Ann. Mag. Nat. Hist.* (9) 13, p. 613, (1924)[1]. Wolfrum, *Col. Cat.* 102, p. 111, (1929). — [1]Gopaldhara: Sikkim.

minutus Jordan

Melanopsacus minutus Jordan, *Ann. Mag. Nat. His.* (9) 13, p. 613, (1924)[1]. Wolfrum, *Col. Cat* 102, p. 111, (1929). — [1]Gopaldhara ; Sikkim

nanellus Jordan

Melanopsacus nanellus Jordan, *Nov. Zool,* 40, p. 204, (1937)[1]. Beeson, *Forest insects,* p. 48, (biology), (1941). [Food-Plant: *Michelia champaca*]. — [1]Dahura, Bagdogra, Kurseong: Bengal.

specialis Jordan

Melanopsacus specialis Jordan, *Ann. Mag. Nat. Hist.* (9) 13, p. 614, (1924)[1]. Wolfrum, *Col. Cat.* 102, p. 111, (1929).

[1]Wallardi: Travancore.

subfasciatus Jordan

Melanopsacus subfasiatus Jordan, *Ann. Mag. Nat. Hist.* (9) 13, pp. 164-615, (1924)[1]. Wolfrum, *Col. Cat.* 102, p. 111 (1929).

[1]Kurseong: Bengal. Nilgiri Hills: Madras.

33. **Merarius** Fairmaire

Merarius, Fairmaire, *Ann. Soc. Ent. France,* (6) 9, p. 56, (1889). Bovie, *Ann. Soc. Ent. Belg.* 49, p. 267, (1905). Wolfrum, *Col. Cat.* 102, p. 53, (1929).

davidis Fairmaire

Merarius devidis Fairmaire, *Ann. Soc. Ent.* France, (6) p. 56, (1889)[1]. Bovie *Ann. Soc. Ent. Belg.*49, p. 267, (1905). Jordan, *Nov. Zool,* 23, p. 361, (1916)[2]; *Opusc. Inst. Scient. Indochine, Fn. Ent.* 1, p. 83, (1923)[3]. Wolfrum, *Col. Cat.* 102, p. 53, (1929).

[1]Moupin.
[2, 3]Tonkin.
[3]Ruby Mines: Burma.

*__quadrituberculatus__ Hope

Meriarius quadrituberculatus Hope, In Gray, *Zool Misc.* p. 27, *(Anthribus)*, (1831)[1].

[1]Nepal.

Merarius glabriceps Jordan, *Nov. Zool.* 18, pp. 106-107, (1911)[1]. Wolfrum, *Col. Cat.* 102, p. 53, (1929). Beeson, *Forest Insects,* p. 48, (biology), (1941).

[1]Kurseong: Bengal.

[Food-Plant: *Mallotus roxburghiansu*].

[Distribution: Darjeeling Div.: Bengal].

34. **Misthosima** Pascoe

Misthosima, Pascoe, *Ann. Mag. Nat.* Hist. (3) 4, pp. 434-435, (1859). Lacordaire, *Gen. Col.* 7, p. 591, (1866). Blackburn, *Tran.s R. Soc. South Austral.* 24, p. 161, (1900). Bovie, *Ann. Soc. Ent.* 49, p. 324, (1905). Jordan, *Nov. Zool.* 34, p. 94, (1928). Wolfrum. *Col. Cat.* 102, p. 109, (1929).

---

* Suggested by Dr. Jordan.

separ Jordan

Misthosima separ Jordan, *Nov. Zool.* 40, pp. 203-204 (1937)[1]. Beeson, *Forest Insects,* p. 48, (biology), (1941).

[Food-plants: *Michelia champaca, Tectona grandis*].

[1]Kandy ; Colombo: Ceylon. Kankath, Nilambur: Madras. [Distribution: — Kurseong Division: Bengal].

35. Nessiara Pascoe

Nessia, Pascoe, *Ann. Mag. Nat. Hist.* (3) 4, p 329, (1859).

Nessiara, Pacoe, *Jour. Ent.* 1, p. 60, (1860). Lacordaire, *Gen. Col.* 7, p. 537, (1866). Jordan, *Nov. Zool.* 1, p. 630, (1894). Bovie, *Ann. Soc. Ent. Belg* 49, p. 256, (1905). Wolfrum, *Col. Cat.* 102, p. 40, (1929).

olivacea Jordan

Nessiara olivacea Jordan, *Nov. Zool.* 19, pp. 88-89, (1912)[1]. Wolfrum, *Col. Cat.* 102, p. 41, (1929).

[1]Anamalai and Karkur Ghat; Nilgiri Hills: Madras.

36. Nessiodocus Heller

Nessiodocus, Heller, *Ent.. Blatt.* 21, p. 106, (1925). Jordan, *Nov. Zool.* 34, p. 83, (1928). Wolfrum, *Col. Cat.* 102, p. 31, (1929)

celsus Jordan

Nessiodocus celsus Jordan, *Nov. Zool.* 38, pp. 308-309, (1932)[1]. Beeson, *Forest Insects,* p. 48, (biology), (1941), [food-plant: *Hopea wightiana*].

[1]Sunksal. S. Kanara: Bombay.

37. Ozotomerus Perroud

Ozotomerus, Perroud, *Ann. Soc. Linn. Lyon,* (2) 1, p. 406, (1853). Lacordaire, *Gen. Col.* 7, p. 572, (1866). Bovie, *Ann. Soc. Ent. Belg.* 49, p. 300, (1905). Jordan, *Rec. Ind. Mus.* 9, p. 215, (1913). Wolfrum, *Col. Cat.* 102, p. 84, (1929).

Dipieza Pascoe, *Ann. Mag. Nat. Hist.* (3) 4, P. 431, (1859).

Oedecerus, Montrouzier, *Ann. Soc. Agr. Lyon,* 7, p. 46, (1855). (Reprint: 1857).*

---

* Suggested by Dr. Jordan.

maculosus Perroud

Ozotomerus maculosus Perroud, *Ann. Soc. Linn. Lyon.* (2)1, p. 407, (1853)[1]. Bovie, *Ann. Soc. Ent. Belg.* 49, p. 300, (1905). Jordan, *Rec. Ind. Mus.* 9, p. 215, (1913)[2]; Beeson, *Indian Forester,* 45, (2), p. 51, (1919)[3]; *Rept. Proc. Fifth Ent. Meet Pusa,* p. 170, (1923). Jordan, *Opusc. Inst. Scient. Indochine, Fn. Ent.* 1, p. 107, (1923)[4]. Wolfrum, *Col, Cat.* 102, p. 84, (1929). Gardner, *Indian For. Rec.* 16 (11), pp. 331-332. Pl. 1, figs. 16—18, (larval characters), (1932); *Indian For. Rec.* 2 (2), pp. 102, 110, (1936). Beeson, *Forest Insects,* p. 48, (biology), (1941).

[1]Calcutta: Bengal. [2]Andamans. [3]Sunderbans, Tista: Bengal. Goalpara: Assam., [4]Laos.

[Food-plants: *Acrocarpus fraxinifolius, Heritiera, fomes, Hopea odorata, Mangife a indica, Mimusops littoralis, Shorea robusta,* unknown wood].

[Distribution: Samsing, Kalimpong Div.: Bengal. Shahapur (East Thana Div.): Bombay. Rengali (Sam-balpur Div.): Orissa. Dehra Dun, Lachiwala, Nagsidh (Dehra Dun Div.): Uttar Pradesh. Siang Yane (N. Toungoo Div.): Burma. Palugama 250 ft., Anuradhapura: Ceylon].

38. **Peribathys** Jordan

Peribathys, Jordan, *Nov. Zool.* 40, p. 333, (1937).

beesoni Jordan

Peribathys beesoni Jordan, *Nov. Zool.* 40, p. 333, (1937)[1].

[1]Sukna. Kurseong: Bengal.

[Food-plant: Unknown wood].

39. **Phaeochrotes** Pascoe

Phaeochrotes Pascoe, *Ann. Mag. Nat. Hist.* (3) 5, p. 42, (1860). Lacordaire, *Gen. Col.* 7, p. 523, (1866). Bovie, *Ann. Soc. Ent. Belg.* 49, p. 247, (1905). Wolfrum, *Col. Cat.* 102, p. 33, (1929).

cinereomaculatus (Motchulsky)

Acorynus cinereomaculatus Motschulsky, *Bull. Mosc.* 36, p. 499, (1863)[1]; 37, p. 399, (*Macrocephalus*)

[1]Ceylon.

(1866)*. Bovie, *Ann. Soc. Ent. Belg* 49, p. 236, (1905). Wolfrum, *Col. Cat.* 102, p. 24, (1929).

* *Phaecohrotes cinereomaculatus* Motschulsky.

Beeson, *Forest Insects* p. 48, (biology), (1941).

[Food-plants: *Cinnamomum cecvidodaphne, Ficus religiosa, Mangifera indica*].

[Distribution: Sukna (Kurseong Div.), Sunderbans: Bengal]. Koina, Singhbum: Orrisa. Mandalay District: Burma].

phorcas Jordan

Phaeochrotes phorcas Jordan, *Ann. Mys. Genova,* 41, p. 80, (1904)[1]. Bovie, *Ann. Soc. Ent. Belg.* 49, p. 247, (1905). Jordan, *Nov. Zool.* 31, p. 234, (1924). Wolfrum, *Col. Cat.* 102, p. 33, (1929).

[1]Carin Cheba: Burma

porcellus Pascoe

Phaeochrotes porcellus Pascoe, *Ann. Mag. Nat. Hist.* (3) 5, p. 42, t. 1, (1860)[1]. Jordan, *Nov. Zool.* 1, p. 629, (1894); *Ann.Mus. Genova,* 41, p. 80, (1904). Bovie, *Ann. Soc. Ent. Belg.* 49, p. 247, (1905). Jordan, *Tijdschr. Ent.* 58, p. 50, (1915)[2]; *Tijdschr. Ent.* 59, p. 161, (1916); *Nov. Zool.* 31, p. 234, (1924). Wolfrum, *Col. Cat.* 102, p.33, (1929). Jordan, *Treubia,* 7 (4), p. 336, (1929)[3]. Beeson, *Forest insects,* p. 48, (biology), (1941).

[Food-plants: *Sideroxylon longepetiolatum* unknown wood].

[1]Mocassar.

[2]Java: Philippine Simalur.

[3]Molacca.

[Distribution: — Andaman].

40. Phaulimia Pascoe

Phaulimia, Pascoe, *Ann. Mag. Nat. Hist.* (3), 4, p. 437, (1859). Lacordaire, *Gen. Col.* 7. p. 551, (1866). Bovie, *Ann. Soc. Ent. Belg.* 49, p. 264, (1905). Wolfrum, *Col. Cat.* 102, p. 49, (1929).

caena Jordan

Phaulimia caena Jordan, *Nov. Zool.* 40, p. 200, (1937)[1]. Beeson, *Forest Insects,* p. 48, (biology), (1941).

[Food-plants: *Calophyllum walkeri, Dipteroarpus zeylanicus, Gordonia zeylanica*].

[1]Mawagankanda, Ratnpura; Ohiya, Uva; Ceylon.

* Suggested by Dr. Jordan.

dissensa Jordan

Phaulimia dissensa Jordan, *Nov. Zool.* 30, p. 172, (1923)[1]. Wolfrum, *Col. Cat.* 102, p. 49, (1929).

[1]Madura District: S. India.

lineata nilgiriensis Jordan

Phaulimia lineata nilgiriensis Jordan, Nov. Zool. 16, p. 308, (1909)[1]. Wolfrum, *Col. Cat.* 102, p. 50, (1929).

[1]Nilgiri Hills: Madras.

priva nigrina Jordan

Phaulimia priva nigrina Jordan, *Nov. Zool.* 19, p. 89, (1912)[1]. Wolfrum, *Col. Cat.* 102, p. 50, (1929).

[1]Karkur Ghat, Nilgiri Hills: Madras.

schaumi

Phaulimia schaumi Pascoe, *Ann. Mag. Nat. Hist.* (4), 8, p. 260, (1871)[1]. Jordan, *Nov. Zool.* 1, p. 634, (1894). Bovie, *Ann. Soc. Ent. Belg.* 49, p. 264, (1905). Jordan, *Nov. Zool.* 30, p. 172, (1923). Wolfrum, *Col. Cat.* 102, p. 50, (1929).

[1]Ceylon.

41. **Phloeobius** Schönherr

Phloeobius Schönherr, Disp. Meth. p. 36, (1826). Lacordaire, *Gen. Col.* 7, p. 576, (1866). Jordan, *Deutsche Ent. Zeitschr.* p. 78, (1902). Bovie, *Ann. Soc. Ent. Belg.* 49, p. 301, (1905). Jordan. *Rec. Ind. Mus.* 9, p. 213, (1913); *Trans. Linn. Soc. London,* 16, p. 254, (1914); *Opusc. Inst. Scient, Indochine, Fn. Ent.* 1, p. 105, (1923).Wolfrum, *Col. Cat.* 102, p. 85, (1929). Zimmerman, *B. P. Bishop Mus.* Occ. Papers, 14, (13), pp. 223-224, (1938).

albescens Jordan

Phloeobius albescens Jordan, *stett. Ent. Ziet.* 56, pp. 198-199, (1895)[1]. Bovie, *Ann. Soc. Ent. Belg* 49, p. 301, (1905). Jordan, *Trans. Linn. Soc. Lond.* 16, p. 254, (1914); *Opusc. Ins. Scient. Indochine, Fn. Ent.* 1, pp. 102, 105, (1923)[2]. Wolfrum, *Col. Cat.* 102, p. 86, (1929). Jordan, *Indian For. Rec.* 18 (3), p. 4, (1933)[3].

[1]Habitat not mentioned in original description, but the specimeans were obtained on Mt. Sukahumi, W. Java, according to Jordan (1933). [2]Indochina. Tonkin. [3]Fraserpet: Coorg. Calcutta: Bengal. Hainan. Java. Sumatra. Indochina.

[Food-plant: Unknown wood].

[Distribution: Cachar Division: Assam. Chilauni, Kalimpong Division: Bengal].

albimaculatus Allard

Phloeobius albimaculatus Allard, *Bull. Soc.Ent. France*, p. 204, (1895)[1]. Bovie, *Ann. Soc. Ent. Belg.* 49, p. 301, (1905). Jordan, *Rec. Ind. Mus.* 9, p. 214, (1913)[2]; *Trans Linn. Soc. Lond.* 16, p. 254, (1914). Wolfrum, Cal. Cat. 102, p. 86, (1929).

[1]Tharrawaddy: Burma.
[2]Rangoon: Burma.

alternans (Wiedemann).

Anthribus alternans Wiedemann, *Zool. Mag.* 1 (3), p. 172, (1819)[1]. Fahraeus, in *Schh. Gen. Curc.* 5, p. 240, (1839).

[1]Bengal.

Anthribus apicalis Walker, *Ann. Mag. Nat. Hist.* (3) 3, p. 262, (1859)[2]. Sharp. *Trans, Ent. Soc. Lond.* p. 319, (1891)[3]. Stebbing, *Indian Forest Insects,* p. 387, (biology), (1914)[4]. Beeson, *Indian Forester,* 45 (2), p. 51 (1919).

[2]Ceylon.
[3]Japan.
[4]Tharrawaddy: Burma. Ceylon.

Philoeobius alternans Jordan, *Stett, Ent. Zeit.* 56, pp. 198-199, (1895); *Ann. Genova* 38, p. 643, (1897)[5]; *Deutsche Ent. Zeitschr.* p. 78, (1902); *Nov. Zool.* 11, p 235, (1904). Boyle, *Ann. Soc. Ent. Belg.* 49, p. 301, (1905). Lefroy, *Indian Insect Life,* 1, p. 380, (1909). Shar & Mur, Trans. Ent. Soc. Lond. p. 570, t. 76, 225, 225a (Morphology), (1912). Jordan, *Rec. Ind. Mus.* 9, pp. 213, 214, (1913)[6]; *Trans. Linn. Soc. Lond.* 16, p. 254, (1914); *Opusc. inst. Scient. Indochine. Fn. Ent.* 1, pp. 102, 105, (1923); [7] *Nov. Zool.* 30, p. 173, (1923); *Nov. Zool.* 34, p. 103, (1938)[8]. Wolfrum, *Col. Cat.* 102, p. 86, (1929). Jordan, *Indian For. Rec.* 18 (3), p. 4 (1933)[9]. Beeson, *forest Insects,* p. 48, (biology), (1941). Mathur, *Indian For. Rec.* 8 (3), p. 15, (larval charactes) (1950)[10].

[5]Kaju Tanam, Padang: Sumatra.
[6]Sibsagar: Assam. Calcutta, Dacca, Maldah: Bengal. Kandy, Kalutara: Ceylon Andaman Islands.
[7]Tonkin. Laos Cambodia.,
[8]Perak.
[9]Fraserpet: Coorg. Extends to Borneo, Celebes and Java.
[10]Ramanguli, Kanara Bombay.

[Food-plant: *Bambusa arundinacea, Bambusa polymorpha, Erythrina, lithosperma, Xylia dolabriformis*].

[Distribution: Samsing: Bengal. Pollibetta: Coorg. Nilambur: Madras. Dehra Dun: Uttar Pradesh. Maymyo; Rangoon: Burma].

brevitarsis Jordan

Phloeolius brevitarsis Jordan, *Indian For. Rec.* 18 (3), pp. 5, 6, (1933)[1].

[1]Fraserpet: Coorg. Aiyur, North Salem: Madras.

ceylonicus Jordan

Phlaeobius ceylonicus Jordan, *Deutsche Ent. Zeitschr.* p. 78, (1902)[1]. Bovie, *Ann. Soc. Ent. Belg.* 49, p. 302, (1905). Jordan, *Trans. Linn. Soc. Lond.* 16, p. 254, (1914). Wolfrum, *Cal. Cat* 102, p 86, (1929). Jordan, *Indian For. Rec.* 18 (3), p. 7, (1933)[2].

[1]Colombo: Ceylon. [2]Fraserpet: Coorg. Kanara: Bombay. Madura, Juranda Madras.

crassicollis Jordan

Phloeobius crassicollis Jordan, *Indian For. Rec.* 18(3), p. 6, (1933)[1]. Beeson, *Forest Insects,* p. 48, (biology), (1941).

[1]Fraserpet: Coorg. Aiyur, N. Salem: Madras.

[Food-plants: *Dendrocalamus strictus* (dry)]

[Distribution: Dehra Dun: U. P.].

gigas (Fabricius)

Ptinus gigas Fabricius, *Syst. Ent.* p. 63, (1775)[1].

[1]Hollandia.

Anthribus griseus Fabricius, *Ent. Syst.* 1 (2), p. 377, (1792). Olivier, *Ent.* 4 (80), p. 9, t. f. 10, (1795). Fabricius *Syst. El.* 2, p. 410, (1801). Boisduval, *Voy. Astrol.* p. 398, (1835). Montrouzier, *Ann. Soc. Agr. Lyoh.* 7(1), p. 45, (1857).

Anthribus longicornis Fabricius, *Ent. Syst. Suppl.* p. 160, (1798)[2]. Jordan, *Stett. Ent. Ziet.* 56, pp. 130, 198, 199 (1895).

[2]India.

Phloeobius gigas, Jordan, *Stett. Ent. Zeit.* 56, pp. 198-199, (1895); *deutsche Ent. Zeitschr.* p. 78, (1902). Bovie, *Ann. Soc. Ent. Belg.* 49, p. 302, (1905). Jordan, *Nova Guinea,* 5, p. 352, (1908)[3]; *Nov. Zool.* 19, p. 145, (1924)[4]; *Trans. Linn. Soc. Lond.* 16, pp. 254, 255, (1914); *Opusc. Inst. Scient. Indochine, Fn. Ent.* 1, pp. 103—105 (1923). Wolfrum, *Col. Cat.* 102, p. 86, (1929). Gardner, *Indian For. Rec.* 16 (II), p. 330, Pl. 1, fig. 12, (larval character), (1932). Jordan, *Indian For. Rec.* 18(3), p. 6, (1933); *Nov. Zool.* 39, p. 275, (1936)[5]. Beeson, *Forest insects,* p, 48, (biology), 1941).

[3]New Guinea. [4]Formosa.

[5]Widely distributed in the Oriental Region, from India to the pacific.

[Food-plants: *Kydia calycina* unknown climber and wood]

[Distribution: Kirwatti, E. Kanara; Manor, N. Thana: Bombay. Dehra Dun: U.P.].

gigas nigroungulatus Gyllenhal

Abthribus nigroungulatus Gyllenhal, *in Schhj, Gen. Curc.* 1, p. 133, (1833)[1].

Phloeobius gigas nigroungulatus, Jordan, *Rec. Ind. Mus.* 9, p. 214, (1913)[2]; *Trans. Linn. Soc. Lond.* 16, p. 255, (1914); *Opusc, Inst. Scient. Indochine, Fn. Ent.* 1, p. 102, (1923)[3]. Wolfrum, *Col. Cat.*, 102, p. 87, (1929).

[1]India and Indo-china. China.
[2]Kurseong: Bengal.
[3]Tonkin. Hanoi. Laos.

lutosus Jordan

Phaloebius lutosus Jordan, *Rec. Ind. Mus.* 9, p. 213-214, (1913)[1]; *Opusc. Inst. Scient. Indochine, fn. Ent.* 1, pp. 102, 105, (1923)[2]. Wolfrum, *Col. Cat.* 102, p. 87, (1929). Jordan, *Treubia,* 7(4), p. 337, (1929)[3]. Beeson. *Forest Insects*, p. 48, (biology), (1941).

[Food-Plant: *Dendrocalamus strictus* (*dry*)]

[1]Java—Andamans. Distributed from Sikkim to Sumba and Buru.,
[2]Laos.
[3]North India to Ceram Buru.
[Distribution: Kanara: Bombay. Dehra dun, Najibabad: Uttar Pradesh. Zigon Division: Burma].

pallipes Jordan

Phloeobius pallipes Jordan, *Stett. Ent., Zeit.* 56, pp. 197-198, (1895)[1]; *Deutsche Ent. Zietschr.* p. 78, (1902). Bovie, *Ann. Soc. Ent. Belg.* 49. p. 302, (1905). Jordan, *Rec. Ind. Mus.* 9, p. 214, (1913)[2]; *Trans. Linn. Soc.* Lond. 16, p. 254, (1914); *Nov. Zool.* 23, p. 363, (1916)[3]; *Opusc. Inst. Scient. Indochine, Fn. Ent.* 1, pp. 102, 106, (1923)[4]. Wolfrum, *Col. Cat* 102, p. 87, (1929).

[1]Sumatra. Perak.
[2]Andamans
[3]Cambodia. Tonkin Vientiane: Indo-china.
[4]Tonkin. Laos. Cambaodia. Cochinchn. Siam.

pilipes Jordan

Phloeobius pilipes Jordan, *Stett. Ent. Zeit.* 56, p. 198, (1895)[1]. Bovie *Ann. Soc. Ent. Belg.* 49, p. 302,

[1]Sumatra.

(1905). Jordan *Rec. Ind. Mus.* 9, p. 214, (1913)[2]; *Trans. Linn. Soc. Lond.* 16, p. 254, (1914); *Opusc. Inst. Scient. Indochine, Fn. Ent.* 1, pp. 103, 106, (1923)[3]; *Nov. Zool.* 34, p. 94, (1928)[4]. Wolfrum, *Col. Cat.* 102 p. 87, (1929). Garnder, *Indian For. Rec.* 16 (11), pp. 329-330, pl. 1, figs. 6—11 (larval characters), (1932). Beeson, *Forest Insects*, p. 48, (biology), (1941).

[2]Andamans. [3]Tonkin. Laos, Cochin-china. [4]Tonkin.

[Food-plants: *Gmelina arborea, Lagerstroemia flos-reginae,* unknown wood].

[Distribution: Cachar, Sibsagar: Assam. Chittagong Hill Tracts; Kurseong: Bengal. Tithimatti: Coorg. Namtu: Burma.].

santalinus Jordan

Phloeobius santalinus Jordan, *Indian For. Rec.* 18(3), pp. 6-7, (1933)[1].

[1]Aiyur, N. Salem: Madras. Fraserpet: Coorg.

[Food-plant: Unknown climber].

[Distribution: Sappal, palghat: Madras, Dehra Dun: U.P.].

42. Phloeomimus Jordan

Phloeomimus Jordan, *Nov. Zool.* 13, p. 409, (1906). Wolfrum, *Col. Cat.* 102 p. 88, (1929).

griseus Jordan

Phloeomimus griseus Jordan, *Nov. Zool.* 13, p. 409, (1906)[1]. Wolfrum, *Col. Cat.* 102, p. 88, (1929).

[1]Tharrawaddy: Burma. [Distribution: Bagdogra, Kurseong: Bengal.

43. Phloeopemon Schönherr

Phloeopemon Schönherr, *Gen. Curc.* 5, p. 159, (1839). Lacordaire, *Gen. Col.* 7, p. 490, (1866). Bovie, *Ann. Soc. Ent. Belg.* 49, p. 219, (1905). Wolfrum, *Col. Cat.* 102, p. 5, (1929).

acuticornis continentalis Jordan

Phloeopemon acuticornis continentalis Jordan, *Nov. Zool.* 30, p. 167, (1923)[1]; *L' Inst. Scient. Saigon. Faune Ent.* 1, p. 72, (1923)[2]. Wolfrum, *Col. Cat.* 102, p. 5, (1929).

[1]Khasis Hills: Assam. Annam. Burma. Tonkin. [2]Indochina.

44. **Physopterus** Lacordaire

Physopterus Lacordaire, *Gen. Col.* 7, p 498 (1866). Boive, *Ann. Soc. Ent. Belg.* 49, p. 227, (1905). Jordan, *Nov. Zool.*20, pp. 261-262, (1913); *Rec. Ind. Mus.* 9, p. 205, (1913). Wolfrum, *Col. Cat.* 102, p. 15, (1929).

agrestis (Boheman)

[1]Bengal.

Phloeophilus agrestis Boheman in *Schonh. Gen.* Curc. 1, p. 157, (1833)[1]. Bovie, *Ann. Soc. Ent. Belg.* 49, p. 230, (1905).

Physopterus agrestis Jordan, *Rec. Ind. Mus.* 9, p. 205, (1913)[2]; *Nov. Zool.* 20, p. 261, (1913). Beeson, *Indian Forestor* 45 (2), p. 51, (1919)[3]; *Rept. Proc. Fifth Ent. Meet. Pusa,* p. 170, (1923). Wolfrum, *Col. Cat.* 102, p. 15, (1929). Beeson, *Forest Insects,* p. 49 (biology), (1941).
[Food-plant: *Shorea robusta*].

[2]Calcutta: Bengal. [3]Siwaliks: Gorakhpur: U.P. Jalpaiguri: Bengal.

[Distribution: Sylhet: Assam. Pusa:Bihar. Dehra Dun: Haldwani; Rikbikesh: U.P.].

analis Jordan

Physopterus analis Jordan, *Nov. Zool.* 13, p. 408, (1906)[1]. Wolfrum, *Col. Cat.* 102, p. 15, (1929). Jordan, *Indian For. Rec.* 18 (3), p. 4, (1933)[2].

[1]Anamalai Hills: Madras. [2]Fraserpet: Coorg. Denkanikota: Madras.

callocerus Jordan

Physopterus callocerus Jordan, *Nov. Zool.* 19, p. 87, (1921)[1]. Wolfrum, *Col. Cat.* 102, p. 15, (1929).

[1]Nilgiri Hills: Madras.

gibbosus Guerin

Physopterus gibbosus Guèrin, *Voy. Deless. Ent.* 2, p. 51, (1843)[1]. Bovie, *Ann. Soc. Ent. Belg.* 49, p. 227, (1905). Jordan, *Nov. Zool.* 19, p. 87, (1912); *Nov. Zool.* 20, p. 261, 262 (1913). Wolfrum, *Col. Cat.* 102, p. 15, (1929).

[2]S. India.

45. **Rawasia** Roelofs

Rawasia, Roelofs, *Notes Leyd. Mus.* 2, p. 203, (1880). Ritsema, *in Snelleman, Midden-Sumatra, 4 (Col).* p.

27, (1887). Bovie, *Ann. Soc. Ent. Belg.* 49, p. 292, (1905). Jordan, *Tijdschr. Ent.* 55, p. 140, (1912); *Rec. Ind. Mus.* 9, p. 212, (1913). Wolfrum, *Col. Cat.* 102, p. 76, (1929).

annnulipes Jordan

Rawasia annulipes Jordan, *Stett. Ent. zeit.* 56, pp. 188-189, (1895)[1]; *Nov. Zool.* 10, p. 129, (1903); *Ann. Mus. Genova,* 41, p. 84, (1904). Bovie, *Ann. Soc. Ent. Belg.* 49, p. 292, (1905). Jordan, *Rec. Ind. Mus.* 9, p. 212, (1913)[2]; *Opusc. Inst. Scient. Indochine, fn. Ent.* 1, p. 92, (1923)[3]; *Nov. Zool.* 34, pp. 93, 103, (1928)[4]. Wolfrum, *Col. Cat.* 102, p. 76, (1929). Mathur, *Indian For. Rec.* 8 (3), pp. 13—15, (larve characters), (1950)[5]. [Food-plant: *Bambusa arundinacea*].

[1]Carin Cheba: Burma.
[2]N. India ,
[3]Laos.
[4]perak. Selangor, Tonkin.,
[5]Ramanguli, N. Kanara, Bombay.

communis communis Jordan

Rawasia communis Jordan, *Stett. Ent. Zeit.* 56, pp. 188, 189, (1895)[1]. Bovie, *Ann. Soc. Ent. Belg.* 49, p. 292, (1905). Jordan, *Rec. Ind. Mus.* 9, p. 212, (1913); *Nov. Zool.* 34, p. 103, (1928). Wolfrum, *Col. Cat.* 102, p. 77, (1929).

[1]Khasis Hills: Assam.

ritsemae Roelofs

Rawasia ritsemae Roelofs, *Notes Leyd. Mus.* 2, p. 204, (1880)[1]. Ritsema in *Snelleman, Midden, Sumatra (Col).* 4, p. 128, (1887). Jordan, *Stett. Ent.* Zeit, 56, pp. 188, 189, (1895); *Nov. Zool.* 10, p. 129 (1903); *Ann. Mus. Genova,* 41, p. 84, (1904). Bovie, *Ann. Soc. Ent. Belg.* 49, p. 292, (1905). Jordan, Rec. Ind. Mus. 9, p. 212, (1913)[2]; *Tijdschr. Ent.* 59, p. 161, (1916)[3]; *Nov. Zool.* 23, p 362, (1916)[4]; *Opusc. Inst. Scient. Indochine, Fn. Ent.* 1, p. 92, (1923)[5]; *Nov. Zool.* 34, p. 103, (1928)[6]. Wolfrum, *Col. Cat.* 102, p. 77, (1929). Beeson, *Forest Insects,* p. 49, (biology), (1941).
[Food-plants: *Bambusa polymorpha* (dry) and dry bamboo].

[1]Sumatra.
[2]Andamans. North India to Java. Boreo.
[3]Bekassa, Deli: Sumatra.
[4]Annam. Upper Tonkin. Common in the Indo-Malayan Countries.
[5]Tonkin. Laos. Annam., [6]Perak.
[Distribution: Bengal. Lansdowne: Uttart Pradesh.

Insein, Myityina, Pyinmana, Shewbo, Toungoo: Burma].

46. **Rhaphitropis** Rietter.

Rhaphitropis, Rietter, *Fn.* Germ., 5, p. 5, (1916). Jordan, *Ann. Mag. Nat. Hist.* (9) 16, p. 257, (1925). Wolfrum, *Col. Cat.* 102, p. 59, (1929).

Rhaphidotropis, Jordan, *Ent. Mitt.* 11, p. 147, (1922).

carbo Jordan

Rhaphitropis carbo Jordan, *Nor. Zool.* 39, p. 316, (1936)[1]. Beeson, *Forest Insects,* p. 49, (biology) (1941). [Food-plants: *Artocarpus lakoocha, Bombax malabaricum, Broussonetia papryifera, Dalbergia latifolia, Euphorbia pulcherrima,* unknown dry twigs]. [1]Dehra Dun: U.P.

Centrimaculatus (Allard)

Araecerus centrimaculatus Allard, *Bull. Soc. Ent. France,* p. 204, (1895)[1]. Bovie, *Ann. Soc. Ent. Belg.* 49, p. 320, (1905). Wolfrum, *Col. Cat.* 102, p. 106, (1929). [1]Belgaum: Bombay.

Rhaphitropis centrimaculatus (Allard).

cor Jordan

Rhaphitropis car Jordan, *Nov. Zool.* 32, pp. 249-250, (1925)[1]. Wolfrum, *Col. Cat.* 102, p. 60, (1929). [1]Nilgiri Hills: Madras.

cosmia Jordan

Rhaphiropis cosmia Jordan, *Nov. Zool,* 40, pp. 200-201, (1937)[1]. Beeson, *Forest Insects,* p. 49, (biology), (1941). [Food-plant: *Michelia champaca*]. [1]Dahura, Bagdogra, Kurseong: Bengal.

discus Jordan

Rhaphitropis discus Jordan, *Nov. Zool.* 39, pp. 315-316 (1936)[1]. Gardner. *Indian For. Rec.* (N.S.) 2 (2), pp. 107-108, figs. 15—17, (larval characters), (1936). [1]Konain, Chakrata: U.P.

[Food-plant: Unknown dry sticks].

incanus Jordan

Rhaphitropis incanus Jordan, *Nov. Zool.* 38, pp. 311-312, (1932)[1]. Beeson, *Forest Insects,* p. 49, (biology), [1]Jhajra, Dehra Dun: U.P.

(1941). [Food-plants; *Dalbergia latifolia, Michelia champaca Morus indica,* Unknown dry twigs and creeper]. Distribution: — Bagdogra, Kurseong: Bengal].

indicus Jordan

Rhaphitropis indicus Jordan, *Nov. Zool.* 32, p. 248, (1925)[1]. Wolfrum, *Col. Cat.* 102, p. 60, (1929). Jordan, *Nov. Zool.* 38, p. 312, (1932); *Nov. Zool.* 40, p. 201, (1937). Beeson, *Forest Insects,* p. 49, (biology), (1941). [Food-plant: *Acacia gageana*]. [4]Kulu: Punjab. [Distribution: — Dehra Dun: Uttar Pradesh].

marchicus Herbst

Rhaphitropis marchicus Herbst, *Naturgesch. Ins. Käf.* 7, p. 167, t. 106, f. 8, (1797)[1]. Bedel, *Fn. Col. Bassin Seine,* 6, pp. 10, 15, (1882—88). Stierlin, *Col. Helret,* p. 423, (1892). Rey, *Echange,* 9, p. 49, (1893). Everts, *Col. Neerland,* 2, p. 535, (1903); 3, p. 502, (1922). Shilsky, *Käf. Eur.* 44, p. 90, (1907). C. Schaufuss, in Calwer, *Käferb.* ed. 6, (1907); 16, p. 1027, (1913). Reitter, *Fauna Germ.* 5, p. 7, t. 153, f. 5, (1916). Kuhnt, *Illustr. Best.-Tab. Käf. Deutschl.* p. 901, (1913). Wolfrum, *Col. Cat.* 102, p. 60, (1929); *Ent. Blätt.* 32 (3). p. 133, (biology), (1936). Jordan, *Ark. Zool.* 41A (6), p. 7, (1949)[2]. [1]Germany: Middle Europe. [2]Kambaiti: Burma.

***marmoratus** (Motschulsky)

Platyrrhinus marmoratus Motschulsky, *Bull. Mosc.* 39, p. 399, (1866)[1]. Bovie, *Ann. Soc. Ent. Belg.* 49, p. 252, (1905). Wolfrum, *Col. Cat.* 102, p. 37, (1929). [1]Ceylon.

Rhaphitropis marmoratus Motschulsky.

stevensi Jordan

Rhaphitropis stevensi Jordan, *Nov. Zool.* 32, pp. 250-251, (1925)[1]. Wolfrum, *Col. Cat.* 102, p. 60. (1929). Jordan, *Nov. Zool.* 40, pp. 200, 201, (1937). Beeson, *Forest Insects,* p. 49, (biology), (1941). [Food-plant: *Acacia gageana*]. [1]Gopaldhara: Assam. Nepal-Sikkim frontier. [Distribution: — Dehra Dun: U. P.].

tamilis Jordan

Rhaphitropis tamilis Jordan, *Nov. Zool.* 32, pp. 250-251, (1925)[1]. Wolfrum, *Col. Cat.* 102, p. 60, (1929). [1]Kandy: Ceylon.

---

* Dr. Jordan writes : *marmoratus* Motsch., was originally described as *Macrocephalus marmoratus. Macrocephalus* Oliver 1789 is preoccupied.

47. **Sintor** Schönherr

Sintor, Schönherr, *Gen. Curc.* 5, p. 148, (1839). Lacordaire, *Gen. Col.* 7, p. 510, (1866). Bovie, *Ann. Soc. Ent. Belg.* 49, p. 234, (1905). Jordan, *Rec. Ind. Mus.* 9. p. 207, (1913). Wolfrum, *Col. Cat.* 102, p. 21, (1929).

Rhinanthribus, Motschulsky, *Bull. Moscou,* 48, p. 241, (1874).

andrewesi Jordan

Sintor andrewsi Jordan *Nov. Zool.* 13, p. 408, (1906)[1]. Wolfrum, *Col. Cat.* 102, p. 21, (1929).

[1]Nilgiri Hills: Madras.

biplaga Jordan

Sintor biplaga. Jordan, *Nov. Zool.* 10, pp. 416-417, (1903)[1].Bovie *Ann. Soc. Ent. Belg.* 49, p. 334, (1905). Jordan, *Rec. Ind. Mus.* 9. p. 207, (1913)[2]; *Nov. Zool.* 34, p. 88, (1928)[3]. Wolfrum, *Col. Cat.* 102, p. 21, (1929).

[1]Khasis Hills: Assam.
[2]Sikkim.
[3]Tonkin.

suturalis Jordan

Sintor suturalis Jordan, *Stett. Ent. Zeit.* 56, pp. 374-375, (1895)[1]. *Ann. Mus. Genova,* 38, p. 627, (1897). Bovie, *Ann. Soc. Ent. Belg.* 49, p. 234, (1905). Wolfrum, *Col. Cat.* 102, p. 22, (1929). Jordan, *Nov. Zool.* 36, p. 297, (1931)[2].

[1]Khasis Hills: Assam.

[2]Java. Cambodia. [Distribution: Haflong (Cachar District): Assam. Maublaing (Shwebo Division) Burma].

48. **Stiboderes** Jordan

Stiboderes, Jordan, *Nov. Zool.* 32, p. 243 (1925). Wolfrum, *Col. Cat.* 102, p. 73, (1929).

Impressus impressus Jordan

Xylindades Impressus Jordan, *Stett. Ent. Zeit.* 56, pp. 253-254, (1895)[1]. Bovie, *Ann. Soc. Ent. Belg.* 49, p. 288, (1905). Jordan, *Opusc. Inst. Scient. Indochine, Fn. Ent.* 1, p. 97, (1923)[2].

[1]Khasis Hills: Assam. Sikkim.

[2]Laos. Tonkin.

Stiboderes impressus Jordan, *Nov. Zool.* 32, p. 243, (1925). Wolfrum, *Col. Cat.* 102, p. 74, (1929). Gardner, *Indian For. Rec.* (N.S.) 2 (2), p. 109-110, figs. 10—12, (larval characters), (1936) Beeson, *Forest Insects,* p. 49, (biology). (1941). [Food-plants: *Phoebe lanceolata,* unknown wood].

[Distribution: — Bagdogra, (Kurseong Division) Samsing, (Kalimpong Division), Bengal].

49. **Straboscopus** Lacordaire

Straboscopus, Lacordaire, *Gen. Col.* 7, p. 533 (1866). Bovie, *Ann. Soc. Ent.* Belg. 49, p. 256, (1905). Jordan, *Rec. Ind. Mus.* 9. p. 207, (1913). Wolfrum. *Col. Cat.* 102, p. 40, (1929)

riehli Lacoradaire

Straboscopus riehli Lacordaire, *Gen. Col.* 7, p. 534, (1866)[1]. Bovie *Ann. Soc. Ent. Belg.* 49, p. 256, (1905). Jordan, *Rec. Ind. Mus.* 9, p. 207, (1913)[2]; *Nov. Zool.* 23, p. 361, (1916). Wolfrum, *Col. Cat.* 102, p. 40, (1929).

[1]Ceylon.
[2]Matale Ceylon.

[Distribution: — Anamalai Hills, 2400 ft. Madras].

sanguinipes sanguinipes Lacordaire

Straboscopus sanguinipes lacordaire, *Gen. Col.* 7, p. 534 (1866)[1]. Bovie, *Ann. Soc. Ent.* Belg. 49, p. 256, (1905). Wolfrum *Col. Cat.* 102, p. 40, (1929)

[1]Assam.

50. **Sympaector** Kirsch.

Sympaector, Kirsch, *Mith. Mus. Dresd.* 1, p. 53, (1875). Bovie, *Ann. Soc. Ent. Belg.* 49, p. 266, (1905). Wolfrum, *Col. Cat.* 102, p. 52, (1929)

revocans (Walker)

Xenocerus revocans Walker, *Ann. Mag. Nat. Hist.* (3) 3, p. 262, (1859)[1].

[1]Ceylon.

staboscopus revocans Jordan, *Nov. Zool.* 1, p. 636, (1894). Bovie, *Ann. Soc. Ent. Belg.* 49, p. 256, (1905). Wolfrum, *Col. Cat.* 102, p. 40, (1929).

Sympaector revocans Jordan, *Nov. Zool.* 1 p. 605, (1894) ("*angulifer*" eratum, should read *revocans*).

51. *__Tropideres__ Schönherr

Tropiders, Schönherr, *Isis von Oken,* 10, p. 1135, (1823); *Curc. Disp. Meth.* p. 35, (1826). Imhoff, *Gen. Curc.* 1, p. 63, (1842). Redtenbacher, *Fauna. Austr.* p. 473, (1849); p. 674, (1858); p. 385, (1874). Jacq. du

* Dr. Jordan writes : Most of these references under *Tropideres* cover each species belonging to different genera.

Val, *Gen. Col. D'Eur.* 1854—68, 4, p. 4, (1854). Bose-Gutfl., *Käf. Deutschl.* p. 457, (1859). C.G. Thoms., *Skand. Col.* 7, p. 19 (1865). Lacordaire, *Gen. Col.* 7, p. 535, (1866)., Seidl, *Fn. Balt.* 1872—75, p. 468, (1875). Bedel, *Fn. Col. Bassin Seine,* 1882—88, 6, pp. 8, 14, (1882). C. Schenki, *Deutsche Käferw.* p. 223, (1885). Seidl, *Fn. Transsylv.* 1888-91, p. 746, (1891). Sharp, *Trans. Ent. Soc. London,* p. 302, (1891). Strierlin *Col. Helvet,* p. 422, (1892). Jordan *Nov. Zool.* 1, p. 608, (1894). Acloque, *Fn. France Col.* p. 354, (1896). Blackburn, *Trans. Soc. S. Austral.* 24, p. 150, (1900). Lameere, *Fn. Belg.* 2, p. 472, (1900). Everts, *Col. Neerland,* 2, p. 534, (1903). Kuhnt *Illustr. Best-Tab. Käf. Deutschl.* p. 901, (1903). Bovie, *Ann. Soc. Ent. Belg.* 49, p. 267, (1905). C. Schaufuss. *Calwer, Käferb.* ed. 6, 1907-16, p. 1027, (1913). Reitter, *Fn. Gem.* 5, p.4, (1916). Henriksen, *Danm.* Fauna. 22, p. 10, (1918). Everts, *Col. Neerland,* 3, p. 501, (1922). Wolfrum, *Col. Cat.* 102, pp. 53-54, (1929); *Ent. Blätt.* 26, p. 90, (1930). Gardner, *Indian For. Rec.* 2 (2), pp. 105-106, (larval characters) (1936).

Amblycerus, Thunberg, *Nova Acta Ups.* 7, p. 146, (1815).

Macrocephalus, Olivier, *Ent.* 4, p. 80, (1795).

Tropidoderes, Gemminger & Harold, *Cat. Col.* 9, p. 2733, (1872). Jordan, *Ent. Mitt.* 11, p. 147, (1922).

asparsus Jordan

Acorynus aspersus Jordan, *Nov. Zool.* 32, p. 290, (1925)[1]; *Nov. Zool.* 34, p. 80, (1928). Wolfrum, *Col Cat.* 102, p. 23, (1929).

[1]Haflong, Cachar: Assam.

Tropideres aspersus Jordan.

bolinus Jordan

Tropideres bolinus Jordan, *Nov. Zool.* 31, p. 233, (1924)[1]. Wolfrum, *Col. Cat.* 102, p. 55, (1929). Beeson, *Forest Insects,* p. 49, (biology), (1941). [Food-plants: *Artocarpus hirsuta, Gmelina arborea, Mallotus philippinensis, Quercus incana, Shorea robusta, Tectona grandis, Terminalia tomentosa*].

[1]Thano Range and Nagsidh Forests, Dehra Dun: U.P. [Distribution: Kasaragod (S. Mangalore Div.): Madras. Saing Yane (N. Tonghoo): Burma].

comes Jordan

Tropideres comes Jordan, *Nov. Zool.* 36, pp. 144-145, (1930)[1].

[1]Shillong: Assam.

***lateralis** Motschulsky

Tropideres leteralis Motschulsky, *Bull. Mosc.* 48 (2), p. 235, (1874)[1]. Bovie, *Ann. Soc. Ent. Belg.* 49, p. 269, (1905). Wolfrum, *Col. Cat.* 102, p. 56, (1929).

[1]E. India.

luteago Jordan

Tropideres luteago Jordan, *Nov. Zool.* 36, pp. 143-144, (1930)[1]. Gardner, *Indian For. Rec.* (N.S.) 2 (2), pp. 106-107, figs. 18—20, (larval characters), (1936). Beeson, *Forest Insects,* p. 49, (1941). [Food-plants: *Albizzia* sp. (dry sticks), *Girardinia heterophylla, Michelia champaca, Tectona grandis,* unknown wood].

[1]Asan, Dehra Dun: U.P.

[Distribution: — Dahura, Bagdogra (Kurseong Division): Bengal. Konakutch (Nilambur Div.): Madras].

marmoreus Jordan

Tropideres marmoreus Jordan, *Nov. Zool.* 1, p. 628, (1894)[1]. Bovie, *Ann. Soc. Ent. Belg.* 49, p. 270, (1905). Wolfrum, *Col. Cat.* 102, p. 56, (1929).

[1]Ruby Mines: Burma [Distribution: Haflong (Cacher District): Assam].

notabilis Jordan

Tropideres notablilis Jordan, *Nov. Zool.* 34, p.82, (1928)[1]. Wolfrum, *Col. Cat.* 102, p. 56, (1929). Beeson, *Forest Insects,* p. 49, (biology), (1941). [Food-plants: *Dolichandrone rheedii, Millettia pendula, Tectona grandis, Terminalis belerica*].

[1]Hoa Binh: Tonkin. [Distribution: Ngayatgyan, Saign yane (N. Toungoo Division); Yanan-ngmyin (Pyinmana Division): Burma].

paviei Lesne

Litocerus paviei Lesne, *Bull.Soc.Ent. France,* 60, p. 91, (1891)[1]; *Nouv. Arch Mus. Paris,* 3, p. 226, (1891). Jordan, *Nov. Zool.* 1, p. 612, (1894); *Stell. Ent. Zeit.* 56, p. 147, (1895); *Nox. Zool.* 10. p. 424, (1903). Bovie, *Ann. Soc. Ent. Belg.* 49, p. 242, (1905). Jordan, *Nov. Zool.* 19, p. 88, 138, (1912)[2]; *Rec. Ind.*

[1]Siam.

[2]Formosa. Kobo: Abor.

---

* Probably not *Tropideres,* writes Dr. Jordan.

Mus. 8, p. 197, (1913)[3]; 9. p. 207, (1913)[4]; *Tijdschr. Ent.* 59, p. 162, (1916)[5]. Beeson, *Indian Forester,* 45 (2), p. 51, (1919)[6]. Jordar, *Opusc. Inst. Scient. Indochine, Fn. Ent,* 1, p. 81, (1923)[7]. Beeson, *Report Proc. Fifth Ent.* Meet. Pusa, p. 170 (1923).

[3]kawkareik: Burma. [4]Burma. [5]Tigembong: Java. [6]Siwaliks, Ramnagar: U.P. [7]lLaos. Vientiane. Annam. Siam.

Tropideres paviei Lesne

Jordan, *Nov. Zool.* 34, p. 82, 99, (1928)[8]. Wolfrum, *Col. Cat.* 102, p. 56, (1929)[9]. Beeson, *Forest Insects,* p. 49, (biology), (1941). [Food-plants: *Ailanthus malabaricus, Bombax malabricum, Butea superba, Dalbergia latifolia, Millettia* sp.,*Shorea robusta*].

[8]Tonkin. Pahang. [9]Java. Halbinsel. Philippine. [Distribution: Assam. Sitapahar, Chittagong: Bengal. Kanara: Bombay. Barkote, Dehra Dun, Nakronda, Thano (Dehra Dun Div.); Lakhanmandi, Chakrata (Haldwani Div.) U.P., Amherst; Byame, Prome; Ikkan, Insein: Burma. Badagamuwa, Kurune gala: Ceylon.]

scitus Jordan

Tropideres scitus Jordan, *Nov. Zool.* 38, pp. 372-373, (1933)[1].

[1]Ruby Mines : Burma.

Tropideres securus Bohemna, *in Schh. Gen. Curc.* 5, p. 207, (1839)[1]. Bovie, *Ann. Soc. Ent. Belg.* 49, p. 271, (1905). Jordan, *Opusc. Inst. Scient. Indochine, Faune Ent.* 1, p. 82, (1923)[2]; *Nov. Zool.* 34, pp. 82, 99, (1928)[3]. Wolfrum, *Col Cat.* 102, p. 56, (1929). Jordan, *Indian For. Rec.* 18 (3), p. 2. (1933)[4]. Gardner, *Indian For. Rec.* (N. S.) 2 (2), pp. 106-107, (larval characters). (1936). Anon., *Indian For. Rec.* (N. S.) 4 (1), p. 32, (1938). Beeson, *Forest Insect,* p. 49, (biology) (1941). [Food-plants: *Albizzia* sp. (dead sticks) *Bombax malabaricum, Dalbergia sissoo, Euphorbia pulcherrima* (drystems), *Lannea grandis, Quercus incana, Shorea assamica, S. Robusta, Tectona grandis, Terminalia tomentosa*].

[1]Calcutta: Bengal.
[2]Annam.
[3]Tonkin. Common and widely distributed in Indo-Malayan Countries. Kuala Lumpur.
[4]Fraserpet: Coorg. Assam. Darjeeling: Bengal. Coimbatore, Madras: Madras, Java.

[Distribution: Haflong (Cachar Division ; U. Dihing, Lakhimpur: Assam. Amarampalam (Nilambur Division): Madras. Barkote, Dehra Dun, Jhajra, Nakronda, Thano (Dehra Dun Div.) ; Lalkua, Chakrata (Haldwani Div.): U.P. Daunge (Meiktila Division); Sing Yane (N. Toungoo Div.) Ruby Mines. Burma].

Litocerus rufescens Roelofs, *C. R. Soc. Ent. Belg.* 22, p.55, (1879)[5]. — [5]Japan.

Litocerus anna Jordan, *Nov. Zool.* 10, p. 424 (1903)[6]. — [6]Borneo.

vigens Jordan

Tropideres vigens Jordan, *Nov. Zool.* 32, p. 239, (1925)[1]. Wolfrum. *Col. Cat.* 102, p. 57, (1929). Beeson, *Forest Insects*, p. 50, (biology), (1941). [Food-plant: *Juglans regia*]. — [1]Naga Hills: Assam.

[Distribution: — Hpemaw, Myitakyina : Burma.]

52. **Tropiderinus** Reitter

Tropiderinus, Reitter *Fn. Germ.*, 5, p. 5, (1916), Wolfrum, *Col. Cat.* 102, p. 58, (1929).

munieri Bedel

Tropideres munieri Bedel, *Faune Col. Bassin Siene*, 1882—88. 6, p. 8, nota 1, (1882)[1]. Bovie, *Ann. Soc. Ent. Belg.*, 49, p. 270, (1905). Schilsky, *Käf. Eur.* 44, p. 88, (1907). — Algier: Spain.

Tropiderinus munieri Bedel. Wolfrum, *Col. Cat.* 102, p. 58, (1929). Beeson, *Forest Insects*, p. 49, (biology), (1941). [Food-plant: *Populus euphrotica*].

[Distribution: — Gazighat (Multan Div.): W. Pakistan].

53. **Tropidobasis** Jordan

Tropidobasis, Jordan, *Nov. Zool.* 30, p. 178, (1923). Wolfrum, *Col. Cat.* 102, p. 89, (1929).

gemella Jordan

Tropidobasis Genella Jordan, *Nov. Zool.* 30, p. 179, (1923)[1]. Wolfrum, *Col. Cat.* 102, p. 89, (1929).

[1]Ceylon. [Distribution: Amarampalam, Nilambur (Nilambur Division): Madras].

54. **Ulorhinus** Sharp

Ulorhinus, Sharp, *Trans. Ent. Soc. London* pp. 300-301, (1891). Bovie, *Ann. Soc. Ent. Belg.* 49, p. 274, (1905). Jordan, *Nov. Zool.* 34, p. 121, (1928). Wolfrum, *Col. Cat.* 102, p. 51, (1929).

brachystomus Jordan

Ulorhinus brachystomus Jordan; *Nov. Zool.* 38, p. 311, (1932)[1]. Beeson, *Forest Insects,* p. 50, (biology), (1941). [Food-plant: *Erythrina indica*].

[1]N. Thana: Bombay

distichus Jordan

Ulorhinus distichus Jordan, *Nov. Zool.* 34, p. 122, (1928)[1].

[1]Ceylon.

55. **Uncifer** Jordan

Uncifer, Jordan, *Ann. Mus. Genova,* 14, p. 88, (1904). Bovie, *Ann. Soc. Ent. Belg.* 49, p. 259, (1905). Jordan, *Ann. Mag. Nat. Hist.* (9) 16, p. 259, (1925). Wolfrum, *Col. Cat.* 102, p. 44, (1929).

alienus Jordan

Ormiscus alienus Jordan, *Tijdschr. Ent.* 55, pp. 137-138 (1912)[1].

[1]Preanger: Java.

Uncifer alienus Jordan, *Ann. Mag. Nat. Hist.* (9) 16, p. 250, (1925). Wolfrum, *Col.* 102, p. 44, (1929). Beeson, *Forest Insects,* p. 50, (biology), (1941). [Food-plants: *Acacia gageana, Ficus glomerata, Ougeinia dalbergioides*].

[Distribution: Dehra Dun: Uttar Pradesh].

basalis Jordan

Uncifer basalis Jordan, *Ann. Mag. Nat. Hist.* (9) 16, pp. 259-260, (1925)[1]. Wolfrum, *Col. Cat.* 102, p. 44, (1929). Jordan, *Nov. Zool.* 36, p. 293, (1931); *Indian For. Rec.* 18 (3), p. 3, (1933)[2].

[1]Medan: Sumatra.
[2]Fraserpet: Coorg.

diffinis Jordan

Uncifer diffinis Jordan, *Nov. Zool.* 32, pp. 253-254, (1925)[1]. Wolfrum, *Col. Cat.* 102, p. 44, (1929). Jordan, *Nov. Zool.* 39, p. 319, (1936).

[1]Kandy: Ceylon.

Myodes Jordan

Uncifer Myodes Jordan, *Nov. Zool.* 40, p. 201 (1937)[1]. Beeson, *Forest Insects*, p. 50. (biology), (1941). [Food-plant: *Michelia champaca*].

[1]Dahura, Bagdogra, Kursoeng: Bengal.

pissodes Jordan

Uncifer Pissodes Jordan, *Indian For. Rec.* 18 (3), p. 3, (1933)[1].

[1]Fraserpet: Coorg.

sticticus Jordan

Uncifer sticticus Jordan, *Ann. Mus. Genova,* 41, p. 88-89, (1904)[1]. Bovie, *Ann. Soc.* Ent. Belg. 49, p. 259, (1905). Wolfrum, *Col. Cat.* 102, p. 44, (1929).

[1]Burma.

stigmosus Jordan

Uncifer stigmosus Jordan, *Nov. Zool.* 39, p. 319, (1936)[1]. Gardner, *Indian For. Rec.* (N.S.). 2 (2), p. 108, fig. 21, (Larval characters), (1936). Jor dan, *Nov. Zool.* 40, p. 201 (1937). Beeson, *Forest Insects* p. 50, (biology), (1941). [Food-Plants: *Artocarpus lakoocha, Mangifera, indica* (dead branches)].

[2]Dehra Dun: U.P.

56. **Xenocerus** Schönherr

Xenocerus, Schönherr, *Gen. curc.* 1, p. 117, (1833). Imhoff, *Gen. Curc.* 1, p. 36, (1842). Lacordaire, *Gen. Curc. 7,* p. 559, (1866). Jordan, *Nov. Zool.* 1, p. 636, (1894); *Stett. Ent. Zeit.* 56, p. 182, (1895), *Ann. Soc. Ent. Belg.* 49, p. 282 (1905). Jordan, *Nov. Zool.* 20, p. 268, (1913); *Rec. Ind. Mus.*9. p. 209 (1913); *Opusc. Inst. Scient. Indochine, Fn. Ent.* 1, p. 101, (key for Indochina species), (1923). Wolfrum, *Col. Cat.* 102, p. 67, (1929). Jordan, *Proc. R. Ent. Soc. Lond. B,* 14 (1-2), p. 10, (1945).

andamanensis Jordan

Xenocerus andamanensis Jordan, *Nov. Zool.* 1, pp. 637-638, (1894)[1]; *Stett. Ent. Zeit.* 56, pp. 184-185,

[1, 2, 3]Andaman Islands.

(1895). Bovie, *Ann. Soc. Ent. Belg.* 49, p. 282, (1905). Jordan, *Nov. Zool.* 20, p. 268, (1913);[2] *Rec. Ind. Mus. 9.* p. 209, (1913)[3]; *Nov. Zool.* 23, P. 361, (1916). Wolfrum, *Col. Cat.* 102, p. 67, (1929).

birmanicus Jordan

Xenocerus birmanicus Jordan, *Nov. Zool.* 10. p. 429 (1903)[1]. Bovie *Ann. Soc. Ent. Belg.* 49, p. 283, (1905). Jordan, *Opusc. Inst. Scient. Indo-chine, Fn. Ent.* 1, pp. 100-101, (1923)[2]. Wolfrum, *Col. Cat.* 102, p. 68, (1929).

[1]Burma.
[2]Laos.

callimus Jordan

Xenocerus callimus Jordan, *Nov. Zool.* 18, pp. 94-95, (1911)[1]; *Rec. Ind. Mus.* 9, p. 210, (1913)[2]. Wolfrum, *Col. Cat.* 102, p. 68, (1929).

[1, 2]Andaman Islands.

khasianus khasianus Jordan

Xenocerus khasianus Khasianus Jordan, *Stett. Ent. Zeit.* 56, pp. 184-185, (1895)[1]. Bovie, *Ann. Soc. Ent. Belg.* 49, p. 284, (1905). Jordan, *Rec. Ind. Mus.* 8. p. 198, (1913)[2]; *Nov. Zool.* p. 361, (1916); *Opusc. Inst. Scient. Indochine, Fn. Ent.* 1, pp. 100-101, (1923)[3]. Wolfrum, *Col. Cat.* 102, p 69, (1929). Beeson, *Forest Insects,* p. 50, (biology), (1941). [Food-plant: *Mesua ferrea*].

[1]Khasis Hills. Assam.
[2]Dejoo: Assam.
[3]Tonkin.
[Distribution: Nambor, Sibsagar: Assam. Ceylon].

lineatus Jordan

Xenocerus Lineatus Jordan, *Nov. Zool.* 1, pp. 647-648, (1894)[1]; *Stett. Ent. Zeit.* 56, p. 187, (1895). Bovie, *Ann. Soc. Ent. Belg.* 49, p. 284, (1905). Wolfrum, *Col. Cat.* 102, p. 69, (1929).

[1]Assam.

rectilineatus Jordan

Xenocerus rectilineatus Jordan, *Nov. Zool.* 1, p. 638-639, 648, (1894)[1]. Bovie, *Ann. Soc. Ent. Belg.* 49, p. 285, (1905). Jordan, *Nov. Zool.* 20, p. 269, (1913)[2]; *Rec. Ind. Mus.* 8, p. 198, (1913)[3]; 9, p. 210, (1913)[4]. Wolfrum, *Col. Cat.* 102, p. 70, (1929). Beeson, *Forest Insects,* p. 50, (biology), 1941. [Food-plants: *Artocarpus lakoocha, Buchanania latifolia, Delbergia panicula, Pterocarpus marsupium*].

[1]Ruby Mines: Burma. Khasis Hills: Assam.
[2]Darjeeling.
[3]Dejoo, Khasis Hills: Assam. Burma, Ceylon.
[4]Cachar, Sibsagar. Sylhet: Assam. Kandy: Ceylon.

[Distribution: Haflong: Assam. Tista Valley 700 ft. Bagdogra, Kurseong; Bengal Dandeli, N. Kanara; Nagargalli, Bengal. Dandeli, N. Kanara; Nagargalli, Belgaum: Bombay. Amarampallam, Nedungayam. Nilambur (Nilambur div.); Madras. Mahaoya, Batticaloa: Ceylon.]

57. **Xylinades** Latreille

**Xylinade,* Latreille, *Families naturelles,* p. 387, (1825).

Ø*Xylinada,* Berthold, *Naturl. Fam. Theirr.* p. 380, (1827), Thon, *entom Archiv.* 1 (3), p. 64, (1828).

Xylinades, Schönherr, *Gen. Curc.* 1, p. 177, (1833); *Gen. Curc.* 5, p. 232, (1839). IMhoff, *Gen. curc.* 1, p. 41, (1842). Lacordaire, *Gen. Col.* 7, p. 561, (1866). Jordan, *stett. Ent. Zeit.* 56, p. 252, (1895). Bovie, *Ann. Soc. Ent. Belg.* 49, p.287, (1905). Jordan, *Rec. Ind. Mus.* 9, p. 210, (1913); *Ent. Mitt.* 11, p. 150, (1922); *Opusc. Inst. Scient. Indo-Chine, Fn. Ent.* 1, p. 99, (19230. Wolfrum, *Col. Cat.* 102, p. 71, (1929).

andamanensis Jordan

Xylindades andamanensis Jordan, *stett. Ent. Zeit.* 56, p. 255, (1895)[1]. Bovie, *Ann. Soc. Ent. Belg.* 49, p. 287, (1905). Jordan, *Rec. Ind. Mus.* 9, p. 210, (1913)[2]. Wolfrum, *Col. Cat.* 102, p.71, (1929). Gardner, *Indian For. Rec.* 16(11), pp. 332-333, pl. 1, figs. 19—21, (larval characters), (1932); *Indian For. Roc.* 2 (2), pp. 109-110, (1936). Beeson, *forest Insects* p. 50, (biology), (1941), [Food-plant: *Mysristica andamanica*].

[1, 2]Andaman Islands.

---

* *Xylinade* : French form nopt accepted in nomenclature.

Ø *Xylinada* : Latin form, translation of the preceding. The International Commission on Zoological Nomenclature has been asked to suppress this name, writes Dr. Jordan.

annulipes Jordan

Sylinades annulipes Jordan, *Stett. Ent. Zeit.* 56, pp. 255-256, (1895)[1]. Bovie, *Ann. Soc. Ent. Belg.* 49, 287, (19050. Jordan, *Rec. Ind. Mus.* 8, p. 198, (1913)[2]; 9, p. 210, (1913)[3]; *Opusc. Inst. Scient Indochine, Fn. Ent.* 1, pp. 98, 99, (1923)[4]. Wolfrum, *Col. Cat.* 102, p. 72, (1929).

[1]Khasis Hills: Assam Calcutta: Bengal. [2]Rotung: Abor. Siam. Tonkin. Khasis Hills: Assam. [3]Shan States. [4]Tonkin. Loas [Distribution: Lopchu 5000 ft. Darjeeling: Bengal].

beesoni Jordan

Xylinades beesoni Jordan, *Nov. Zool.* 32, pp. 240-241, (1925)[1]. Wolfrum *Col. Cat.* 102, p. 72, (1929), Beeson, *Forest Insects*, p. 50, (biology), (1941). [Food-plant: *Pterocarpus marsupium*].

[1]Nedungayam, Nilambur: Madras.

foveatus Jordan

Xylinades foveatus Jordan, *Stett. Ent. Zeit.* 56, pp. 257-258, (1895)[1]. Bovie, *Ann. Soc. Ent. Belg.* 49, p. 287, (1905). Jordan, *Rec. Ind. Mus.* 9, p. 210, (1913)[2]; *Tijaschr. Ent.* 61, p. 47, (1915); *Opusc. Inst. Scient. Indoschine, Fn. Ent.* 1, pp. 98-99, (1923)[3]; *Nov. Zool.* 32, pp. 240-241, (1925). Wolfrum, *Col. Cat.* 102, p. 72, (1929).

[1]Khasis Hills: Assam.

[2]Sibsagar: Assam. [3]Laos.

[Distribution: Jiri Forest, Cachar: Assam].

indignus Walker

Xylinades indignus Walker, *Ann. Mag. Nat. Hist.* (3) 3, p. 261, (1859)[1]. Bovie, *Ann. Soc. Ent. Belg.* 49, 288, (1905). Wolfrum, *Col. Cat.* 102. p. 72, (1929). [Food=palnt:"Dambu", unknown wood].

[1]Ceylon.

[Distribution: Anshi, Sunksal, Kanara: Bombay, Nilambur: Madras. Mahaoya, Batticaloa: Ceylon].

parumsignatus Jordan

Xylinades parumsignatus Jordan, *Nov. Zool.* 38, pp. 306 307, (1922)[1].

[1]Bhutan.

plagiatus Jordan

Xylinades plagiatus Jordan, *Stett. Ent. Zeit.* 56, p. 257, (1895)[1]. Bovie *Ann. Soc. Ent. Belg.* 49, p. 288, (1905) Jordan, *Rec. Ind. Mus.* 8. p. 198, (1913)[2]; 9, p. 210 (1913)[3]. Stebbing *indian Forest insects*, pp. 386-387, (biology), 1914)[4]. Jordan, *Nov. Zool.* 23, p. 362, (1916)[5]. Beeson, *Indian Forester*, 45 (2), p. 51 (1919)[6]. Jordan, *Opusc. Inst. Indoschine, Fn. Ent.* 1, pp. 97, 99, (1923)[7]. Wolfrum, *Col. Cat.* 102, p. 73, (1929). Anon., *Indian For. Rec.* (N.S.) 4(1), p. 32, (biology), (1938). Beeson, *forest Insects*, p. 50, (biology), (1941).

[Food-plants: *Dalbergia sissoo, Lannea grandis, Shorea robusta, Xylia dolabriformis*].

[1]Khasis Hills: Assam. [2]Dejoo, N. Lakhimpur: Assam. Rotung: Abor. Burma. [3]Dikrang Valley, Sibsagar: Assam. [4]Tharrawaddy: Burma. [5]Annam. Vientiane. [6]Buxa: Bengal. [7]Tonkin. Loas. Annam. Cochinchina. Siam.

[Distribution: Haflong Cachar; Sylhet; Assam. Kashmir. Dehra Dun: U.P. Yauaungmyin, Pyinmana Burma].

striatifrons Jordan

Xylinades striatifrons Jordan, *stett. Ent. Zeit.* 56, pp. 260-261, (1895)[1]. Bovie, *Ann. Soc. Ent. Belg.* 49, p. 289, (1905). Jordan, *Nov. Zool.* 19, p. 143, (1912); *Opusc. Inst. Scient. Indochine,* Fn. Ent. 1, pp. 98-99, (1923)[2]. Wolfrum, *Cal. Cat.* 102, p. 73, (1929).

[1]Khasis Hills: Assam.
[2]Tonkin.

sulcifrons Jordan

Xylindades Sulcifrons Jordan, *Stett. Ent. Zeit.* 56, pp. 257-258, 263, 264, (1895)[1]. Bovie, *Ann. Soc. Ent. Belg.* 49, p. 289, (1905). Jordan, *Rec. Ind. Mus.* 9, p. 211, (1913)[2]; *Nov. Zool.* 32, p. 245, (1925). Wolfrum, *Col. Cat.* 102, p. 73, (1929).

[1]Khasis Hills: Assam.
[2]Maldah: Bengal. Khasis Hills: Assam.

tamilanus Jordan

Xylinades tamilanus Jordan, *Nov. Zool.* 38, pp. 305-306, (1932)[1].

[1]Ceylon.

PART XXIX – ASILIDAE: DIPTERA

Family **Asilidae**

Asilidae Loew, *Linn. Ent.,* 2:384, (1847). Schiner, *Verh. zool. bot. Ges. Wein,* 16: 649, (1866); 17: 355. (1867); *Dipt. Novara Reise,* (1868). Wulp, *Catalogue Diptera South Asia,* Hague: 75-100, (1896). Williston,-*Manual of North American Diptera,* New Haven: 192-204. (1908). Kertesz, *Catalogus Dipterorum,* Budapest, 4: 49-348, (199). Hermann, *Zool. jab.,* 43: 161-94, (1920). Melin, *Zool. Bidr. Uppsala,* 8: I-317, (1923). Currans, *The families and genera of N. American Diptera,* NewYork: 167-84, (1934). Bromley, *Ann. ent. Soc. Amer.,* 27: 74-113, (1934); *Lingnam Sci. f.,* 21: *(1-4):* 87-105, (1945). Efflatoun, *Mem. Soc. R. Ent. Egypte,* 4 (2): 31, (1934). Hardy, *Ann. Mag. nat. Hist.,* (10), 13: 498-525, (1934); 14: I-35, (1934); 16: 161-87, 405-26, (1935); *Ent. mon. mag.,* 84: 116-19, (1948). Clements. *proc. R. ent. Soc.* London, (B) 20: 10-14, (1915). Martin, C.H., *Bull. Amer. Mus. nat. Hist.,* 112 (1): 343-86, (1957).

Asilidae was divided by Loew (1847) into three sub-families, Dasypogoninae (including *Leptrogaster* Meigen), Laphriinae (including *laphystia* Lowe) and Asilinae. Wulp (1896) and Kertesz (1909). followed this plan in their catalogues Williston 1908 pointed out the unsatifacotry distinction between Dasypogoninae and Laphriinae on the one hand and between Asilinae and Leptogastrinae on the other; on the basis of palpi he suggested the merger of the former two sub-families under Dasypogoninae and that of the latter two sub-families under Asilinae. This grouping was endorsed by Hardy (1934, 1935, 1948) as a result studies of the terminalia of both sexes. Hardy (1948) further divided the two sub-families into various tribes on the basis of the structure of prosternum, which was shown by Clements (1951) to be highly variable even in the same tribe and, therefore, unreliable. This is confirmed by the present author as a result of a critical sudy of the literature and the examination of specimens of this family lying in the National Pusa Collection of insects. It is however evident from hardy's investigations that some phylognetic relationship exists between Asilianae and Leprogastrinae and between Dasypogoninae and Laphriinae. This indicates that the former two sub-families probably had common ancestors and the latter two sub-families descended from other common ancestors.

In 1868 Schiner enacted sub-family Leptogastrinae for including *leptogaster* and related genera, which was approved by Hermann (1920),

Efflatoun (1934), Bromely (1934, 1935) and more recently by Martin (1957), Schiner thus considered Asilidae to be consisting of four sub-families *t.e.,* Asilsing Leptogastrinae. This arrangement was Challanged by Hermann (1920) but at the same time he accepted the four sub-families. Melin (1923), studied the female genitialia relative to habits, and advanced certain views which were not acceptable to workers, Bromley (1934, 1935), Efflatoun (1934) and others accepted these four subfamilies. In addition to giving keys of the sub-families the present author has drawn the keys to the Indian genera and incorporated in the manuscript.

Key to sub-families

1.	Palpi one-jointed; antennae without terminal style: the arista is terminal and varies from long to short	2
-	Palpi two-jointed; antennae may or not bear a terminal style: the arista when present is short to very minute, but rarely long	3
2.	With Ist and 2nd longitudinal veins (R_1 and R_2 +$_3$) almost always reaching the wings margin separately (cell R_1 almost always open) or if Cell R_1 just closed vein R_2 + $_3$ is sharply curved up terminally so as to form a blunt end to the cell. Bristles and pubescence reduced to a minimum	**Leptogastrinae**
-	With these verins meeting before the wing margin (cell R_1) always closed and with a short petiole, viens R_1 and R_2 + $_3$ meeting at an almost equal curve. Usually etongate bristly but not hairy flies	**Asilinae**
3.	With cell R_1 almost always open, or if just closed then vein R_2 + $_3$ is sharply curved up terminaly so as to form a blunt end to the cell. Antennae may or may not have terminal style; the arista when present is short, very rarely long. Bristles or pubescence always obvious.	**Dasypogoninae**
-	with cell R_1 always closed and with a short petiole, veins R_1 and R_2 + $_3$ meeting a terminal style; the arista when present is microscopic and is contained in a pit. Usually stout coarsely or wooly-haired files	**Laphriinae**

Subfamily I.-**LEPTOGASTRINAE**

leptogastrinae Schiner, *Dipt. Novara Reise,* (1868). Willston, *manual of N. American Diptera,* New Haven: 197, (1908). Hermann, *Zool. jb.,* 43: 164-65, (1920).

Bromley, *Ann. Ent. Soc. Amer.*, 27: 78, (1934); *Lingnan Sci. F.*, 21 (1-4): 89, (1945). Efflatoun, *Mem. Soc. R. Ent. Egypte* 4 (2): 38, (1934). Martin, Bull. *Amer, Mus. nat. Hist.*, 112 (I): 343-86, (1957).

Leptogastrini Hardy, *Proc. Linn. Soc. N.S.W.*, 52: 389, (1927); *Ann. Mag. Nat Hist.*, (10), 16: 161, (1935); *Ent. mon. Mag.* 84: 118, (1948).

This sub family consists of slender and fragile bare species with elongate abdomen and reduced alula on the wings, the latter giving them the pelicellated wing structure that is reminscent of dragon of sub-order Zygoptera. Venation simple and practically all the veins running to the wings border without unduly curving and absence or extreme reduction of interradial corss veins. This sub family resembles Asillnae in having one-jointed palpi, and resembles Dasypogoninae in having cell R_1 almost always open. The pulvilli are absent on all legs, and elongate slender empodium between the very long claws is present. Female genitalia simple and devoid of ovipositor. It consists of one genus and six species from the Indian region.

Genus 1. **Leptogaster** Meigen

leptogaster Meign, *Illiger's Mag. Ins.*, 2: 269 (1803) (Type for the genus: *Asilus cylindrica* Degreer). Wulp, *Catalogue Diptera South Asia Hague:* 75 (1896). Kertez, *Catalogus Dipterorum*, Budapest, 4: 49, (1909), (*Syn. Gonypes* Latreille). Ricard, *Ann. Mag. nat. Hist.*, (8), 9: 474 (1912). Enderlein, *Wien, ent. zig.*, 23: 153 (1914), (*Syn. Euscelindia* Westwood) Meijere, *Tijdschr. Ent*, 56 suppl: 34, 42 (1914). White., *Proc. roy. Soc. Tasmania;* 149. (1916). Hardy., *Proc. Linn. Soc. N.S.W.*, 52: 389 (1927); *Ann. Mag. Nat. Hist.* (10), 16: 161-62, (1935). Frey, *Notual. ent. Helising.*, 17: 39, (1937).

Gonypes Latreille, *Hist. nat. des. curst et des insect.*, 14: 309 (1804).

Euscelidia Westwood, *Trans. R. Ent, Soc. Lond.*, 5: 232 (1849).

Psilonyx Aldrich, *Proc. U. S. nat. Mus.*, 62: 5, (1923).

1. albimann Walker

Leptogaster albimanus Walker, *Proc. Linn. Soc. Lond.*, 3: 89 (1859). Röder, *Ins. Aru, Ent. Nachricht.* 19: 234, (1893). Wulp, *Catalogue Deptera South Asia*, Hague: 76 (1896). — [1]Aru Is., India, Burma and Ceylon

leptogaster albimana Kertesz, *Catalogus Dipterorum*, Budapest, 4: 49, (1909).[1]

2. cheriani Bromley

Leptogaster chaeriani Bromley, *Indian J. agric. Sci.*, 8: 863 (1938). — Coimbatore (S. India)

3. marion, Walker.

Leptogaster marion Walker, *List Ditera* British *Museum*, London, 2: 484 (1849); 7, Suppl. 3: 773 (1855). Wulp, *Catalogues Diptera, South Asia*, Hague: 75 (1896). Kertesz, *Catalogus Dipterorum*, Budapest, 4: 54 (1909)[1]. — [1]Bengal

4. ophinoea Frey

Leptogaster ophionea Frey, *Notul, ent. Helisingf.*, 17: 44 (1937). — India

5. refuescens Frey

Leptogaster rufescens Frey, *Notul. ent. Helisinfg.*, 17: 47, (1937). — India

6. simpelx Bigot.

leptogaster simplex Bigot, *Ann. Soc. ent. Fr.*, (5)., **8:** 444, (1878). Wulp. *Catalogue Diptera South Asia, Hague:* 76 (1896). Kertesz, *Catalogus Dipterorum*, Budapest, **4:** 56 (1909).[1] — [1]India, Burma and Ceylon

Subfamily 2. ASILINAE

Asilinae Loew, *Linn. Ent.*, 2: 384 (1847). Schiner, Verh. Zool.-*bat. Ges. Wein*, 16: 649 (1866); 17: 355 (1867); *Dipt. Novara Reise*, (1868). Wulp, *N. American Diptera, New Haven:* 203 (1908). Kertesz, *Catalogus Dipterorum, Budapest*, **4:** 208 (1909). Hermann, *Zool. jb.*, **43:** 164-165 (1920). Hardy, *Proc. Linn. Soc.*

N.S.W., 52: 389-90 (1927); *Ann. Mag. nat.* (10), **16**: 161-83, 405-26 (1935); *Ent. mon. Mag.*, **84:** 116-19, (1948). Bromely *Ann. ent. Soc. Amer.*, 27: 78 (1934); *Lingnan Sci. J.*, **21** (1-4): (1945). Efflatoun, *Mem. Soc. R. ent. Egypte*, **4** (2): 31 (1934). Clements, *Proc. R. ent. Soc. London*, (B), **20:** 10-14, (1951).

This subfamily resembles leptogastrinae in having one-jointed plapi. It consists of usually elongate bristly flies. The genera *Emphysomera, Merodontiana Michotamia* and *Ommatus* are characteristic in having pectinate antennae (*i.e.*, the arista is long and provided with filaments). These genera resemble Leptogastrinae in having simple female devoid of ovipositor. Their affinities with rest of the genera of Asilianae are due to the joining of vein $R_2 + {}_3$ with R_1 before reaching the wing border and M_3 similarly runs to M_4. The rest of the genera of Asilinae are more advanced than the above mentioned four genera in having well formed ovipositor with the female genitalia, and that is why in subsequent pages the account of the former is given after that of the latter. Asilinae is represented in the Indian region by 17 genera and 143 species. The following is the key to the Indian genera.

Key to the Genera

1. Arita long and provided with filaments	2
- Arista devoid of filaments	4
2. Third joint of antennae always considerably longer than the first two joints together; arista with pectinate hairs clustered at tip	*Michotamia* Macquart
- Third joint of antennae as long as both the first and second together; arista long and pectinate: the hind femur some-what thickened and provided with a round tooth beneath near the base	*Merodontima* Enderlein
- Third joint of antennae short, usually shorter than the firot two joints together; arista pectinate below	3
3. Face with the tubercle	*Ommatius* Wiedermann
- Face quite flat, without tubercles	*Emphysomera* Schiner

4.	The discal cell (Ist M_2) bulged into posterior cell (5th R); marginal vein thickened, or it not thickened it is moderately widened	5
-	Not with this combination of character	6
5.	Face moderatley produced mouth margin; each metanotal calosity with a truft of hairs; Epiygium very large and foot like	*Clephydroneura* Becker
-	Facial protuberance and metanotal hairs completely absent; Epipygium not large & not foot-like.	**Synolcus* Loew
6.	Slopes or lateral swellings of the metanotum bare Slopes or lateral swelling of the metanotum pilose	7
7.	Third antennal segment very long, narrow, tapering and with a short thick arista; face very narrow	12
-	Third antennal segment short, with long arista: face not usually narrow	*Senoprosopis* Macquart
8.	The posterior branch of the third vein (R^5) curves forward to meet the costa at or before the apex of the wings; two or three submarginal cells.	8
-	The posterior branch of the third vein (R_3) terminates beyond the tip of the wing	9
9.	Ovipositor cylinderical with a terminal circlet of spines; two submarginal cells (3rd R and 4 th R); large species.	10
-	Ovipositor elongate, without a terminal circlet of spines; three submarginal cells	*Proctacanthus* Macquart
10.	Two submarginal cells.	*Erax* Scopoli
	[The Ist posterior cell (5th R) normal (open); pubesence on disc of thorax long, and on the centre of the posterior half tufted and conspicuously mane-like presutural dorso-central hairs present].	*Neolophonotus* Engel
-	Three submarginal cells.	11
11.	Short spines on cerci; head not particularly broadened; the second submarginal cell (4th R) as long or only a little shorter than the first submarginal cell (3rd R); bare species	*Philodicus* Loew.

* *Synolcus* Loew, an African genus is included in the key as it closely resembles *Clephydroneura* Becker.

- No spines on cerci; head broadened; the second submarginal cell (4th R) at most half as long as the first submarginal cell (3rd R) more hairy species.	*Bactra* Meigen
- With long spines on cerci; the third submarginal cell (5th R) usually closed, but may be very much narrowed or wide open; pubescence on disc of thorax minute all over and only post-sutural dorso-central hairs present.	*Apoclea* Macquart
12. Facial gibbosity shining black	*Rhadiurgus* Loew
- Facial gibbosity pollinose, weakly or strongly developed	13
13. Ovipositor, may be laterally compressed, elongate and slender, giving the abdomen a 'rat tailed' appearance	14
- Ovipositor conical, not elongate and its tip without spines	*Asilus* Linnaeus
14. Ovipositor only slightly produced and its tip usually without spines; occipital bristles not proclinat; ovipositor when long, the first joint of the posterior tarsus normal, and the occipital bristles proclinate (bristles rather long and curved at almost right angle at or near the apical third); dorsocentral hairs extending to the suture or a little distance beyond; thoracic bristles may or may not be disposed in man-like fashion; legs prevailing black or light and opaque-coloured	*Cerdistus* Loew
- Ovipositor only slightly produced and without bristles; dorsocentral hairs not extending in front of the sutures; male genitalia compact,never leaving a large open space from dorsal view (a difference from **Philonicus Loew** the latter has strongly curved male genitalia which leave a large opne space on the apical half); legs prevailing shining yellow	*Heligmoneura* Bigot
- Ovipositor very long; first joint of posterior tarsus thickened and spured; head broadened	*Astochia* Becker

Genus 1. **Emphysomera** Schiner

Emphysomera Schiner, *Verh. zool-bot. Ges. Wien,* 16: 665 (1866), (Type for the *genus: Ommatius conopsoides* Wiedermann). Wulp. *Catalogue Diptera*

South Asia, Hague: 97 (1896). Kertesz, *Catalogus Dipterorum.* Budapest, 4: 313 (1909).

1. conopsoides (Wiedermann)

Ommatius conopsoides Wiedermann, Aussereuopaische zweiflugeligc Insecten, Hamm, I: 422 (1828). Macquart, *Suit. a Buffon,* I: 314 (1834); *Dipteres Exotiques,* Paris, I (2): 132 (1838). Guerin, *Iconogr. d. Regne Anim. Ins.,* 7: 536, PI. XCIV, fig. 3, (1835). Walker, *List Diptera British Museum,* London 7, Suppl. **3**: 757 (1855). Rondani, *Ann. Mus. Civ.* Genova, 7: 450 (1875).

Emplhysomera conopsoides Schiner, *verh. zool.-bot. Ges. Wien,* **16:** 682 (1866). Wulp, *Catalogue Diptera South Asia,* Hague: 97 (1896). Kertesz, *Catalogus Dipterorum,* Budapest, 4: 313 (1909)[1]. Meijere, *Tijdschr. Ent.,* **54:** 314 (1911); **58,** *Suppl.:* 24, 77 (1915).

[1]Bengal (India) Sumatra and Borneo

2. femorata Bigot

Emplhysomera femorata Bigot, *Ann. Soc. Ent. Fr.,* (5), 5: 243 (1875). Wulp, *Catalogue Diptera South Asia,* Hague: 98 (1896). Kertesz, *Catalogus Dipterorum,* Budapest, 4: 313 (1909)[1].

[1]Ceylon.

3. nigra Schiner

Emplhysomera nigra Schiner, *Dipt. Novara Reise* 195 (1868). Wulp, *Tijdschr. Ent.,* (2), 7 (15); 255 (1872); *Smatra Exped. Diptera: 26* (1881); *Catalogue Diptera South Asia,* Hague: 98 (1896). Kertesz, *Catalgous Dipterorum* Budapest, **4:** 313 (1909)[1].

[1]Nicobar Is. (India) and Gilolo (Sumatra).

4. spathulata (Doleschall)

Ommatius spathulata Doleschall, *Naturk. Tijdschr. Ned. Ind.,* (4), 3 (17) 89 (1858).

Emphysomera spathulata Schiner, *Verh. zool.-bot. Ges. Wien.* 16; 682 (1866); *Dipt. Novara Reise:* 195, Pl. III, fig. I, (1868). Wulp, *Catalogu Diptera South Asia,* Hague: 97 (1896). Kertesz, *Catalogus Dipterorum,* Budapest, **4:** 313 (1909).[1]

omnatius platymelas Walker, *Proc. Linn. Soc. Lond.,* **6** (8): 21 (1862)[2].

[1,2]Nicobar Is. (India), Gilolo (Sumatra) and Amboina Is.

Genus 2. **Merodontina** Enderlein

Merodontina, Enderlein, *Zool. Ans.*, **46**: 262 (1914), (Type for the genus: *Merodontina sikkiments* Enderlein).

1. **Sikkimensis** Enderlein

Merodontina sikkimensis Enderlein, *Zool. Anz.*, **46:** 262 (1914). Sikkim.

Genus 3. **Michotamia** Macquart

Michotamia macquart, *Dipteres* Exotiques, Paris, I (2): 72 (1838), (Type for the genus: *Michotamia analis* Macquart). Wulp, *Catalogue Diptera South Asia*, Hague: 89 (1896). Ketresz, *Catalogus Dipterorum*, Budapest, **4:** 207 (1909). Bromley, *Lingnan Sci. J.*, **21** (1-4): 97 (1945), (Syn. Allocotosia Schiner).

Allocotosia Schiner, *Verh. Zool-bot. Ges. Wien.* **16:** 845 (1866). Wulp, *Tijdschr. Ent.* (2), 7 (15): 249 (1872); *Catalogue Diptera South Asia* Hague: 97 (1896). Kertesz. *Catalogus Dipterorum*, Budapest, **4:** 312 (1909).

1. **analis** Macquart

Michotamia analis Macquart, *Dipteres Exotiques*, Paris, **I** (2): 72, PI. VI, fig. 6, (1838). Walker, *List Diptera British Museum*, London, 7, Suppl. 3: 565 (1855). Wulp, *Catalogou Diptera South Asia*, Hague: 89 (1896); *Tijdschr. Ent.*, **41**: 131 (1898). kertesz, *Catalogus Dipterorum*, Budapest, **4:** 207 (1909).[1]

[1]Java and Bengal.

2. **annulata** *Bigot*

Michotamia annulata Bigot, *Ann. Soc. ent. Fr.*, (5), **8**: 239 (1878). Wulp, *Catalogue Diptera South Asia*, Hague: 89 (1896). Kertesz, *Catalogus Dipterorum*, Budapest, **4:** 205 (1909)[1]

[1]Burma.

3. **Aurata** (Fabricius)

Asilus aurata Fabricius, *Ent. Syst.*, **4**: 387 (1794). *Dasypogon aurata* Fabricius, *systema Anthliat:* 167 (1805).

Ommatius aurata Wiedermann, Dipteres Exotiques, Paris: 213 (1821); *Ausseeuropaische zweifluglige insecten,* Hamm **I:** 420 (1828). Macquart, *Suit,* a Buffon, **I**: 314 (1934); *Dipteres Exotiques,* Paris, **I** (2): 133 (1838). Walker, *List Diptera British Museum,* London, **2:** 479 (1849): 7, Suppl. 3: 759 (1855). Schiner, *Verh. zool.-bot Ges. Wien,* 16: 681 (1866). Wulp. *Tijdschir Ent.,* (2), 7 (15): 249, PI. XII, fig. 7 (1872); *Catalogue Diptera South Asia,* Hague: 97 (1896); *Tijdschr. Ent.,* **41:** 150 (1898). Kertesz, *Catalogus Dipterorum,* Budapest, **4:** 312 (1909)[1]. Bromely, *Rec. Indian Mus.,* **37** (2): 230 (1935); *Indian J. Agric. Sci.,* **8** (6): 868 (1938).[2] — [1]Asia, Celebes [2]Darjeeling (Bengal)

4. compedita (Wiedermann)

Ommatius compedita Wiedermann, *Aussereuro-paische zweiflugelige Insecten,* Hamm **I:** 419 (1828). Walker, *List Diptera British Museum,* London, 7, Suppl. **3:** 758 (1855). Schiner, *Verh, zoool.-bot. Ges. Wien,* **16:** 681 (1866).

Allocotosia compedita Wulp, *Catalogue Diptera South Asia,* Hague: 97 (1896). Kertesz,

Catalogus Dipterorum, Budapest, **4**: 312 (1909).[1] — [1]India or East India.

Genus **4. Ommatius** Wiedermann.

Ommatius Wiedermann, *Dipteres Exotiques, Paris:* 213 (1821); *Aussereuropaische Exotiques, Paris:* 213 (1821); *Aussereuropaische zweiflugelige Insecten,* Hamm **I:** 418 (1828), (Type for the genus *Asilus chinensis* Fabricius). Wulp, *Catalogue Diptera, Catalogus Diperorum,* Budapest, **4:** 306 (1909). Meijere, *Tijdschr. Ent.* **54:** 321 (1911). Ricardo, *Ann. Mag. nat. Hist.,* (8), II: 162 (1913). White, *Proc. roy. Soc. Tasmania:* 167 (1916); 89 (1917). Hardy, *Proc. Linn. Soc. N. S.W.,* **46:** 296 (1921); *Ann. Mag.nat. Hist.,* (10), **16:** 168 (1935). Malloch, *Proc.* Linn. Soc. N.S.W., **53:** 609 (1928).

1. ater Bromley

ommatius ater Bromley, *Rec. Indian Mus.,* 37: 227 (1935). — Simla (Punjab).

2. carmichaeilis Bromley

Ommatius carmichaelis Bromley, *Rec. Indian Mus.,* **37:** 225 (1935). — Darjeeling (Bengal)

3. **chinensis** (Fabricius)

Asilus chinennsis Fabircius, *Ent. Syst.,* **4:** 383 (1794). Wiedermann, *Dipteres Exotiques,* Paris: 185 (1821); *Aussereuropaische Zweiflugelige insecten,* Hamm **I:** 434 (1828).Walker, *List Diptera British Museum,* London 7. Suppl. 3: 726 (1855). Schiner, *Verh. zool.-bot Ges Wien,* **16**: 683 (1866).

Dasypogon chinensis Fabricius, *Systema Anthat:* 169 (1805).

Ommatius chinensis Walker, *List Diptera British Museum,* London, **2**: 470 (1849). Roder, *Ent. Nach.,* **16**: 88 (1890). Wulp *Catalopgue* Diptera South Asia, Hague: 98 1896)[1]. Kertesz, *Catalogue Dipterorum,* Budapest, **4:** 306-7 (1909)[2], (*Syn. Ommatius androcles* Walker, *O. concinnens* Wulp, *O.coryphe* Walker, *O. flavescens* Fabricius, *O. fulvidus* Wiedermann, *O. fulvus* Doleschall, *O. inextricatus* Walker, *O. pennus* Walker and *Asilus garmotii* Guerin). Ricardo, *Ann. Mag. nat. Hist.* (8), 11: 163 (1913). Meijere, *Tijdschr. Ent.,* **58**: Suppl.: 78 (1915). Edwards *J. F.M.S. Mus.,* **8**(3): 35 (1919). Hardy, *Proc. roy. Soc. Qd.,* **40:** 62 (1928). Malloch, *proc. Linn. Soc. N.S.W.,* **54**: 408 (1929). Hardy, *Ann. Mag. nat. Hist.,* (10), **16**: 168 (1935)[3], (probably not conspecific with that of Fabricius). Bromley *Indian j. agric. Sci.,* **8**(6): 868 (1938)[4].

[1]China.
[2]Asia, Japan and India.
[3]Australia.
[4]India.

Dasypogon flavescens Fabricius, *Systema Antliat:* 169 (1805).

Asilus flavescens Wiedermann, *Aussereuropaische zweiflugelige Insecten,* Hamm **I:** 437 (1828). Walker, *List Diptera British Museum* London, 7 Suppl. 3: 725 (1855).

Ommatius flavescens Schiner, *Verh. zool.-bot. Ges.Wien,* **16:** 683 (1866). Wulp, *Catalogue Diptera South Asia,* Hague: 98 (1896)[5].

[5, 6]Java and Sumatra.

Ommatius fulvidus Wiedermann, *Dipteres Exotiques,* Paris: 214 (1821); *Aussereuropaische zweifluglige Insecten,* Hamm **I**: 420 (1828). Macquart, *Suit a Buffon,* **I:** 313, Pl. VII, fig. 9a, b, (1834): *Dipteres Exotiques,* Paris, **I**(2): 132 (1838); Suppl. **3**: 29, Pl. III., fig. 2 (1948). Walker, *List Diptera British*

Museum, London 7, Suppl. **3**: 757 (1855). Doleschall *Natuurk. Tijdschr. Ned. Ind.,* (n), 7 (10): 408 (1856); (3) **4,** (14): 394 (1857).Schiner, *Verb. zool.-bot. Ges. Wien,***16:** 681 (1866); *Dipt. Novara Reise:* 193 (1868). Wulp, *Tijdschr. Ent.,* (2) **7** (15): 259 (1872); *Sumatra Exped.* Diptera: 25, Pl. I. fig. 15-16 (1881); *Catalogue Diptera South Asia,* Hague: 98 (1896)[6]; *Tijdschr. Ent.,* **41**: 152 (1898): **42:** 53 (1899). Osten-Sacken, *Ann. Mus. Civ.* Genova, **16**: 424 (1882); *Ent. Z.,* **26:** 111 (1882).

[6]Malaya-Peninsula.

Asilus garnotii Guerin, *Voyage de la Coquillett Zool. Atlas,* Pl. XX, fig. 8 (1830).

Ommatius androcles Walker, *List Diptera British Museum,* London, **2:** 470 (1849).

Ommatius corphe Walker, *List Diptera British Museum,* London, **2:** 469 (1849).

Ommatius pennsu Walker, *List Diptera British Museum,* London, **2**: 469 (1849); 7, suppl. **3:** 758 (1855); *proc. Linn. Soc.* Lond.,: 14 (1856). Coquillet, *Proc. U.S. nat. Mus.,* 21: 312 (1898).

Ommatius fulvus Doleschall, *Naturrk. Tijdschr. Ned. Ind.,* (3), **4** (14), Pl. VI, fig. 5 (1857).

Ommatius inextricatus Walker, *Proc. Linn. Soc. Lond.,* **6**: 21 (1862).

Ommatius concinnens Wulp, *Tijdschr. Ent.,* (2), 7 (15): 260 (1872). Wulp, *Catalogue Diptera South Asia,* Hague: 99 (1896).

4. **frauenfeldi** Schiner

Ommatius frauenfeldi Schiner, Dipt. *Novara Reise:* 193 (1868). Wulp, *Catalogue Diptera South Asia,* Hague: 99 (1896). Kertesz, *Catalogues Dipterorum,* Budapest, **4**: 308 (1909)[1].

[1]Nicobar Is. (India).

5. **leucopogon** Widermann

Ommatius Leucopogon Wiedermann, *Analecta Ent.,* **25**: 21 (1824); *Aussereuropaische zweiflugelige Insecten* Hamm I: 423 (1828). Walker, *List Diptera British Museum,* London 7, Suppl. 3: 758 (1855), Schiner, *Verh. zool-bot. Ges. Wien,* **15:** 682 (1866). Wulp, *Catalogue Diptera South Asia,* Hague: 98 (1896).

Kertesz. *Catalogue Dipterorum,* Budapest, **4:** 309 (1909)[1].

[1]india or East India.

6. lividipes Bigot.

Ommatius lividipes Bigot, *Proc. Asiat. Soc. Beng*: 138 (1890). Kertesz. *Catalogus Dipterorum,* Budapest, **4**: 309 (1909)[1].

[1]Simila (Punjab)

7. minor Doleschall

Ommatius minor Doleschall, *Natuurk. Tijdschr. Ned. Ind.,* (3) **4** (14); 394 Pl. VI, fig. 4 (1857)). Wulp, *Tijdschr. Ent.,* (2), 7 (15): 266 (1872); *Catalogue Diptera South Asia,* Hague: 99 (1896)[1]; *Teremszetr. Fuz.,* 21: 420 (1898); *Tijdschr. Ent.,* **41:** 153 (1898). Kertez., *Termeszetr. Fuz.,* **20**: 611, (1897); *Catalogus Dipterorum,* Budapest, **4**: 310 (1909)[2], (*Syn. Ommatius noctifer* Walker and *O. Spinnibarbis* Wulp.)

[1]Amboina Is.

[2]India and New Guinea

Ommatius noctifer Walker, *Proc. Linn. Soc. Lond.,***3:** 88, 129 (1859); **5:** 148, 237, 281 (1861); **6:** 7 (1862); **7**: 208 (1864).Schiner, *Verh. zool-bot. Ges. Wien.* **17**: 410 (1867). wulp, *Catalogue Diptera South Asia,* Hague: 99 (1896).

Ommatius spinibarbis Wulp, *Tijdschr. Ent.,* (2), 7 (15); 265, Pl. XII, fig. 16 (1872). osten-Sacken, *Ann. Mus. civ. Genova* **16**: 425 (1882).

8. nanus Walker

Ommatius nanus Walker, *Insecta Saunersiana Diptera,* London, I, Suppl. (1851): *List Diptera British Museum,* London, 7, Suppl. 3: 20 (1855); *Proc. Linn. Soc. lond.,* **5:** 237 (1861)[1];: 208 (1864). Wulp, *Catalogue Diptera South Asia,* Hague: 99 (1896)[2]. Kertes, *Catalogus Dipterorum, Budapest,* **4**: 310 (1909).

[1]New Guinea

[2]Mysol Is. and India of East India.

9. nigrifemorata Bornley

Ommatius nigrifemorata Bromley, *Rec. Indian Mus.,* 37: 226 (1935).

Darjeeling (Bengal)

10. peregrinum (Wulp)

Emphysomera Pergrina Wulp. *Tijdschr. Ent.,* (2), 7 (15): 253, PL. XII, fig.s 9-11 (1872); *Sumatra Exped. Diptera:*

[1]Sumatra, Borneo and Ternate

26 (1881); *Catalogue Diptera South Asia,* Hague: 98 (1896). Kertesz, *Catalogues Dipterorum,* Budapest, **4:** 313 (1909)[1]. Meijere, Tijdschr. Ent., **54:** 314 (1911). Edwards, *J.* F. M. S. *Mus.,* **8** (3): 36 (1919).

Ommatius peregrinum, Brobley, *Rec. Indian* Mus., **37** (2): 230 (1935)[1]. — [1]India

11. **retrahens** Walker

Ommatius retrahens Walker, *Proc. Linn. Soc. Lond.,* **3:** 88 (1859); **6:** 8, 18 (1862); **7:** 208, 233 (1864). Wulp, *Catalogue Dipterorum,* Budapest, **4:** 311 (1909)[2]. — [1]Is. of Aru, Gilolo Ternate, Ceram and Misol; [2]India

12. **reubicundus** Wulp

Ommatius rubicundus Wull, Tijdschr. Ent., (2), 7 (15): 276 (1872): *Sumatra Exped. Diptera:* 25 (1881); *Catalogue Diptera South Asia, Hague* 100 (1896). *Kertesz, Catalogus Dipterorum,* Budapest, **4:** 311 (1909)[1]. Meiere, *Tijdschr. Ent.,* **54:** 316 (1911)[2]; **58,** Suppl.: 23 (1915)[3]. — [1]India; [2]Java; [3]Sumatra

13. **Subgracilis** Bromley

Ommatius subgracilis Bromley, *Rec. Indian Mus.,* 37: 226 (1935). — Darjeeling (Bengal).

14. **tenellus** Wulp

Ommatius tenellus Wulp, *Trans. R. ent. Soc. Lond.,* **97,** Tab. III, fig. 14-15 (1899). Bezzi, *Bull. Soc. ent.,* Ital, **37:** 294 (1906). Kertesz, *Catalogus Dipterorum,* Budapest, **4:** 312 (1909)[1]. Engel, *Lindner's Die fliegen,* Stuttagart, **24:** 38 (1930)[2]. Charian, *Indian. agric. Sci.,* **8**(6): 868 (1938)[3]. — [1]Arabia and Erythraea; [2]Arabia and Egypt; [3]South India

Genus 5. **Apoclea** Macquart

Apoclea Macquart, *Dipteres Exotigues,* paris, **I** (2): 119 (1838), (Type for the genus: Asilus *algira* Linnaeus). Kertesz, *Catalolgus Dipterorum,* Budapest, **4:** 231 (1909). Engel, *Lindner's Die Fliegen,* Stuttgart, **24:** 40 (1930).

1. **indica** Bromley

Apoclea indica Bromley Rec. *Indian Mus.,* **37:** 225 (1935). — Madras.

Genus. 6. **Asilus** Linnaeus

Asilus Linnacus, *Systema Naturae*, ed., 10: 605 (1758), (Type for the genus: *Asilus crabroniformis Linnaeus): Faun. Suec:* 469 (1761). Wulp, *Catalogue diptera South Asia, Hague:* 93 (1896). Kertesz, *Catalogues Dipterorum,* Budapest, **4**: 250 (1909).

Key to Subgenera

1. Tergitres on each side of the abdomen near the incisures with a transverse row of bristles which are distinct from the hairs: facial gibbosity strongly developed and reaching at least slightly above the middle of the face; ground colour of the body dark. — *Machimus* Loew

Tergites without such bristles; bright coloured species, with colour more or less on wings; facial gibbosity pollinose, weakly developed. — *Asilus* Linnaeus sens. strict.

Subgenus 1. **Asilus** Linnacus *sens. strict.*

(Type for the subgenus: *Asilus crabronformis* Linnaeus)

1. atripes (Fabricius)

Dasypogon atripes Fabricius, *Systema Anliat:* 170 (1805). Wiedermann, *Dipteres Exotiques,* Paris: 195, (1821); *Aussereuropaische zweiflugelige Insecten,* Hamm **I**: 455 (1828). Walker, *List Diptera British Museum,* London, 7, Suppl. **3**: 695 (1855).

Asilus atripes Schiner, *Verh. zool.-bot. Ges. Wien,* **16**: 685 (1866). Aldrich, *Catalogue N. American Diptera,* New Haven: 282 (1905)[1]. Kertesz, *Catalogus Dipterorum,* Budapest, **4:** 251 (1909)[2]. Ricardo, *Ann. Mag. nat. Hist.,* (9), **10:** 52 (1922).

[1]West Indies
[2]India

2. aurimystax Bromley

Asilus aurimystax Bromley, *Amer. Mus. Novit.,* **336:** 3 (1928)[1]; *Rec. Indian Mus.,* **37** (2): 230 (1935)[2].

[1]China
[1]Darjeeling (Bengal)

3. bifasciatus Oliver

Asilus bifasciatus Oliver, *Encylop. Method,* **4**: 263 (1789). Kertesz, *Catalogus Dipterorum* Budapest, **4:** 252 (1909)[1]. Ricardo, *Ann. Mag. nat. Hist.,* (9), **10:** 52 (1922)[2].

[1, 2]India or East India

4. **ephippium** Macquart

Asilus ephippium Macqurt, *Dipteres Exotiques,* Paris, Suppl. 5: 62, (1855). Wulp, *Catalogue Diptera South Asia,* Hague: 93 (1896). Kertesz, *Catalogus Dipterorum,* Budapest, **4**: 255 (1909)[1]. Ricardo, *Ann. Mag. Nat.* Hist., (9), **10:** 52 (1922)[2].

[1,2]India or East India

5. **jamensus** Walker

Asilus jamensus Walker, *List Diptera British Museum* London, **2:** 428 (1849); 7, Suppl. 3: 728 (1855) Wulp, *Catalogue Diptera South Asia,* Hague: 93 (1896). Kertesz, *Catalogus Dipterorum,* Budapest, **4**: 258 (1909)[1].

[1]Bengal

6. **laetus** Wiedermann

Asilus laetus Wiedermann, *Analecta Ent.*: 24 (1824): *Aussereuropaische zweiflugelige insecten,* Hamm **I:** 435 (1828). Walker, *List Diptera British Museum,* London, 6, Suppl. **3:** 728 (1855). Schiner, *Verh. zool-bot. Ges. wien,* **16:** 683 (1866). Wulp, *Catalogue Diptera South Asia,* Hague: 93 (1896). Kertesz, *Catalogus Dipterorum,* Budapest, **4**: 258 (1909)[1]. Ricardo *Ann. Mag. Nat. Hist.* (9), **10:** 52 (1922).

[1]India or East India

7. **nigrimysaceus** Macquart

Asilus nigrimystaceus Macquart, Dipteres Exotiques, Paris, Suppl. 4: 91 (1849). Wulp, *Cataloue Diptera South Asia,* Hague: 93 (1896). Kertesz, *Catalogues Dipterorum,* Budapest, **4:** 261 (1909)[1]. Ricardo, *Ann. Mog. Nat. Hist.* (9), **10**: 52 (1922)[2].

[1, 2]Pondichery.

8. **nudipes** Macquart

asilus nudipes Macquart, *Dipteres Exotiques,* paris, Suppl. **2:** 42 (1847), Walker, *List Diptera Britist Museum,* London, 7, Suppl. **3**: 725 (1855). Wulp, *Catalogue Diptera South Asia,* Hague: 93 (1896). Kertesz, *Catalogue Dipterorum,* Budapest, **4:** 261 (1909)[1]. Ricardo, *Ann. Mag. Nat. Hist.,* (9), **10:** 52 (1922).

[1]India or East India

9. **opulentus** Walker

Asilus opulentus Walker, *Insecta Saundersian Diptera,* London, **I**: 150 (1851); *List Diptera British Museum,* London, 7, Suppl. 3: 724 (1855). Wulp, *Catalogue Diptera South Asia,* Hague: 94 (1896) Kertesz *Catalogue Dipterorum,* Budapest, **4**: 261 (1909)[1]. Ricardo, *Ann. Mag. nat. Hist.* (9), **10:** 44 (1922). Bromley, *Rec. Indian Mus.,* **37** (2): 230 (1935)[2].

[1]India or East India

[2]Darjeeling (Bengal)

10. **orientalis** Ricardo.

Asilus orientalis Ricardo, *Ann. Mag. nat./ Hist.,* (9), 10: 46 (1922)[1]. Bromley, *Rec. Indian Mus.,* **37** (2): 230 (1935).

[1]Khasi hills (Assam)

11. **paterculus** Walker

Asilus paterculus Walker, *Insecta Saundersiana Diptera,* London, **I:** 146 (1851); *List Diptera British Museum,* London, 7, Suppl. 3: 7288, (1855). Wulp, *Catalogues Dipterorum,* Budapest, **4:** 262 (1909)[1]. Ricardo, *Ann. Mag. Nat. Hist.* (9), **10:** 52 (1922).

[1]India or East India.

12. **pracefiniens** Walker

Asilus praefiniens Walker, *Insecta Saundersian Diptera,* London, **I:** 146 (1851); *List Diptera British Museum,* London, 7, Suppl. 3: 728 (1855). Wulp, *Catalogue Diptera South Asia. Hague:* 94 (1896). Kertesz, *Catalogus Dipterorum,* budapest, **4**: 262 (1901)[1]. Ricardo, *Ann. Mag. nat. Hist.,* (9) **10:** 52 (1922), (Type not traceable and should be deleted from the list).

[1]India or East India

13. **pusio** Wiedermann

Asilus pusio Wiedermann, *Zoo. Mag.,* **I** (3): 36 (1819); *Dipteres Exotiques,* Paris: 25 (1821); *Ausseru-ropaische zweiflugelige Insecten,* Hamm **I:** 455 (1828). Macquart, Dipteres *Exotiques,* Paris, **I** (2): 143 (1838). Walker, *List Diptera British Museum,* London, 7, suppl. 3: 727 (1855). Schiner, *Verh.zool-bot. Ges Wien,* **16:** 685 (1866). Wulp, *Catalogue Diptera South Asia,* Hague: 93 (1896), Kertsz. *Catalogue Dipterorum,* Budapest, **4**: 262 (1909)[1]. Ricardo, *Ann. Mag. nat. Hist.,* (9) **10:** 52 (1922)[2].

[1, 2]Java and India or East India

14. **pulcher** Ricardo

Asilus pulcher Ricardo, *Ann. Mag. nat. Hist.*, (9) **10:** 51 (1922). Dehra Dun (U.P.).

15. **ricardoi** Bromley

Asilus ricardoi Bromley, *Rec. Indian Mus.*, **37:** 219 (1935). Simla (Punjab)

16. **trifarius** Macquart

Asilus trifarius Macquart, *Dipteres Exotiques*, Paris, **I** (2): 142 (1838). Walker, *List Diptera British Museum*, London, 7, Suppl. 3: 728 (1855). Wulp *Catalogue Diptera south Asia*, Hague: 93 (1896). Kertesz, *Catalogues Dipterorum*, Budapest, **4:** 265 (1909)[1]. Ricardo, *Ann. Mag. nat. Hist.*, (9) **10:** 52 (1922). [1]Pondichery

Subgenus 2. **Machimus** Loew

Machimus Loew, *Linn. Ent.*, **4:** I (1849). Kertesz, *Catalogus Dipterorum*, Budapest, **4:** 299 (1909). Ricardo, *Ann. Mag. nat. Hist.*, (9) **3:** 44 (1919). Engel, *Lindner's Die Fliegen*, Stuttgart, **24:** 152 (1930), (Type for the subgenus: *Asilus atricapillus* Fallen). Bromely, *Lingnan Sci. j.*, **21** (1-4): 104 (1945), (*Syn. Tolmerus* Loew).

Tolmerus Loew, *Lim. Ent.*, **4:** 94 (1849). Wulp *Catalogue Diptera South Asia*, Hague: 96 (1896). Kertesz. *Catalogus Dipterorum*, Budapest, **4**: 297 (1909). Ricardo, *Ann. Mag. nat. Hist.*, (9), **10:** 61 (1922). Engle, *Lindner's Die Fliegen*, Stuttgart, **24:** 152 (1930), *(Syn.* of *Machimus* Loew*)*.

1. **angularis** (Ricardo)

Tolmerus angularis Ricardo, *Ann. Mag. nat. Hist.*, (9) **10:** 64 (1922). Mussoorie (U.P.)

2. **assamensis** Ricardo

Machimus assamensis Ricarod, *Ann. Mag. nat. Hist.*, (9), **3**: 46 (1919)[1]. [1]Assam

3. **coerulescens** Ricardo

Machimus coerulescens Ricardo, *Ann. Mag. nat. Hist.*, (9) **3:** 47 (1919)[1]. [1]Kumaon (U.P.)

4. dubius Ricardo

Machimus dubius Ricardo, Ann. Mag. nat. Hist., (9), **3**: 47 (1919). — [1]Kumaon (U.P.)

5. excelsus Ricardo.

Machimus excelsus Ricardo *Ann, Mag. Nat. Hist.,* (9) **10:** 68 (1922). — Sikkim and Tibet

6. hirsuta Bromley

Machimus Hirsuta Bromley, *Rec: Indian Mus.,* **37**: 220 (1935). — Khasi Hills (Assam).

7. hirtipes Ricardo.

Machimus hirtipes Ricardo, Ann. Mag. Nat. Hist., (9) **3**: 49 (1919). — Khasi Hills (Assam)

8. Incisularis Bromley

Machimus incisularis Bromley, *Rec. Indian Mus.,* **37**: 220 (1935). — Darjeeling (Bengal)

9. indianus Ricardo

Machimus indiansu Ricardo. *Ann. Mag. nat. Hist.,* (9), **3**: 50 (1919)[1]. — [1]Kumaon and Mussoorie (U.P.) and West Himalaya.

10. infrafemoralis Bromley

Machimus infrafemoralis Bromley, *Rec. Indian Mus.,* **37** 221: (1935). — Darjeeling (Bengal).

11. inutilis Bromely

Machimus inutilis Bromely, *Rec. Indian Mus.,* **37:** 222 (1935). — Dalhousie (Punjab).

12. Khasiensis Ricardo

Machimus khasiensis Ricardo, *Ann. Mag. Nat. Hist.,* (9), **3**: 51 (1919). — Khasi HIill (Assam)

13. montanus Ricardo

Machimus montanus Ricardo, *Ann. Mag. nat. Hist.,* (9) **3**: 52 (1919)[1]. — [1]Nilgiri Hills

14. nicobarensis (Schiner) — Nicobar Is. (India)

Tolmerus nicobarensis Schiner, *Dipt. Novara Reise:* 192 (1868). Wulp, *Catalogue Diptera South Asia,* Hague:

96 (1896). Kertesz, *Catalogus Dipetrorum*, Budapest, **4:** 297 (1909).

15. nigrinus Ricardo

Machimus nigrinus Ricardo, *Ann. Mag. Nat. Hist.*, (9), **3:** 53 (1919)[1]. [1]Kumaon (U.P.)

16.. pallipes Ricardo

Machimus Pallipes Ricardo, *Ann. Mag. Nat. Hist.*, (9), **10:** 65 (1922). Kumaon

17. parvus Ricardo

Machimus parvus Ricardo, *Ann. mag. Nat. Hist.*, (9), **3:** 54 (1919)[1]. [1]Nilgiri Hills (India) Ceylon and Siam

Tolmerus jparvus Ricardo, *Ann. Mag. nat. Hist*, (9), **10:** 64 (1942)[1]. [2]Ceylon

18. pubescens Ricardo

Machimus pubescens Ricardo, *Ann. Mag. nat. Hist.*, (9), **10:** (1922). Sikkim and Tibet

19. punjabensis Bromley

Machimus punjabensis Bromely, *Rec. Indian. Mus.*, **37:** 220 (1935). Kangra Valley (Punjab)

20. rufipes Ricardo

Machimus rufipes Ricardo, *Ann. Mag. nat. Hist.*, (9), **10:** 69 (1922)[1]. Bromerly, *Rec. Indian Mus.*, **37:** 230 (1935)[2]. [1]Dehra Dun (U. P.) and Kangra (Punjab)[2] [2]Darjeeling (Bengal)

21. subgenitalis Bromley

Machimus subgenitalis Bromley, *Rec. Indian Mus.*, **37:** 221 (1935). Dalhousie (Punjab)

22. *tibialis* Ricardo

Machimus tibialis Ricardo, *Ann. Mag. nat. Hist.*, (9), **3:** 55 (1919) Dehra Dun and Kumaon (U.P.).

Genus 7. **Astochia** Becker

Astochia Becker, *Ann. Mus. zool, Petersb.*: 538 (1913), (Type for the genus: *Astochia metatarsata* Becker). Hermann, *Arch. Naturgesch.*, (A) 82, **5:** 23 (1917). Engel, *Lindner's Die fliegn,* **24:** 49 (1930).

1. **canis** Bromely

Astochia canis Bromley, *Rec. Indian Mus.*, **37:** 223 (1935). Upper Burma

2. **determinatus** (Walker)

Asilus determinatus Walker, *Proc. Linn. Soc. Lond.*, **4:** 107 (1860): **5:** 265 (1861). Kertesz, *Catalogus Dipterorum,* Budapest, **4:** 225 (1909)[1].

Itamus determinatus Wulp, *Catalogue Diptera South Asia,* Hague: 97 (1896).

Astochia determinates Bromley, *Rec. Indian Mus.*, **37:** 230 (1935)[2]. [1, 2]Darjeeling (Bengal)

3. **Maculipennsi** (Walker)

Trupanea maculipennis Walker, *List Diptera British Museum,* London, 7, Suppl. **3**: 605 (1855).

Promachus maculipennis Wulp, *Catalogue Diptera South Asia,* Hague: 91 (1896). Kertesz, *Catalogus Dipterorum,* Budapest, **4:** 223 (1909)[1]. [1]Hongkong (China)

Astrochia maculipennis Bromely, *Rec. Indian Mus.*, **37:** 230 (1935)[2]. [2]Khasi Hills (Assam)

4. **trichura** Hermann,

Astochia trichura Hermann, *Arch. Naturgesch.*, A 82, **5:** 31 (1917). Ceylon

Genus 8. **Bactria** Meigen

Bactria meigen, syst. Beschr. zweifl. Ins. Aachen, **2:** 307-08, Pl. 21, figs 6, 8 (1820), (Type for the genus: *Aslius maculatus* Fabricius). Hobby. *Ent. Mag.*, **72:** 182, 84 (1936), [*Syns. Trupanea* Macquart (*praeocc*), *Promachus* Loew, *Telejoneura* Rondani, *Philomachus* Karsch. *Trypanoides* Becker and *Enagaedium* Engel].

Trupanea Macquart, *Dipteres Exotiques,* Paris, **I** (2): 91 (1838) (*Praeocc.*).

Promachus Loew, *Linn. Ent.*, 3: 390 (1848). Wulp, *Catalogue Diptera South Asia,* Hague: 90 (1896). Kertesz, *Catalogues Dipterorum,* Budapest, **4:** 216 (1909). Ricardo, *Ann. Mag. Nat. Hist.,* (8), II: 412 (1913). White, *Proc. Roy, Soc. Tasmania:* 171 (1916). Hardy, *Proc. Linn Soc. N.S.W.,* **45:** 250 (1920): **46:** 239 (1921); **52:** 391 (1927); *Ann. Mag. nat. Hist.,* (10), **16:** 175 (1935). Malloch, *Proc. Linn. Soc. N.S.W.,* **53**: 610 (1928).

Telejoneura Rondani, *Archivio per la Zool.,* **3**: 48 (1863).

Philomachus Karsch, *Ent. Z.,* **31:** 375, (1887). Kertesz, *Catalogus Dipterorum,* Budapest, **4:** 228 (1909), (*praeocc*).

Trypanoides Backer, *Ent. Mitt.,* **14:** 71 (1925).

Enagaedium Engel, *Konowia,* **8:** 459 (1930).

1. apivorus (Walker)

Trupanea apivorus Walker, *Trans. R. ent. Soc. Lond.,* (n) **5**: 282 (1860).

Promachus opivorus Osten-Sacken, *Catalogue N. American Diptera,* new Haven, ed. **2**: 234 (1878). Wulp, *Catalogue Diptera South Asia,* Hague: 91 (1896). kertesz, *Catalogus Dipterorum* Budapest, **4**: 217 (1909). Ricardo *Ann. Mag. nat. Hist.,* (9), **5:** 217 (1920)[1].

[1]Burma

2. beesoni (Ricardo).

promachus beesoni Ricardo, *Ann. Mag. nat. Hist.,* (9) **8**: 175 (1921)[1].

[1]Burma

3. binghamensis (Ricardo)

promachus binghamensis Ricardo, *Ann. Mag. nat. Hist.,* (9) **5**: 215 (1920)[1]. Bromley, Rec. *Indian Mus.,* **37:** *230* (1935)[2].

[1]Sikkim
[2]Darjeeling (Bengal)

4. calanus (Walker)

Trupanca calanus Walker, *Insecta Saundersiana Diptera,* London, **I:** 122 (1851); *List Diptera British Museum,* London, 7. Suppl. **3:** 607 (1855).

Promachus calanus Wulp, Catalogue Diptera South *Asia,* Hague: 91 (1896). Kertesz, *Catalogus Dipterorum,* Budapest, **4:** 218 (1909)[1]. Ricard, *Ann. Mag, nat. Hist.,* (9). **5:** 216 (1920).

[1]India or East India

5. **ceylonicus** (Macquart)

Trupanea ceylonicus Macquart, *Dipteres Exotiques*, Paris **I**:(20; 100, 207 (1938); *List Diptera British Miseum*, London, 7: Suppl. 3: 606 (1855).

Promachus ceylonicus Wulp, *Catalogue Diptera South Asia*, Hague: 91 (1896). Kertesz, *Catalogus Dipterorum*, Budapest, **4:** 91 (1909)[1]. Ricardo , *Ann. Mag. Nat. Hist.*, (9), **5**: 213 (1920). — [1]Ceylon

6. **contractus** Walker

Trupanea contractus Walker, *Insecta Saundersiana Diptera*, London, **I:** 120 (1851): *List Diptera British Museum*, London, 7, Suppl. **3:** 606 (1855).

Promachus contractus Wulp. *Catalogue Diptera South Asia*, Hague: 91 (1896). Kertesz, *Catalogus Dipterorum*, Budapest, **4:** 219 (1909)[1]. Ricardo, *Ann. Mag, nat. Hist.*, (9), **5:** 217 (1920). — [1]India or East India.

7. **duvaucelii** (Macquart)

Trupanea duvaucelli Macquart, *Dipteres Exotiques, paris*, **I:** (2): 97 (1838). Walker, *List Diptera British Museum London*, 7, Suppl. **3:** 602 (1855).

Promachus duvaucelii Wulp, *Catalogue Diptera South Asia*, Hague: 91 (1896). Kertesz, *Catalogus Dipterorum Budapest*, **4:** 219 (1909)[1]. Ricardo, *Ann. Mag. nat. Hist.*, (9), **5:** 214. (1920). Bromley, *Indian, j. agric., Sci.* **8** (6) 868 (1938). — [1]Bengal

8. **ghumtiensis** (Bromley)

Promachus ghumtiensis Bromley, *Rec. Indian Mus.*, **37:** 224 (1935). — Darjeeling (Bengal)

9. **herteropterus** (Macquart)

Trupanea Heteropterus Macquart, *Dipters Exotigques*, Paris, **I (2)**: 96 Pl. IX, fig. 3 (1838). Walker, *List Diptera British Museu*, London, Suppl. **3:** 603 (1855).

Promachus heteropterus Wulp, *Catalogue Diptera South Asia*, Hague: 90 (1896). Kertesz, *Catalogus Dipterorum*, Budapest, **4:** 221* (1909)[1]. Ricardo, *Ann. Mag. nat. Hist.*, (9), **5:** 216, (1920)[2]. Bromley, *Indian, J. agric. Sci.*, **8**(6): 868, (1938). — [1]Malabar coast (S. India). [2]Bellary district (S. India).

10. **leoninus** (Loew)

Promachus leoninus Loew, *Linn. Ent.*, **3:** 404, (1848); Ent., Z., **12:** 378 (1868). Wulp, *Tijdschr. Ent.*, **41:** 131, Pl. IV, fig. 6-8, (1898): **42:** 45 (1899). Kertesz, *Catalogus Dipterorum*, Budapest, **4:** 222 (1909)[1]. Ricardo., *Ann. Mag. nat. Hist.*, (9), **5:** 219 (1920)[2]. Bromley, *Rec. Indian Mus.*, **37** (2): 230 (1935)[3].

[1]Asia minor and Himalaya (India).
[2]Upper Burma
[3]Darjeeling (Bengal).

Trupanea leoninus Walker, *List Diptera British Museum* London, 7, Suppl. **3**: 585 (1855).

11. **leucotrichodes** (Bigot)

Promachus leucotrichodes Bigot, *Bull. Soc. zool. Fr.*, **17**: 210 (1892). Wulp, *Catalogue Diptera South Asia*, Hague: 92 (1896). Kertesz. *Catalogus Dipterorum*, Budapest, **4:** 222 (1909)[1]. Ricardo, *Ann. Mag. nat. His.*, (9), **8:** 183 (1921).

[1]India or East India

12. **maculatus** (Fabricius)

Asilus maculatus Fabricius, *Syst. Ent.*: 794 (1775): **4**: 383 (1794); *Species Insecta*, **2:** 264 (1781); *Mantissa insect.*, **2:** 359 (1787). Oliveir, *Encyclop. Method.*, **4:** 267 (1789) Gmel., *Systema natutae* **5:** 2898 (1792). Meigen, *System Beschreib.*, **2:** 307 (1820). Wiedermann, *dipteres Exotiqus*, Paris: 202 (1821); *Aussereuropasiche zweiflugelige Insecten*, Hamm **I:** 483 (1828). Macquart, *Suit. a Buffon*, **I:** 311 (1843).

Dasypogon maculatus Fabricius, *Systoma Antiliat:* 169 (1805).

Trupanea maculatus Macquart, *Dipteres Exotiques* Paris, **I:** (2): 99 (1838). Walker, *List Diptera British Museum*, London, 7, Suppl. **3**: 385 (1855).

Promachus machulatus Loew, *Linn. Ent.* **3**: 406 (1848). Schiner, *Verh. zool.-bot. Ges. Wein*, **16:** 688 (1866). Wulp, *Tijdschr. Ent.*, (2), 7(15): 226 (1872); *Catalogue Diptera South Asia*, Hague: 90 (1896). Röder, *Ent. Nachrichit.*, **19:** 235 (1893). Kertesz, *Catalogue Dipterorum*, Budapest, **4:** 222 (1909)[1]. Ricardo, *Ann. Mag. nat. Hist.*, (9), **5**: 218 (1920), (*Syns. Promachus flavibarbis* Macquart and *P. Copillus* Walker). Bromley, *Indian J. agric Sci.*, **8** (6): 868 (1938).

[1]Europe, Orient, Asia, Java and Ceylon.

Trupanea flavibarbis Macquart, *Dipteres Exotiques,* paris, **I**(2): 96 (1838). Walker, *Iist Diptera British Museum,* London, 7, Suppl. 3: 607 (1855).

Promachus flavibarbis Wulp, *Catalogue Diptera South* Asia, Hague: 90 (1896). Kertesz, *Catalogus Dipterorum,* Budapest, **4:** 220 (1909)[2]. [2]Pondicherry (India)

Asilus copilus Walker, *List Diptera British Museum,* London, **2:** 389 (1849).

Trupanea copillus Walker, *List Diptera British Museum.* London, 7, Suppl. 3: 607 (1855).

promachus copillus Wulp, *Catalogue Diptera South Asia,* Hague: 91 (1896). Kertesz, *Catalogus Dipterorum,* Budapest, **4:** 219 (1903)[3]. [3]India or East India

13. marcii (Macquart)

Trupanea marcii Macquart, *Dipteres Exotiques,* Paris, **I**(2): 97 Pl. IX, fig. 2 (1838). Walker, *List Diptera British Museum,* London, 7, Suppl. **3:** 602 (1855).

Promachus Marcii Wulp, *Catalogue Diptera South Asia,* Hague: 90 (1896). Kertesz, *Catalogus Dipterorum,* Budapest, **4:** 223 (1909)[1]. Ricardo *Ann. Mag. nat. Hist.,* (9), **10**: 214 (1920)[2]. [1]India or East India [2]Bhandani (India)

14. nicobarensis (Schiner)

promachus nicobarensis Schiner, *Dipt. Novara Reise:* 177 (1868). Wulp, *Catalogue Diptera South Asia,* Hague: 92 (1896), Kertesz, *Catalogus Dipterorum,* Budapest, **4:** 223 (1909)[1]. Ricardo, *Ann. Mag. nat. Hist.,* (9), **5:** 213 (1920). [1]Nicobar Is. (India)

15. orientalis (Macquart)

Trupanea orientalis Macquart, *Dipteres Exotiques,* Paris, **I**(2): 96 (1838). Walker, *List Diptera British Museum,* London, 7 Suppl. **3:** 603 (1855). Wulp, *Catalogu Diptera South Asia,* Hague: 90 (1896). Kertesz, *Catalogus Dipterorum, Budapest,* **4:** 223 (1909)[1]. [1]India or East India

16. pallidus (Ricardo)

Promachus pallidus Ricardo, *Ann. Mag. nat. Hist.,* (9), **8:** 177 (1921)[1]. [1]Burma

17. Pseudomaculatus (Ricardo)

Promachus pseudomaculatus Ricardo, *Ann. Mag. nat. Hist.,* (9), 5: 218 (1920)[1]. — [1]Ceylon

18. ramakrishnai (Bromley)

Promachus ramakrishnai Bromley, *Indian j. agric. Sci.,* **8:** 867 (1938). — N. Malabar (S. India).

19. rufibarbis (Macquart)

Trupanea rufibarbis Macqurt, Dipteres Exotiques, Paris, Suppl. **3**: 25, Pl. II, fig. 9 (1848). Walker, *List Diptera British Museum,* London, 7, suppl. **3**: 603 (1855).

Promachu rufibarbis Wulp, *Catalopuge Diptera South Asia,* Hague: 91 (1896)[1]. Kertesz. *Catalogus Dipterorum,* Budapest, **4:** 225 (1909)[2]. — [1]East India [2]Java.

20. rufipes (Macquart)

Trupanea rufipes Macquart, *Dipteres Exotiques,* Paris, Suppl. 4: 80, Pl. VIII, fig. I (1849).

Promachus rufipes Wulp, *Catalogue Diptera South Asia,* Hague: 91 (1896). Kertesz, *Catalogues Dipterorum,* Budapest, **4:** 225 (1909)[1]. Ricardo, *Ann Mag. nat. Hist.,* (9), **5**: 213 (1920)[2]. — [1]Asia [2]India, Burma and Cochin-China

21. tristis (Bigot)

Promachus tristis Bigot, *Bull. Soc. zool.* 17: 211 (1892). Wulp, *Catalogue Diptera South Asia,* Hague: 92 (1896). Kertesz, *Catalogus Dipterorum,* Budapest, **4:** 226 (1909)[1]. Ricardo, *Ann. Mag. nat. Hist.* (9) **8:** 185 (1922). — [1]Mahe, Madras (S. India) and East India

22. varipes (Macquart)

Trupanea varipes Macquart, *Dipteres Exot. ques,* Paris, **I**(2): 97 (1938); Suppl. 1: 79 (1848). Walker, *List Diptera British Museum,* London, 7, Suppl. **3:** 605 (1855).

Promachus vasripes Wulp, *Tijdschr, Ent.,* (2), 7 (15): 229 (1872); *Catalogue Diptera South Asia,* Hague: — [1]Bengal (India) and Manilla (Phillipines).

90 (1896). Kertesz, *Catalogus Dipterorum,* Budapest. **4:** 227 (1909)[1]. Ricardo, *Ann. Mag. nat. Hist.,* (9), **5:** 213 (1920).

23. Westermani (Macquart)

Trupanea westermanni Maquart, *Dipteres Exotiques,* Paris, **I**(2): 98 (1838). Walker, *List Diptera British Museum,* London, 7, Suppl. **3:** 601 (1855).

promachus westermanni Wulp. *Catalogue Diptera South Asia. Hague:* 91 (1896). Kertesz. *Catalogus Dipterorum,* Budapest, **4:** 227 (1909)[1]. Ricardo. *Ann. Mag. Hist.,* (9) **5:** 213 (1920).

[1]Bengal (India) and Java

24. yerburiensis (Ricardo)

Promachus yerburiensis Ricardo, *Ann. Mag. nat. Hist.,* (9), **5:** 220 (1920)[1]. Bromley, *Rec. Indian Mus.,* **37** (2): 230 (1935)[2]; *Indian J. agric. Sci.,* **8**(6): 868 (1938).

[1]Ceylon Madras and Coimbatore
[2]Tuticorin (Madras)

Genus 9. **Cerdistus** Loew

Lophonotus Macquart, *Dipteres Exotiques,* **I** (2): 125 (1829), (*Praeocc.*)

Cerdistus Loew, *Linn. Ent.* **4:** 74 (1849), (Type for the genus: *Asilus Erythrurus* Meigen). Hardy, *Ann. Mag. nat. Hist.,* (10), **16:** 405 (1935) (*Syns. Dyswachus* Loew and *Neoitamus* Osten-sacken).

Itamus Loew, *Linn. Ent.,* **4:** 84 (1849), (Praeocc). Wulp, *Catalogue Diptera South Asia,* Hague: 96. (1896). Ricardo, *Ann. Mag. nat. Hist.,* (9), **3:** 58 (1919).

Dysmachus Loew, *Dipt. Fauna Sudafrika,* **I**: 143 (1860). Kertesz, *Catalogus Dipterorum,* Budapest, **4**: 274 (1909), (*Syn. Lophonotus* Macquart).

Neoitamus Osten-Sacken, *Catalogue N. American. Diptera,* new Haven, ed. 2: 82, 235 (1878). Kertesz, *Catalogus Dipterorum,* Budapest, **4**: 290, (1909) (*Syn. Itamus* Loew).

Asilus terebratus Macquart, *Dipteres Exotiques,* Paris, **I** (2): 149 (1838). Walker, *List Diptera British Museum,* London, 7, Suppl. **3:** 752 (1855).

Asilus latro Doleschall, Natuurk. Tijdschr. Ned. Ind., (3) **4,** (14): 394 (1857): (4), **3**(17): 89 (1858) Schiner, *Dipt. Novara Reise: 189* (1868).

Asilus involutus Walker, *Proc. Linn. Soc. Lond.*, 5: 281(1861): **6:** 7 (1862).

Itamus involutus Osten-Sacken, *Ann. Mus. Civ. Genova,* **16:** 423 (1882). Wulp, *Catalogue Diptera South Asia,* Hague: 97 (1896).

Neoitamus involutus Kertesz, *Catalogus Dipterorum,* Budapest, **4:** 292 (1909)[4].

Asilus normalis Walker, *proc. Linn. Soc.* Lond., **6:** 18 (1862).

Itamus normalis Wulp, *Catalogue Diptera South Asia,* Hague: 97 (1896).

Neoitemus mormal's Kertesz, Catalogus Dipterorum, Budapest, **4:** 293 (1909).

Itamus dipygus Schiner, *Dipt. Novara Reise: 188, Pl.* II, fig. 12 (1868).Wulp, *Catalogue Diptera South Asia,* Hague: 97 (1896); *Tijdschr. Ent.,***41:** 144 (1898).

Neoitamus dipygus Kertesz, *Catalogues Dipterorum,* Budapest, **4:** 292 (1909)[5].

Itamus dentipes Wulp, *Tijdschr. Ent.,* (2), 7 (15): 248 (1872); *Catalogue Diptera South Asia,* Hague: 97 (1896)[6].

Asilus vertebratus Wulp, *Tijdschr. Ent.,* **41:** 145 (1898).

[4, 6]Islands of Batjan Gilolo, Ternate and Salawatti

8. philus (Walker)

Asilus philus Walker, *List Diptera British Museum,* London, **2**: 393 (1849); 7, Suppl. **3:** 725 (1855). Wulp, *Catalogue Diptera South Asia,* Hague: 93 (1896).

Itamus philus Wulp, *Tijdschr. Ent.,* **41:** 145, (1898); **42:** 55 (1899).

Neoitamus philus Kertesz, *Catalogus Dipterorum,* Budapest, **4**: 293 (1909)[1]. Ricardo, *Ann. Mag.nat. Hist.,* (9), **3**: 58-59 (1919)[2].

[1]Silhet (Assam)
[2]Khasi Hills (Assam), Sikkim and Tannaserim (Burma)

9. pulcher (Ricardo)

Neoitamus Pulcher Ricardo, *Ann. Mag. nat. Hist.*, (9) **3**: 65 (1919). [1]Ceylon

10. siamensis (Ricardo)

Neoitamus siamensis Ricardo, *Ann. Mag. nat. Hist.*, (9) **3:** 68 (1919). India

11. tarsalis (Ricardo)

Neoitamus tarsalis Ricardo, *Ann. Mag. Nat. Hist.*, (9). **3:** 70 (1919). Ceylon

12. tolmeroides (Bromley)

Dysmachus tolmeroides Bromley, *Amer. Mus. Novit.*, **336**: 3 (1928); *Rec. Indian Mus.*, **37**(2): 230 (1935)[1]. [1]Khasi Hills (Assam)

13. tropicus (Ricardo)

Neoitamus tropicus Ricardo, *Ann. Mag. Nat. Hist.*, (9), **3:** 69 (1919). India

Genus 10. **Clephydroneura** Backer

Clephydroneura Backer, *Ent. Mitt.*,**14**-68 (1925). Engel, *Ann. Transv. Mus.*,**13**: 143 (1929). Frey, *Rev. Sussie Zool.*, **41**: 317 (1934). Oldroyd, *Ann. Mag. nat. Hist.*, (11), **I:** 450-54 (1938). (Type for the genus: *Asilus xanthopus* Wiedermann redescribed and key to its species given).

1. annulatus (Fabricius)

Asilus annulatus Fabricus, *Syst. Ent.*: 794 (1775); **4**: 384 (1794); *Species Insect.*,**2:** 464 (1781); *Mantissa insect.*, **2:** 360 (1787).Olivier, *encyclop. Method.*, **4:** 268 (1789). Gmel *systema Naturae*, **5:** 2899 (1792). Wiedermann, *Dipteres Exotiqeus*, Paris: 205 (1821); *Aussereuropaische zweiflugelige Insecten*, Hamm **I**: 434 (1828). Walker, *List Diptera British Museum*, London, 7, Suppl. **3:** 727 (1855).

Dasypogon annulatus Fabricius, *Syst.* Analiat: 171 (1805).

Mochtheris annulauts Schiner, *Verh. zoo.-bot. Ges. Wien*, **16:** 683 (1866). Wulp Catalogue Diptera

South Asia, Hague: 96 (1896); Tijdschr. *Ent.*, **41:** 142 (1898); **42:** 55 (1899).

Synolcus annulatus Röder, *Ent. Nach.*,**19**: 235 (1893). Kertesz, *Catalogus Dipterorum*, Budapest, **4:** 282-83 (1909)[1]. Ricardo *Ann. Mag. nat. Hist.*, (9), **10**: 58 (1922)[2], (*Syns. Synolcus flavicornis* Macquart and *Asilus barium* Walker). Bombay, *Indian j. agric. Sci.*, **8**(6): 868 (1938). — [1]India or East India; [1, 3, 6]Ceylon.; [2, 5]Dehra Dun (U.P.); [4, 5]Bengal

Clephydroncura annulatus Oldroyd, *Ann. Mag. nat. Hist.*, (11), **I:** 463-64 (1938)[3].

Asius flavicornis Macquart *Dipteres Exotiques*, Paris, **I**(2): 142 (1838). Walker, *List Diptera British Museum*, London, **2:** 459 (1849); 7, Suppl. **3:** 728 (1855).Wulp, *CatalogueDiptera South Asia*, Hague: 93, (1896)[4].

Clephydroneura flavicornis Oldroyd, *Ann. mag nat. Hist.*, (II), **I:** 464-65 (1938)[5]. — [5]Coimbatore

Asilus barium Walker, *List Diptera British Museum*, London, **2:** 426 (1849); 7, Suppl. **3:** 728 (1855); *proc. Linn. Soc. Lond.*, **I:** 14, 116 (1856). Wulp, *Catalogue Diptera South Asia*, Hague: 93 (1896) Kertesz, *catalogus Dipterorum, Budapest*, **4**: 252 (1909)[6]. — [6]Sidapur (Coorg), Borneo & Sumatra.

Clephydroneura barium oldroyd, *Ann. Mag. nat. Hist.* (II), **I:** 462-63 (1938).

2. **apicalis** Oldroyd

Clephydroneura apicalis Oldroyd, *Ann. mag. nat. Hist.* (II), **I:** 462-63 (1938). — Tinnevelly (S. India) and Ceylon

3. **bengalensis** (Macquart).

Asilus bengalensis Macquart. *Dipteres Exotiques*, Paris, **I**(2): 141, Pl. XII, fig. 2 (1838). Walker, *List Diptera British Museum*, London, 7, Suppl. **3:** 725 (1855). Wulp, *Catalogue Diptera SouthAsia*, Hague: 93 (1895).

Synolcus bengalensis Wulp, *Tijdschr. Ent.*, **41:** 142 (1898); **42: 55** (1899). Kertesz, *Catalogus Dipterorum*, Budapest, **4:** 283 (1809). *Kertesz,* catalogus Dipterorum, Budapest, **4:** 283 (1909)[1]. Ricardo. *Ann. Mag. Nat. Hist.*, (9), **10:** 60 (1922)[2].Meijere, *Tijdschr. Ent.*, 56, Suppl. 64 (1914); **58,** Suppl.: 23, 77 (1915)[3]. — [1]Bengal; [2]Coimbatore; [3]Sumatra; [1,4]Java

Clephydroneura bengalensis Oldryod, *Ann. Mag. nat. Hist.*, (II) **I:** 460-61 (1934)[4]. — [4]Sepanjang Is.

4. **brevipennis** Oldroyd

Clephydroneura brevipennis Oldryd, *Ann. Mag. nat. Hist.*, (II), **I:** 469-70 (1938). — Nelliampathi Hills and Travancore (S. India).

5. **distinctus** Oldroyd

Clephydroneura distinctus Oldroyd, *Ann. Mag. nat. Hist.*, (II) **I:** 467 (1938). — Coonoor (S. India)

6. **duvaucelii** (Macuart)

Asilus durvaucelii Macquart, *Dipteres Exotiques,* Paris, **I:** (2): 141, pl. XII, fig. 1 (1838).Walker, *List Diptera British Museum,* London, 7. Suppl. **3:** 727 (1855). Wulp, *Catalogue Diptera South Asia* Hague: 93 (1896). Kertesz, *Catalogus Dipterorum,* Budapest, **4:** 255 (1909)[1]. — [1]Bengal.

Synolcus duvaucelii Ricardo, *Ann. Mag. nat. Hist.*, (9), **10:** 59 (1922).

Clephydroneura duvaucelii Oldroyd, *Ann. Mag. nat. Hist.*, (II), **I:** 464 (1938)[2]. — [2]Madura district (S. India), North-West India, Jabbalpur (M.P.)

7.**exilis** Oldroyd

Clephydroneura exilis Oldroyd, *Ann. Mag. nat. Hist.*, (II), **I:** 468-69 (1938). — [1]Burliyar, Coonoor (S. India).

8. **gymnura** Oldroyd

Clephydroneuragymnura Oldroyd, *Ann. Mag. nat. Hist.*, (II), **I:** 470-71 (1938). — East India. Khangpur (M.P.).

9. **iamenus** (Walker)

Asilus iamenus Walker, *List Diptera British Museum,* London, **2:** 428 (1849); 7, Suppl. **3**: 728 (1855).

Synolcus iamenus Ricardo, *Ann, Mag. nat. Hist.*, (9), **10:** 59 (1922)[1]. Bromely, *Indian J. agric. Sci.*, 8(6): 868 (1938). — [1]Dehra Dun (U.P.) Kotagiri, Mercara & Madras (S. India).

Clephydroneura iamenus Oldroyd, *Ann. Mag. nat. Hist.*, (II) **I:** 452 (1938).

10. minor Oldroyd

Clephydroneura minor Oldroyd, *Ann. Mag. nat. Hist.*, (II), **I:** 461-62 (1938). — Khasi Hill and Burma

11. pulla Oldroyd

Clephydroneura Pulla Oldroyd, *Ann. Mag. nat. Hist.*, (II), **I:** 467 (1938). — Vizagapatnam (S. India).

12. semirufus Oldroyd

Clephydromeura semirufus Oldroyd, *Ann. Mag. nat. Hist.* (II) **I:** 465-67, (1938). — Kodaikanal, Madukarai and Coimbatore (S. India)

13. xanthopus (Widermann)

Asilus xanthophus Wiedermann, *Zool. mag.,* I: (3): 5 (1819).

Asilus xanthopus Wiedermann, dipteres Exotiques, Paris: 186, (1821); *Aussereuropaische zweiflugelige Insecten,* Hamm **I:** 436 (1828). Walker, *List Diptera British Museum,* London, 7, Suppl. **3:** 726 (1828).

Synolcus xanthopus Wulp, *Tijdschr. Ent.,* (2), 7 (15): 240, Tab. XI, fig. 19. (1872); *Catalogues Diptera South Asia,* Hague: 95, (1896)[1]; *Tijdschr. Ent.,* **41:** 142 (1898). Osten-Sacken *Ann. Mus. Civ.* Genova, **16:** 423 (1882). Kertesz, *Catalogus Dipterorum* Budapest, **4:** 283 (1909)[2] (*Syn. Asilus sundaicus*). Meijere, *Tijdchr.Ent.,* **54:** 310 (1911). Edwars, *J. F. M. S. Mus.,* **8:** (3) 34 (1919). Ricardo, *Ann. mag. Nat. Hist.,* (9), **10: 58** (1922)[3]. — [1-3, 4]Java and Sumatra. [1, 2](Celebes) [3]Selangor

Mochtherus xanthopus Schiner, *Verh. zool.-Bot.Ges. Wien,* **16:** 683 (1866).

Clephydroneure xanthopus Oldroyd, *Ann. Mag. nat. Hist.,* (II), **I:** 455-57 (1938) (Redescribed).

Asilus sundaicus Jaennicke, *Abh.* senckenb. *naturf. Ges.,* **6**: 363 (1867).

Clephydroneura sundaicus Oldroyd, *Ann. Mag. nat. Hist.,* (II) **I:** 454-55 (1938).[4] — [4]Burma.

Genus II. **Erax** Scopoli

Erax Scopoli, *Ent. Carinol:* 359, (1763), Type for the genus: *Erax barbatus* Scopoli). macquart, *Dipteres Exotiques,* Paris, **I** (2): 107 (1838), Wulp, *Catalogue Diptera South Asia,* Hague: 94 (1896). Kertesz, *Catalogus Dipterorum,* Budapest. **4:** 232 (1909), (*Syn. Efferia* Coquillett.)

Efferia Coquillett, *Canad. Ent.,* **25:** 175 (1893).

1. **curiatius** Walker

Grax curiatius Walker, *List Diptera British Museum,* London, **2:** 432 (1849); 7 Suppl. **3:** 642 (1885). Wulp, *Catalogue Diptera South Dipterorum,* Budapest, **4:** 235 (1909)[1]. Ricardo. *Ann. Mag.Nat. Hist.,* (9) **10:** 73 (1922), (Specimen not traceable and may have to be deleted from the list). — Nepal

2. **rufiventris** Macquart

Erax rufiventris Macquart, *Dipteres exotiques,* Paris, **I:** (2): 108 (1838). Walker, *List Diptera British Museum,* London, 7. Suppl **3:** 642, (1855). Wulp, *Catalogue Diptera Southa Asia,* Hague: 94 (1896). Kertesz., *Catalogus Dipterorum,* Budapest, **4:** 242 (1909)[1]. — [1]Bengal.

Genus 12. **Heligmoneura** Bigot

Mochtherus Loew, *Linn. Ent.,* **4:** 58 (1849), (praeocc.).

Hilimoneura Bigot, *Thomson's Achiv. Ent.,* **2:** 356 (1858), (Type for the genus: *Asilus pallipes* Meigen). Kertesz, *Catalogus Dipterorun,* Budapest **4:** 285 (1909). Engel, *Ann. Tranv. Mus.,* 135 (1927), (*Syn. Cinadus* Wulp).

Cinadus Wulp, *Tijdschr. Ent.,* **41:** 139, (1898). Kertesz, Catalogus Dipterorum, Budapest, **4:** 303 (1909). Ricardo, *Ann. Mag. nat. Hist.,* (**9**), **10:** 70 (1922).

1. **armatipes** (Macquart)

Asilus armatipes Macquart, Dipteres Exotiques Paris Suppl. 5: 63 Pl. II. fig. 8 (1855)[1]. Ricardo, *Ann. Mag, nat. Hist.,* (9), **10:** 45 (1922), (*Syn. Asilus shalumus* Walker). — [1]North China Border

Heligmoneura armatipes Bromley, *Rec. Indian Mus.,* 37 (2): 230 (1935)[2]. — [2]Darjeeling (Bengal)

Asilus shalumus Walker, *Trans. R. ent. Soc. Lond.* n, **4:** 131 (1857), Wulp, *Catalogue Diptera South Asia,* Hague: 94, (1896). Kertesz, *Catalogus Dipterorum,* Budapest, **4:** 263, (1909)[3].

[3]China

2. congedus (Walker)

Asilus congedus Walker, *Insecta Saundersiana Diptera,* London, **I:** 138 (1851); *List Diptera British Museum,* London, 7, Suppl. **3:** 726 (1855). Wulp, *Catalogue Diptera South Asia,* Hague: 94 (1896). Kertesz, *Catalogus Dipterorum,* Budapest, **4:** 253 (1909)[1].

[1]India or East India

Heligmoneura congedus Ricardo, *Ann. mag. Nat. Hist.,* (9), **10:** 60 (1922).

3. gnava (wulp)

Mochtherus gnava Wulp, *Tijdschr. Ent.,* (2), 7 (15): 243, Pl. XII, fig. 3 (1872); 19; 174 (1876); *Catalogue Diptera South Asia, Hague:* 96 (1896)[1] Osten-Sacken, *Ann, Mus. Civ. Genova,* **16:** 423 (1882).

[1]Islands of Java, Celebes and Gilolo

Heligmoneura gnava Kertesz, *Catalogus Dipterorum,* Budapest, **4:** 287 (1909)[2]. Ricardo *Ann. Mag. nat. Hist.,* (9). **3:** 73 (1919); **10:** 60 (1922)[3].

[2]India
[3]Kumaon and Dehra Dun (U.P.)

4. indianus Ricardo

Heligmoneura indianus Ricardo, *Ann. Mag. nat. Hist.,* (9), **3:** 74 (1919); 10: 61 (1922).

India

5. montana (Ricardo)

Asilus montana Ricardo, *Ann. Mag. nat. Hist.* (9), **10:** 47 (1922)[1].

[1]Khasi Hills (Assam)

Heligmoneura montanus Bromley, *Rec. Indian Mus.,* 37, (2): 230 (1935)[2].

[2]Darjeeling (Bengal)

6. tenuicornis (Walker)

Asilus tenuicornis Walker, *proc. Linn. Soc.* Lond., **4:** 108 (1860). Wulp, *Catalogue Diptera South Asia,* Hague: 94 (1896). Kertesz, *Catalogus Dipterorum,* BUdapest **4:** 264 (1909)[1].

[1, 3]Celebes

Cinadus tenuicornis Ricardo, *Ann. Mag. nat. Hist.,* (9), **10:** 70 (1922)[2], (*Syn. Cinadus spurius* Wulp).

[2]Dehra Dun (U.P.)

Cinadus spurius Wulp, *Tijdshr. Ent.* **41:** 140, Pl. V fig. 1-4 (1898); **42:** 48 (1899). Kertesz, *Catalogus Dipterorum,* Budapest, **4:** 303 (1909)[3].

7. **trisignata** Ricardo

Heligmoneura trisignata Ricardo, *Ann. Mag. nat. Hist.,* (9), **10:** 61 (1922)[1]. Bromley, *Indian. j agric. Sci.,* **8** (6): 868 (1938), (Wrongly spelled as *tricynata* Ricardo). — [1]S. Arcot and Bellary Dist. (S. India).

8. **yenpingensis** (Bromley)

Asilus yenpingensis Bromley, *Amer. Mus. Novit.,* **336:** 1928)[1]. — [1]China.

Heligmoneura yenpingensis Bromley, *Rec. India Mus.,* **37**(2): 230 (1935)[2]. — [2]Darjeeling (Bengal)

Genus 13. **Neolophonotus** Engel.

lophonotus Engel, *Bull. Soc. R. ent., Egypte:* 346 (1925) *Lindner's Die Fliegen,* Stuttgart,**24:** 38-39 (1930), (Type for the genus: *Ommatius chalconotus* Wiedermann). Efflatoun, *Mem Soc. R. ent. Egypte,* **4** (1): 38 (1934).

1. **indicus** Bromley

Neolophonotus indicus Bromley, *Rec. India Mus.,* **37:** 222 (1935). — Central India.

Genus 14. ***Philodicus** Loew

Philodicus Loew, *Linn. Ent.,* 3: 391 (1848), (Type for the genus: *Asilus javanus* Wiedermann) Wulp *Catalogue Diptera South Asia,* Hague 92 (1896). Kertesz, *Catalogus Dipterorum,* Budapest, **4:** 228 (1909). Ricardo, *Ann. Mag. nat. Hist.,* (9), **8:** 117, 184 (1921) *(Syn., Alcimus* Loew).

Alcimus Loew, *Linn. Ent.,* **3:** 391, 4 (1848), Wulp, *Catalogue Diptera south Asia,* Hague: 90 (1896). Kertesz, *Catalogus Dipterorum,* Bduapest, **4:** 229 (1909).

1. **Cylonicus** Schiner

Philodicus ceylonicus Schiner, *Dipt. Novara Reise:* 179 (1868). Wulp, *Catalogue Diptera South Asia,* Hague:

* *ilodicus insularis* Ricardo and *P. vandesseanus* Ricardo are present in the Coll. of Agric. Res. Inst., Coimbatore (S. India), *vide* Bromely, *Indian f. agric. Sci.,* 8 (6): 868 (1938), but not known otherwise whether found in India.

93 (1896), kertesz, *Catalogus Dipterorum,* Budapest, **4:** 228 (1909)[1]. Ricardo *Ann. Mag. nat. Hist.,* (9), **8:** 192 (1921)[2]. — [1,2]Ceylon

2. **femoralis** Ricardo

Philodicus femoralis Ricardo, *Ann. Mag. nat. Hist.,* (9). **8:** 190 (1921)[1]. — [1]Burma

3. **fuscipes** Ricardo

Philodicus Fuscipes Ricardo, *Ann. Mag. nat. Hist.,* (9), **8:** 188 (1921)[1]. — [1]Siam

4. **fuscus** (Macquart)

Trupanea fuscus Macquart *Dipteres Exotiques,* Paris, **I** (2): 104 (1838). Walker, *List Diptera British Museum,* London, 7, Suppl. **3:** 603 (1855).

Philodicus fucus Schiner, *Dipt. Novara Reise:* 178 (1868). Kertesz, *Catalogus Dipterorum,* Budapest **4:** 228 (1909)[1]. Wulp, *Catalogue Diptera South Asia,* Hague: 93 (1896), Ricardo, *Ann. Mag. nat. Hist,* (9), **8:** 187 (1921). — [1]Bengal (India) and Java.

5. **grandissimus** Ricardo

Philodicus grandissimus Ricardo, *Ann. mag. nat. hist.,* (9), **8:** 186 (1921). — Bombay.

6. **hospes** (Wiedermann).

Asilus hospes Wiedermann, *Zool. Mag.,* **I:** (3): 32 (1819); *Dipteres Exotiques,* Paris: 207 (1821); *Aussereuropatische zweiflugelige Insecten,* Hamm **I:** 495 (1828).

Trupanea hopes Walker, *List Diptera British Museum,* London, 7, Suppl. 3: 603 (1855).

Alcimus hospes Schiner, *Verh. zool.-bot. Ges. Wien,* **16:** 689 (1866). Röder, *Ent. Nach.,* **19:** 235 (1893). Wulp, *Catalogue Diptera South Asia,* Hague: 90 (1896). Kertesz, *Catalogus Dipaterorum,* Budapest, **4:** 229 (1909)[1]. — [1]Ceylon, and Tranquebar, Madras (India).

Philodicus hopes Ricardo, *Ann. Mag.nat. Hist.,* (9) **8:** 189 (1921).

7. javanus (Wiedermann).

Asilus javanus wiedermann, Zool. Mag., **I**(3): 4 (1819); *Dipteres Exotiques,* Paris: 205 (1821); *Aussereuropaische zweiflugelige Insecten,* Hamm **I:** 494 (1828).

Trupanea javanus Macquart, *Dipteres Exotiques* Paris, **I** (2): 98 (1838). Walker, *List Diptera British Museum,* London, 7, Suppl. **3**: 602 (1855).

Philodicus javanus Schiner, *Verh. zool-bot Ges. Wien,* **16:** 689 (1866). Wulp, *Tijdschr. Ent.,* (2), 7 (15): 232, Pl. XI, fig. 15 (1872); **23;** 169 (1880); *Sumatra Exped Diptera:* 24 (1881); *Catalogue Diptera South Asia,* Hague: 92 (1896); *Tijdschr. Ent.,* **41:** 134, 135 (1898). Kertesz, *Catalogus dipterorum, Budapest* **4:** 228-29 (1909)[1]. Meijer, *Tijdschr. Ent.,* **58,** Suppl.: 77 (1915). Ricardo, *Ann. Mag. nat. Hist.,* (9), **8:** 168-88 (1921), (*syns. Asilus agnitus Wiedermann. Philodicus perplexus* Wiedermann, *Trupanea rubritarsatus* Macquart, *Promachus gobares* Walker, *p. telifer* Walker, *P. sagittifer* Walker, P. *innotablis* Walker, *P. confins* Walker, *Asilus melanurus* Dolseschall, *Promachus inserens* Walker, and *Philodicus rubritarsatus* Schiner).

[1]Islands of Java and Sumatra.

[1]Timor Islands.

Asilus agnitus Wiedermann, *Zoo. Mag.,***I** (3): 35 (1819); *Dipteres Exotiques,* Paris: 212 (1821); *Aussereuropaische zweiflugelige Insecten,* Hamm, **I:** 496 (1828).

Asilus perplexus Wiedermann, *Aussereuropasiche zweiflugeligs Insecten,* Hamm, **I:** 495 (1828).

Philodicus perplexus Schiner, *Verh. zool-bot. Ges. Wein,* **16**: 689 (1866).

Trupanea rubritarsatus Macquart, *Dipteres Exotiques,* Paris, **I**(2): 99 (1838). Walker, *List Diptera British Museum,* London, Suppl. **3:** 605 (1855).

Asilus gabares Walker, *List Diptera British Museum,* London, **2:** 420 (1849).

Trupanea gobares Walker, *List Diptera British Museum,* London, 7, Suppl. 3: 604 (1855).

Promachus gobares Wulp, *Catalogue Diptera South Asia,* Hague: 91 (1896).Kertesz, *Catalogus Dipterorum,* Budapest, **4:** 221 (1909)[2].

[2]Slihet (Assam).

Trupanea telifer Walker, *Insecta Saunersiana Diptera*, London, **I:** 115 (1851); *List Diptera British Museum*, London, 7, Suppl. **3:** 606 (1855).

Promachus telifer Wulp, *Catalogue Diptera South Asia*, Hague: 91 (1896). Kertesz, *Catalogus Dipterorum*, Budapest, **I.** 226 (1909)[3]. — [3, 4]East India.

Trupanea sagittifer Walker, *insecta Saundersiana, Diptera*, London, I: 116 (18510; *List Diptera British, Museum*, London, 7, Suppl. 3: 606 (1855).

Promachus sagittifer Wulp, *Catalogue Diptera South Asia*, Hague: 91 (1896). Kertesz, *Catalogues Dipterorum*, Budapest, **4:** 225 (1909)[4].

Trupanea innotabilis Walker, *List Diptera British Museum* London, 7, Suppl. **3:** 604 (1855).

Promachus innotabilis Wulp, *Catalogue Diptera South Asia*, Hague: 91 (1896) Kertesz, *Catalogus Dipterorum*, Budapest, **4:** 221 (1909)[5].

Trupanea confinis Walker, *List Diptera British Museum*, London 7, Suppl. **3:** 656 (1855).

Promachus confinis Wulp, *Catalogue Diptera South Asia*, Hague: 91 (1896). Kertesz, *Catalogus Dipterorum*, Budapest, **4:** 219 (1909).

Asilus meianurus Doleschall, *Naturrk. Tijdschr. Ned. Ind.*, (n), 7: (10): 408, Pl. VI, fig. 2 (1856).

Trupanea inserens Walker, *Proc. Linn. Soc. Lond.*, I: 116 (1857).

Promachus inserens Wulp, *Catalogue Diptera South Asia*, Hague: 91 (1896); Kertesz, *Catalogus Dipterorum*, Budapest, 4: 221 (1909)[6]. — [6]Bornea Island.

Philodicus rubritarsatus Schiner, *Verh. zool-bot. Ges. Wien*, **17:** 390 (1867).

8. **meridionalis** Ricardo

Philodicus meridionalis Ricardo, *Ann. Mag. nat. Hist.*, (9), **8:** 188 (1921). — Ceylon.

9. **pellidipennis** Ricardo

Philodicus Pellidpennis Ricardo, *Ann. Mag. nat. Hist.*, (9), **8:** 191 (1921). — Karachi (Pakistan).

10. propinquus Bromley

Philodicus propinquus Bormley, *Indian,* J. *agric.* **8:** 866 (1938). — Ponani (India).

11. pruthi Bromely

Philodicus pruthi Bromely, *Rec. Indian Mus.,* **37:** 224 (1935). — Madras.

12. thoracicus Ricardo

Philodicus thoracicus Ricardo, *Ann. Mag. nat. Hist.,* (9), **8:** 189 (1921). — Ceylon.

13. univentris (Walker)

Trupanea univentris Walker, *Insecta Saundersiana Diptera,* London, **I:** 114 (1851): *List Diptera British Museum,* London, 7, Suppl. **3:** 602 (1855).

Promachus univentris Wulp, *Catalogue Diptera South Asia,* Hague: 91, (1896). Kertesz, *Catalogus Dipetororum,* Budapest, **4:** 227 (1909)[1]. — [1]India or East India

Philodicus univentris Ricardo, *Ann. mag. nat. Hist.,* (9), **8:** 190 (1920). Bromely, *Rec. India. Mus.,* **37** (2): 230 (1935)[2]. — [2]Upper Burma.

Genus 15. **Procetacanthus** Macquart

Proctaconthus Macquart, *Dipteres Exotiques,* Paris, **I**(2): 120 (1838), (Type for the genus: *Proctacanthus niveus* Macquart). Kertesz. *Catalogus Dipterorum, Budapest,* **4:** 245 (1909), *(syn. Acanthodelphia* Bigot and *Proctacantha* Scudder).

Acanthodelphia Bigot, *Ann. soc.ent, Fr,* (3), **5:** 545 (1857).

Proctacantha Scudder, *Nomend, Zool. Univ. Indlex:* 262 (1884).

I. penultimus (Walker)

Asilus penultimus Walker, *Insecta Saundersiana Diptera,* London, **I:** 134 (1851); *List Diptera British Museum,* London, 7, Suppl. 3: 726 (1855) Wulp, *Catalogue Diptera South Asia,* Hague: 94 (1896). Kertesz, *Catalogus Dipterorum* Budapest, **4:** 262 (1909)[1]. — [1]India or East India

Protacanthus penultimus Ricardo, *Ann. Mag. nat. Hist.,* (9), **10:** 43 (1922).

Genus 16. **Rhadiurgus** Loew

Rhadiurgus Loew, *Linn. Ent.*, **4:** 133 (1849), (Type for the genus: *Asilus variabilis* Zetterstedt). Wulp, *Catalogue Diptera South Asia,* Hauge: 95 (1896). Kertesz, *Catalogus Dipterorum,* Budapest, **4:** 271 (1909).

1. **bifidus** (Fabricuis)

Dasypogon bifidus Fabricius, *Syst. Antliat:* 170 (1805)

Asilus bifidus Wiedermann, *Dipters Exotique,* Paris: 190 (1821); *Aussereuropaisch Zweiflugelige Insecten,* Hamm, **I:** 443 (1828). Walker, *List Diptera British Museum,* London, 7, Suppl. 3: 726 (1855).

Rhadiurgus bifidus Schiner, *Verh. zool.-bot. Ges. Wien,* 16: 684 (1866). Wulp, *Catalogue Diptera South Asia,* Hague: 95 (1896). Kertesz, *Catalogus Dipterorum,* Budapest, **4:** 271 (1909)[1].

[1]Tranquebar (Madras)

Genus 17. **Senoprosopis** Macquart

Senoprospis Macquart *Dipteres Exotiques,* Paris, **I:** (2): 130 (1838), Type for the genus: *Senoprosopis* diardii Macquart). Wulp, Catalogue *Diptera South Asia,* Hague: 96 (1896). Kertesz, *Catologus Dipterorum,* Budapest, **4:** 282 (1909), (Wrongly spelled as *Stenoprosopis: Syn. Senoprosopus* Walker).

Senoprosopus Walker, *List Diptera British Museum,* London, 7, Suppl. **3:** 762 (1855).

1. **diardii** Macquart

Senoprosopis diardii Macquart, *Dipteres* Exotiques, Paris, **I:** (2): 130 Pl. XI, fig. I (1838). Wulp *Catalogue Diptera South Asia,* Hague: 96 (1896). Kertesz, *Catalogus Dipterrorum* Budapest, **4:** 284 (1909)[1]. Ricardo, *Ann. Mag. nat. Hist.,* (9), **10:** 73 (1922), (specimen not traceable in British Museum Collection).

[1]Bengal.

Subfamily 3. **DASYPOGONINAE**

Dasypogoninae (partim) Loew, *Linn Ent.* **2:** 384, (1847). Wulp *Catalogue Diptera south Asia,* Hague: 76 (1896). Kertesz *Catalogues Dipterorum,* Budapest, 4: 57 (1909).

Dasypognoniae Schiner, *Verh, Zool.-bot. Ges. wien.*, **16:** 649 (1866); **17:** 355 (1867); *Dipt. Novara Reise* (1868). Williston, *manual of No. American Dipetra*, New Haven: 197 (1908). Hermann, *Zool. jb.*, **43:** 164-65 (1920). Bromley, *Ann. Ent. Soc. Amer.*, **27:** 78 (1934); *Lingnan Sci.* J., **21** (1-4): 89-90 (1945). Hardy, *Ann.Mag., nat Hist.*, (IO) **13:** 498-525, (1934); **14:** 1-35 (1934); *Ent. mon. Mag.*, **84:** 116-19 (1948). Efflatoun, *Mem. Soc. R. Ent. Egypte,* **4** (3): 199-443 (1937). Clement, *proc. R. ent.*

Soc. London, (B) **20:** 10-14 (1951).

The subfamily Dasypogoninae consists of 14 genera and 69 species from the Indian Region. It resembles Laphriinae in having two jointed palpi, and resembles Leptogastrinae in having cell R_1, almost always open. The following is the key to all the Indian genera of this subfamily.

Genus 1. **Cyrtopogon** Loew

Cyrtopogon Loew, Linn. Ent., 2: 516 (1847), (Type for the genus: *Asilus ruficornis* Fabricius). Kertsez., *Catalogue Dipetrorum*, Budapest, **4**: 66, (1909), (*Syns. Enarmostus* Walker and *Eupalamus* Jaennicke). Hardy. *Ann., Mag., nat. Hist.*, (10), **14**: 2 (1934)

Enarmostus Walker *Insecta Saundersiana Diptera,* London, I: 102 (1851); *List Diptera British Museum,* London, 6 Suppl. **2:** 423 (1854).

Eupalamus Jaennicke, *Ent., Z.*, II: 86 (1867), (praeocc). Wulp, *Catalogue Diptera South Asia,* Hague: 76, (1896).

1. **khasiensis** Bromley.

Cyrtopogen khasiensis Bromley, *Rec. Indian Mus.*, **37** (2): 228 (1925). Assam

2. **laphrides** Walker.

Cyrtopogen laphrides Walker, *Insecta Saundersiana Diptera,* London, **1**: 99, Pl. III, fig. 7 (1851); *List Diptera British Museum,* London, **6**, Suppl. 2: 474 (1854). Wulp, *Catalogue Diptera South Asia,* Hague: 76, (1896). Kertesz, *Catalogus.* [1]India

Dipterorum, Budapest, **4**: 68, (1909)[1]. Bromley, *Rec. Indian Mus.,* **37** (2): 230 (1935)[2] [2]Punjab

3. laxenecera Bromley

Cyrtopogon laxencera Bromley, *Rec. Indian Mus.,* **37** (2): 228, (1935) Dalhosie (Punjab)

Genus 2. **Damalis** Fabricius

Damalis Fabricius, *Systema Antiliat:* 147 (1805), (Type for the genus: *Damalis marginata* (Wulp), Wulp, *Catalogue Diptera South Asia* Hague 77 (1896). Kertesz, *Catalogus Dipterorum,* Budapest, **4**: 93 (1909), (*Syn. Chalcidomorpha* Weswood). Meijere, *Tijdschr. Ent.,* **54:** 300-01 (1911). Pritchard, J. *N. Y., ent., Soc.,* **46***:* 11-12 (1938). Hardy *Ent., mon. Mag.,* **84** 118, (1948)* (*preoccupied*).

Xenomyza Wiedermann, *Zool. Mag.,* **1** (1): 60 (1817). Frey, *Rec. suisse Zool Genova,* **41***:* 311 (1934), (Syn. *Damalis* Fabricius which is wrongly mentioned as *Damalis* Westwood). Hardy, *Ent. mon. Mag.,* **84**: 118 (1948), (*Syn. Damalis Fabricius*).

1. claripennis Bromely

Damalis claripennis Bromley, *Rec. Indian Mus.,* **37** (2): 229 (1935). Darjeeling (Bengal)

2. felderi (Schiner)

Damalis felderi Schiner, *Verh.zool.–bot. Ges. Wien.,* **17**: 365 (1867). Osten-Sacken, Z. *ent.,* **26**: 105 (1882). Wulp, *Catalogue Diptera South Asia,* Hague: 78 (1896). Kertesz, *Catalogus Dipterorum,* Budapest, **4**: 93 (1909)[1]. [1]Ceylon

3. fulvipes (Westwood)

Chalcidomorpha fulvipes Westwood, *Ann. Soc., ent. Fr.,* 685 (1835).

---

* Hardy points out that *Damalis* name is not available to Asilidae, and therefore he kept *Xenomyza* Wiedermann.

Now it has been noticed that *Damalis* was used in Lepidoptera by Huebner (1819) and in mammals by Smith (1827) and Gray (1846) and in Molluscs by Leach (1847). Since Fabricius (1805) used this name for an Asilid much earlier than others, his naming takes precidence and is therefore retained.

Damalis fulvipes Walker, *List Diptera British Museum*, London, 7, Suppl. **3** 765 (1855). Osten-Sacken, *Z. ent.*, 26: 104 (1882). Röder, *Ent. Nach.*, **19**: 234 (1893). Wulp, *Catalogue Diptera South Asia*, Hague: 77 (1896). Kertesz, *Catalogus Dipterorum*, Budapest, **4**: 93-94 (1909)[1]. — [1]Ceylon

4. fusca (Walker)

Damalis fusca Walker, *List Diptera British Museum*, London, **2**: 481 (1849); 7 Suppl. **3**: 764 (1855). Osten-Sacken, *Z. ent.*, **26**: 104 (1882). Wulp. *Catalogue Diptera South Asia*, Hague: 77 (1896).

Damalis fuscus Kertesz, *Catalogus Dipterorum*, Budapest, **4**: 94 (1909)[1]. Bromley, *Rec. Indian Mus.*, **37**(2): 230 (1935)[2]. — [1]Bengal [2]Darjeeling

5. paradoxa Frey

Xenomyza paradoxa Frey, *Rev. suisse Zool. Genova*, **41**: 314 (1934). — India

6. planiceps (Fabricius)

Damalis planticeps Fabricius, *Systema Antliat*: 148 (1805). Wiedermann, *Aussreuropaische zweiflugelige Insecten*, Hamm **1**: 417 (1828). *Macquart, Suit. a Buffon*, **1**: 316 (1834); *Dipteres Exotiques*, Paris, **1** (2): 154 (1838). Walker, *List Diptera British Museum*, London, **2**: 480 (1849); 7, Suppl. **3**: 764, (1855). Schiner, *Verh. zool.--bot. Ges. Wien*, **16**: 681 (1866). Osten-Sacken, *Z. ent.*, **26** 104 (1882). Wulp, *Catalogue Diptera South Asia*, Hague: 77, (1896). Kertesz, *Catalogus Dipterorum*, Budapest, **4**: 95 (1909)[1]. Bromley, *Rec. Indian Mus.*, **37**(2): 230 (1935). — [1]Darjeeling (Bengal) and Tranquebar (Madras)

7. signatus (Walker)

Damalis signatus Walker, *Trans. R. ent. Soc.* London, (n), **5**: 284 (1860). Osten-Sacken, *Z. ent.*, **26**: 105 (1882). Wulp, *Catalogue Diptera South Asia*, Hague: 77 (1896). Kertesz, *Catalogus Dipterorum*, Budapest, **4**: 95 (1909)[1]. — [1]Burma

8. speculiventris (Meijere)

Damalis speculiventris Meijere, *Tijdschr. Ent.*, **54**: 304 (1911)[1]. — [1]Dehra Dun (U.P.)

9. **tibialis** (Macquart)

Damalis tibialis Macquart, *Dipteres Exotiques,* Paris, **1**(2): 154, Pl. XII, fig. 6 (1838). Walker, *List Diptera British Museum,* London, 7, Suppl. **3**: 764, (1855). Osten-Sacken, *Z. ent.,* **26** 104 (1882). Wulp, *Catalogue Diptera South Asia,* Hague: 77, (1896). Kertesz, *Catalogus Dipterorum,* Budapest, **4**: 95 (1909)[1]. — [1]India

10. **vitalisi** Frey

Xenomyza vitalisi Frey, *Rev. sussie Zool. Genova,* **41**: 312 (1934). — India

11. **vitripennis** Osten-Sacken

Damalis vitripennis Osten-Sacken, *Z. ent., 26:* 106 (1882). Wulp, *Catalogue Diptera South Asia,* Hague: 78, (1896). Kertesz, *Catalogus Dipterorum,* Budapest, **4**: 95 (1909)[1]. Meijere, *Tijdschr, Ent.,* **54**: 301 (1911). Bezzi, *Philipp. J. Sci.,* **12:** 124 (1917)[2]. — [1, 2]Philippines Is.

Genus 3. **Dasypogon** Meigen

Dasypogon Meigen, *Illiger's Mag. f. Ins.,* **2**: 270 (1803), (Type for the genus: *Asilus reutonus* Linnaeus). Wulp, *Catalogue Diptera South Asia,* Hague: 80, (1896). Kertesz, *Catalogus Dipterorum,* Budapest, **4**: 125 (1909).

1. **albonotatus** Wiedermann

Dasypogon albonotatus Weidermann, *Dipteres Exotiques,* Paris: 181 (1821); *Aussereuropaische zweiflugelige Insecten,* Hamm **1**: 402 (1828). Walker, *List Diptera British Museum,* London, 6, Suppl. **2**: 475, (1854). Schiner, *Verh. zool.-bot. Ges. Wien,* **16**: 679 (1866). Wulp, *Catalogue Diptera South Asia,* Hague: 80, (1896). Kertesz, *Catalogus Dipterorum,* Budapest, **4**: 125 (1909)[1]. — [1]Bengal

2. **decretus** Walker

Dasypogon decretus Walker, *Trans. R. ent. soc.* London, (n) **5**: 279 (1860). Wulp, *Catalogue Diptera South Asia,* Hague: 81, (1896). Kertesz, *Catalogus Dipterorum,* Budapest, **4**: 127 (1909)[1]. — [1]Burma

3. **dorsalis** Wiedermann.

Dasypogon dorsalis Wiedermann, *Analecta Ent.*: 27 (1824); *Aussereuroparsche zweiflugetige Insectn*, Hamm **1**: 413 (1828). Walker, *List Diptera British* Museum, London, 6, Suppl. **2**: 476 (1854). Schiner, *Verh, zool-bot. Ges. Wien*, **16**: 681 (1866). Kertesz, *Catalogus Dipterorum*, Budapest, **4**: 127 (1909)[1]. [1]India

4. **incisus** Macquart

Dasypogon incisus Macquart, *Dipteres Exotiques*, Paris, Suppl. 5: 49 (1855). Wulp, *Catalogue Diptera South Asia*, Hague: 81 (1896). Kertesz. *Catalogus Dipterorum*, Budapest, **4**: 128 (1909)[1]. [1]India

5. **inopinatus** Walker

Dasypogon inopinatus Walker, *Trans. R. ent. Soc. Lond.*, (n) 5: 278 (1860). Wulp, *Catalogue Diptera South Asia*, Hague: 81 (1896). Kertesz. *Catalogus Dipterorum*, Budapest, **4**: 129 (1909)[1]. [1]Burma

6. **inopportunus** Walker,

Dasypogon inopportunus Walker, *Trans. R. ent. Soc. Lond.*, (n) 5: 278 (1860). Wulp, *Catalogue Diptera South Asia*, Hague: 81 (1896). Kertesz. *Catalogus Dipterorum*, Budapest, **4**: 129 (1909)[1]. [1]Burma

7. **scatophagoides** Walker

Dasypogon scatophagoides Walker, *List Diptera British* Museum, London, 6, Suppl. **2**: 472 (1854). Wulp, *Catalogue Diptera South Asia*, Hague: 81 (1896). Kertesz. *Catalogus Dipterorum*, Budapest, **4**: 131 (1909)[1]. [1]India

Genus 4. **Holcocephala** Jaennicke

Discocephala Macquart, *Dipteres Exotiques*, Paris, **1** (2): 50 (1838), (*praeocc*).

Holcocephala Jacnnicke, *Abh. Senckenb. Naturf. Ges.*, **6**: 359 (1867), (Type for the genus: *Dasypogon abdominolis* Say). Kertesz, *Catalogus Dipterorum*, Budapest, **4**: 90 (1909), (*Syns. Discocephala* Macquart, *Helcocephala* Loew and *Holocephala* Williston).

Helcocephala Loew, *Z. ent.*, **18**: 377 (1874).

Holocephala Williston, *Trans. Amer. ent. Soc.*, **18**: 72 (189). Pritchard, *J. N. Y. ent. Soc.*, **46**: 11–21 (1938); *Ann. ent. Soc. Amer.*, **34**: 350 (1941). Hardy, *Ent. mon. Mag.*, **84**: 118 (1948).

II. prytanis (Walker)

Dasypogon prytanis Walker, *List Diptera British Museum*, London, **2**: 363 (1849).

Discocephala prytanis Walker, *List Diptera British Museum*, London, 6, Suppl. **2**: 499 (1854).

Holcocephala prytanis Wulp, *Catalogue Diptera South Asia*, Hague: 77 (1896). Kertesz. *Catalogus Dipterorum*, Budapest, **4**: 92 (1909)[1]. (*Syns. Dasypogon prytanis* Walker and *Discocephala prytanis* Walker). [1]Bengal

Genus 5. **Laphistia** Loew

Laphistia Loew, *Linn. Ent.*, **2**: 538 (1847), (Type for the genus: *Laphistia sabulicola* Loew). Kertesz. *Catalogus Dipterorum*, Budapest, **4**: 157 (1909). (*Syns. Lophyctis* Loew and *Asicya* Arrib).

Laphyctis Loew, *Ofvers, Kongl. Vet. Akad. Forhandl.*, **15**: 338 (1858); *Dipterenfn, S. Afrika*: 112 (1860). Wulp, *Catalogue Diptera South Asia*, Hague: 77 (1896).

Laphyctis Loew, *Öfvers.Kongl. Vet. Akad. Förhandl.*, **15**: 338 (1858); *Dipterenfn. S. Afrika*: 112 (1860). Wulp, *Catalogue Diptera South Asia*, Hague: 77 (1896).

Asicya Arrib, *Annl. Soc. Cient. Argent.*, **9**: 224 (1880).

1. stignaticalis Bigot.

Laphistia stigmaticalis Bigot, *Ann. Soc. ent. Fr.*, (5), **8**: 430 (1878). Wulp, *Catalogue Diptera South Asia*, Hague: 77 (1896). Kertesz. *Catalogus Dipterorum*, Budapest, **4**: 158 (1909). Ceylon

Genus 6. **Microstylum** Macquart

Microstylum Macquart, *Dipteres Exotiques*, Paris, **1**(2): 263 (1838), (Type for the genus: *Dasypogon venosum* Wiedermann). Wulp, *Catalogue Diptera South Asia*, Hague: 79 (1896). Kertesz. *Catalogus*

Dipterorum, Budapest, **4**: 149 (1909). (*Syns. Megapollion* Bigot and *Megapollyon* Walker).

Megapollyon Walker, *Insecta Saundersiana Diptera,* London, **1**: 85 (1851).

Megapollyon Bigot, *Ann. Soc. etn. Fr.,* (3) 5: 548 (1857).

1. **albolimbatum** Wulp.

Microstylum albolimbatum. Wulp, *Tijdschr. Ent.,* **41**: 118, Pl. III, fig. 1-4 (1898). Kertesz. *Catalogus Dipterorum,* Budapest, **4**: 149 (1909)[1]. (*Syn. M. albolincatum* Wulp). [1]India

Microstylum albolincatum Wulp, *Tijdschr. Ent.,* **42**: 48 (1899).

2. **apicale** (Wiedermann).

Dasypogon apicale Wiedermann, *Dipteres Exotiques,* Paris, 216, Pl. II, fig. 8 (1821); *Aussereuropaische zweiflugelige Insecten,* Hamm **1**: 372, Pl. VI, fig. 4 (1828). Walker, *List Diptera British Musecum,* London, 6, Suppl. **2**: 473 (1854).

Microstylum apicale Macquart, *Dipteres Exotiques,* Paris, **1**(2): 29 (1838). Schiner, *Verh. zool-bot. Ges. Wien.* **16**: 676 (1866). Wulp, *Catalogue Diptera South Asia,* Hague: 79 (1896). Kertesz. *Catalogus Dipterorum,* Budapest, **4**: 150 (1909)[1]. Bromley, *Rec. Indian Mus.,* **37**(2): 230 (1935).[2] [1]India [2]Darjeeling (Bengal)

3. **apiforme** (Walker)

Dasypogon apiforme Walker, *Insecta Saundersiana, Diptera,* London, **1**: 97 (1851); *List Diptera British Museum,* London, 6 Suppl. **2**: 472 (1854).

Microstylum apiforme Wulp, *Catalogue Diptera South Asia,* Hague: 80 (1896). Kertesz. *Catalogus Dipterorum,* Budapest, **4**: 150 (1909)[1]. [1]India

4. **balbillus** (Walker)

Dasypogon balbillus Walker, *List Diptera British Museum,* London, **2**: 307 (1849); 6 Suppl. **2**: 472 (1854).

Saropogon balbillus Walker, *List Diptera British Museum,* London, **2**: 307 (1849).

Microstylum balbillus Wulp, *Catalogue Diptera South Asia,* Hague: 79 (1896). Kertesz. *Catalogus Dipterorum,* Budapest, **4**: 150 (1909)[1]. [1]Nepal

5. basalis Brunetti

Microstylum basalis Brunetti, *Ann. Mag. nat. Hist.,* (10). **2**: 81, (1928). Ceylon

6. basirufum, Bigot.

Microstylum basirufum Bigot, *Ann. Soc. ent. Fr., (5),* **8**: 403 (1878). Wulp, *Catalogue Diptera South Asia,* Hague: 80 (1896). Kertesz. *Catalogus Dipterorum,* Budapest, **4**: 150 (1909)[1]. [1]Assam

7. bicolor, Macquart

Microstylum bicolor Macquart, *Dipteres Exotiques,* Paris, Suppl. **4**: 62, Pl. VI, fig. 3 (1849). Wulp, *Catalogue Diptera South Asia,* Hague: 80 (1896). Kertesz. *Catalogus Dipterorum,* Budapest, **4**: 150 (1909)[1]. [1]Silhet (Assam)

8. brevipennatum, Bigot

Microstylum brevipennatum Bigot, *Ann. Soc. ent. Fr., (5),* **8**: 403 (1878). Wulp, *Catalogue Diptera South Asia,* Hague: 80 (1896). Kertesz. *Catalogus Dipterorum,* Budapest, **4**: 150 (1909)[1]. [1]India

9. brunnipenne Macquart

Microstylum brunnipenne Macquart, *Dipteres Exotiques,* Paris, Suppl. **4**: 62, Pl. VI, fig. 2 (1849). Kertesz. *Catalogus Dipterorum,* Budapest, **4**: 150 (1909)[1], (*Syn. Microstylum brevipenne* Wulp). Enderlein, *Wien. ent. Ztg. Wien,* **33**,: 171 (1914). [1]Silhet (Assam)

Microstylum brevipenne Wulp, *Catalogue Diptera South Asia,:* Hague: 80 (1896).

10. erythropygum Bigot

Microstylum erythropygum Bigot, *Ann. Soc. ent. Fr., (5),* **8**: 402 (1878). Wulp, *Catalogue Diptera South Asia,* Hague: 80 (1896). Kertesz. *Catalogus Dipterorum,* Budapest, **4**: 151 (1909)[1]. [1]Assam

11. **eximium** Bigot.

Microstylum eximium Bigot, *Ann. Soc. ent. Fr.,* (5), **8**: 404 (1878). Wulp, *Catalogue Diptera South Asia,* Hague: 80 (1896). Kertesz. *Catalogus Dipterorum,* Budapest, **4**: 151 (1909)[1]. [1]Burma

12. **flaviventre** Macquart

Microstylum flaviventre Macquart, *Dipteres Exotiques,* Paris, Suppl. **4**: 62, Pl. VI, fig. 4 (1849); Suppl. 5: 48, (1855). Wulp, *Catalogue Diptera South Asia,* Hague: 80 (1896). Kertesz. *Catalogus Dipterorum,* Budapest, **4**: 151 (1909)[1]. [1]Silhet (Assam) and China border.

13. **fulviventre** Wulp

Microstylum fulviventre Wulp, *Tijdschr. Ent.,* **41**: 116 (1898); **42**: 48 (1899). Kertesz. *Catalogus Dipterorum,* Budapest, **4**: 152 (1909)[1]. [1]India

14. **haemorrhoidale** Bigot

Microstylum haemorrhoidale Bigot, *Ann. Soc. ent. Fr.,* (5), **8**: 404 (1878). Wulp, *Catalogue Diptera South Asia,* Hague: 80 (1896). Kertesz. *Catalogus Dipterorum,* Budapest, **4**: 152 (1909)[1]. [1]Burma

15. **imbutum** (Walker)

Dasypogon imbutum Walker, *Insecta Saundersiana Dipetera,* London, **1**: 96 (1851); *List Diptera British Museum,* London,d 6, Suppl. 2: 472 (1854).

Microstylum imbutum, Wulp, *Catalogue Diptera South Asia,* Hague: 80 (1896). Kertesz. *Catalogus Dipterorum,* Budapest, **4**: 152 (1909)[1]. [1]India

16. **libo** (Walker)

Dasypogon libo Walker, *List Diptera British Museum,* London, **2**: 342 (1849); 6 Suppl. **2**: 472 (1854).

Stichopogon libo Walker, *List Diptera British Museum,* London, **2**: 312 (1849).

Microstylum libo Wulp, *Catalogue Diptera South Asia,* Hague: 80 (1896). Kertesz. *Catalogus Dipterorum,* Budapest, **4**: 152 (1909)[1]. [1]India

17. nigricauda (Wiedermann)

Dasypogon nigricauda Wiedermann, *Analecta Ent.*: 26 (1824); *Aussereuropaische zweiflugelige Insecten*, Hamm, **1**: 377 (1828). Walker, *List Diptera British Museum*, London, 6 Suppl. **2**: 475 (1854).

Microstylum nigricauda, Schiner, *Verb. zool.-bot. Ges. Wien*, **16**: 677 (1866). Wulp, *Catalogue Diptera South Asia*, Hague: 79 (1896). Kertesz. *Catalogus Dipterorum*, Budapest, **4**: 153 (1909)[1]. [1]India

18. nitidiventre Bigot

Microstylum nitidiventre Bigot, *Ann. Soc. ent. Fr., (5)*, **8**: 405 (1878). Wulp, *Catalogue Diptera South Asia*, Hague: 80 (1896). Kertesz. *Catalogus Dipterorum*, Budapest, **4**: 153 (1909)[1]. [1]Burma

19. otacilius (Walker)

Dasypogon otacilius Walker, *List Diptera British Museum*, London, **2**: 344 (1949); 6 Suppl. 2: 472 (1854).

Stichopogon otacilius Walker, *List Diptera British Museum*, London, **2**: 344 (1849).

Microstylum otacilius Wulp, *Catalogue Diptera South Asia*, Hague: 80 (1896). Kertesz. *Catalogus Dipterorum*, Budapest, **4**: 153 (1909)[1]. [1]India

20. polygnotus (Walker)

Dasypogon polygnotus Walker, *List Diptera British Museum*, London, **2**: 305 (1849); 6 Suppl. 2: 472 (1854).

Saropogon polygnotus Walker, *List Diptera British Museum*, London, **2**: 305 (1849).

Microstylum polygnotes Wulp, *Catalogue Diptera South Asia*, Hague: 79 (1896).

Microstylum polygnotus Kertesz., *Catalogus Dipterorum*, Budapest, **4**: 153 (1909)[1]. [1]Silhet (Assam)

21. proclive (Walker)

Dasypogon proclive Walker, *Trans. R. ent. Soc. Lond.*, (n), 5: 277 (1860).

Saropogon proclive Bigot, *Ann. Soc. ent. Fr.*, (5), **8** 221 (1878).

Microstyum proclive Wulp, *Catalogue Diptera South Asia*, Hague: 80 (1896). Kertesz. *Catalogus Dipterorum*, Budapest, **4**: 153 (1909)[1]. [1]Burma

22. rhypae (Walker)

Dasypogon rhypae Walker, *List Diptera British Museum*, London, **2**: 305 (1849); 6, Suppl. **2**: 472 (1854).

Saropogon rhypae Walker, *List Diptera British Museum*, London, **2**: 305 (1849).

Microstyum rhypae Wulp, *Catalogue Diptera South Asia*, Hague: 79 (1896). Kertesz. *Catalogus Dipterorum*, Budapest, **4**: 153 (1909)[1]. [1]Silhet (Assam)

23. rufoabdominalis Brunetti

Microstylum rufoabdominalis Brunetti, *Ann. Mag. nat. Hist.*, (10), **2**: 80 (1928). Ceylon

24. strigatum Enderlein

Microstylum strigatum Enderlein, *Wien, ent. Ztg.*, **33**: 171 (1914). Sikkim

25. sura (Walker)

Dasypogon sura Walker, *List Diptera British Museum*, London, **2**: 345 (1849); 6, Suppl. **2**: 472 (1854).

Lasiopogon sura Walker, *List Diptera British Museum*, London, **2**: 345 (1849).

Microstylum sura Wulp, *Catalogue Diptera South Asia*, Hague: 80 (1896). Kertesz. *Catalogus Dipterorum*, Budapest, **4**: 154 (1909)[1]. [1]India

26. trimelas (Walker)

Dasypogon trimelas Walker, *Insecta Saundersiana Diptera*, London, **1**: 97 (1851); *List Diptera British Museum*, London, 6 Suppl. **2**: 473 (1854).

Microstylum trimelos Wulp, *Catalogue Diptera South Asia*, Hague: 80 (1896).

Microstylum trimelas Kertesz. *Catalogus Dipterorum,* Budapest, **4**: 154-55 (1909)[1]. (*Syn. M. trimelos* Wulp), Bromley, *Rec. Indian Mus.,* **37** (2): 230 (1935)[2].

[1]India
[2]Darjeeling (Bengal)

27. vica (Walker)

Dasypogon vica Walker, *List Diptera British Museum,* London, **2** 304 (1849); 6 Suppl. **2**: 472 (1854).

Saropogon vica Walker, *List Diptera British Museum,* London, **2** 304 (1849).

Microstylum vica Walker, *Proc. Linn. Soc. Lond.,* **1** 112 (1856). Wulp *Catalogue Diptera South Asia* Hague: 79 (1896); *Tijdschr. Ent.,* **41**: 116 (1898); **42**: 53 (1899). Kertesz. *Catalogus Dipterorum,* Budapest, **4**: 155 (1909)[1]. (*Syn. Microstylum spinitarse* Macquart). Enderlein, *Wein, ent.: Ztg.,* **33**: 173 (1914).

[1]Borneo and Silhet (Assam)

Microstylum spinitarse Macquart, *Dipteres Exotiques,* Paris, Suppl. 4: 61, pl. VI, fig. 1 (1849). Wulp *Catalogue Diptera South Asia* Hague: 80 (1896).

28. whitei Brunetti

Microstylum whitei Brunetti, *Ann. Mag. nat. Hist.,* (10), 2: 78 (1928).

Genus 7. **Neolaparus** Williston

Laparus Loew, *Proger. Realschule Meseritz,* 1851: 4 (praeocc).

Neolaparus Williston, *Psyche, 5: 255* (1889), (Type for the Genus: *Dasypogon volcutus* Walker). Engel, *Lindner's Die Fliegen,* Stuttgart, **24**: 442-43 (1930). Bromley, *Lingnan Sci. J.,* **21:** (1-4): 95, (1945), (*Syn. Cenopogon* Wulp).

Cenopogon Wulp. *Tijdschr. Ent.,* **41**: 120 (1898). Kertesz. *Catalogus Dipterorum,* Budapest, **4**: 122 (1909).

1. volcatus (Walker)

Dasypogon volcatus Walker, *List Diptera British Museum,* London, **2**: 346 (1849); 6 Suppl.. **2**: 469 (1854). Wulp, *Catalogus Dipetera South Asia,* Hague: 81 (1896).

Lasiopogon volcatus Walker, *List Diptera British Museum,* London, **2**: 346 (1849).

Cenopogon volcatus Wulp, *Tijdschr. Ent.*, **41**: 122 (1898); **42**: 47, 53 (1899), Kertesz. *Catalogus Dipterorum*, Budapest, **4**: 122 (1909)[1]. [1]Himalaya (India), China and Java

Dasypogon hypsaon Walker, *List Diptera British Museum*, London, **2**: 348 (1849); 6 Suppl.. **2**: 469 (1854). Wulp, *Catalogus Dipetera South Asia*, Hague: 81 (1896).

Lasiopogon hypsaon Walker, *List Diptera British Museum*, London, **2**: 348 (1849).

Cenopogon hypsaon Wulp, *Tijdschr. Ent.*, **41**: 122 (1898).

Dasypogon bifidus Wulp, *Tijdschr. Ent.*, **41**: 121 pl. III, fig. 6-7 (1898).

Genus 8. **Oldroydia** Hull

Oldroydia Hull, *Ann. Mag. nat. Hist.*, (12), **9**: 398 (1956), (Type for the genus: *Oldroydia hamata* Hull).

1. hamata

Oldroydia hamata Hull, *Ann. Mag. nat. Hist.*, (12), **9**: 399 (1956). Assam (India)

Genus 9. **Plesiomma** Macquart

Plesiomma Macquart, *Dipteres Exotiques*, Paris, **1** (2): 54 (1838), (Type for the genus: *Plesiomma nigra* Macquart). Kertesz. *Catalogus Dipterorium*, Budapest, **4**: 145 (1909), (*Syn. Plesioma* Schiner).

Plesioma Schiner, *Verh. Zool-bot. Ges. Wien*, **16** 702 (1866).

1. **lineatum** (Fabricius)

Asilus lineatum Fabricius, *Species Insect.*, **2**: 465 (1781); *Mantissa Insect.* **2**: 360 (1787); *Ent. Syst.*, **4**: 386 (1794). Olivier, *Encyclop Method*, **4**: 269 (1789). *Gmel, Systema Naturae*, **5**: 2900 (1792).

Dasypogon lineatum Fabricius, *Systema Antliat*: 167 (1805). Wiedermann. *Dipteres, Exotiques*, Paris: 221 (1821); *Aussereuropaische zweifluegelige Insecten*, Hamm, **1**: 358 (1828). Walker. *List Diptera British Museum*, London 6 Suppl. 2: 428 (1954).

Plesiomma lineatum Schiner, *Verh. zool.-bot. Ges. Wien*, **16**: 677 (1866); **17**: 374 (1867). Aldrisch, *Catalogue*

N. American Diptera, New Haven: 260 (1905). Kertesz. *Catalogus Dipterorum* Budapest, **4**: 146 (1909)[1].

[1]Ins. St. Crux & India.

Genus 10. **Saropogon** Loew

Saropogon Loew, *Linn. ent.,* **2**: 439 (1847), (Type for *genus:* Dasypogon luctuosus *Meigen). Wil-listen, Trans. ent. Soc.,* **18**: 74 (1891). Wulp *Catalogue Diptera South Asia,* Hague: 79 (1896). Kertesz. *Catalogus Dipterorum,* Budapest, **4**: 136 (1909). Hardy, *Ann. Mag. nat. Hist.,* (10), **14**: 2 (1934).

1. **scalaris** Bigot

Saropogon Scalaris Bigot, *Ann. Soc. ent. Fr., (5),* **8**: 413 (1878). Wulp, *Catalogue Diptera South Asia,* Hague: 79 (1896). Kertesz. *Catalogus Dipterorum,* Budapest, **4**: 140 (1909)[1].

[1]India

Genus 11. **Seylaicus** Loew

Scylaticus Loew, *Öfvers. Kongl. Vet. Akad. Förh.,* **14**: 346, 347 (1858), (Type for the genus: *Scylaticus zonatus* Loew). Wulp, *Catalogue Diptera South Asia,* Hague: 78 (1896). Kertesz. *Catalogus Dipterorum,* Budapest, **4**: 103 (1909), (*Syn. Euthrix* Philippi).

Euthrix Phillippi, *Verh. zool.–bot. Ges. Wien,* **15**: 690 (1865), (pracoec.).

1. **indicus** Bromley

Scylaticus indicus Bromley, *India J. agric. Sci.* **8**: 864 (1938).

Tinnevilly Hills (S. India)

Genus 12. **Senobasis** Macquart

Senobasis Macquart, *Dipteres Exotiques,* Paris, **1**(2): 52 (1838), (Type for the genus: *Senobasis anialis* Macquart). Kertesz. *Catalogus Dipterorum,* Budapest, **4**: 123 (1909), (*Syns. Blepharepitum* Rondani and *Planetolestes* Arrib).

Blepharepium Rondani, Truqui, Studi. Ent., **1**: 89 (1848).

Lochites Schiner, *Verh. zool.-bot. Ges. Wien,* **16**: 671 (1866), *(praeoce.).* Wulp, *Catalogue Diptera South Asia,* Hague: 79 (1896). Kertesz. *Catalogus Dipterorum,* Budapest, **4**: 124 (1909)

Planetolestes Arrib., *Anal. Soc. Cient. Argent.*, **7**: *147 (1879).*

1. **testaceus** (Bigot)

Lochites testaceus Bigot, *Ann. Soc. ent., Fr., (5)*, **8**: 425 (1878). Wulp, *Catalogue Diptera South Asia*, Hague: 79 (1896). Kertesz. *Catalogus Dipterorum*, Budapest, **4**: 125 (1909)[1].

[1]Burma

Genus 13. **Stenopogon** Loew

Stenopogon Loew, *Linn. Ent.*, **2**: 453 (1847), (Type for the genus: *Asilus sabaudus* Fabricius). Kertesz. *Catalogus Dipterorum*, Budapest, *4*: 76 (1909). Ricardo, *Ann. Mag. nat. Hist.*, (8), **10**: 155 (1912). White, *Proc. roy. Soc. Tasmania*, 162 (1916); 78 (1917). Harly, *Proc. Linn. Soc. N. S. W., 46:* 291 (1921); *51:* 309 (1926); *55:* 259 (1930); *Ann. Mag. nat. Hist.*, (10), **14**: 13 (1934). Malloch, *Proc. Linn. Soc. N.S.W., 53:* 607 (1926). Bromley, *Lingnan Sci. J.*, **21**: (1-4): 93 (1945), (*Syn. Scleropogon* Loew).

Scleropogon Loew, *Ent. Z.*, **10**: 26 (1866). Wulp, *Catalogue Diptera South Asia*, Hague: 77 (1896). Kertesz. *Catalogus Dipterorum*, Budapest, **4**: 75 (1909).

1. **ambryon** (Walker)

Dasypogon ambryon Walker, List *Diptera British Museum*, London, **2**: 311 (1849); 6, Suppl. **2**: 473 (1854) . Wulp, *Catalogue Diptera South Asia*, Hague: 81 (1896).

Habropogon ambryon Walker, *List. Diptera British Museum*, London, **2**: 311 (1849).

Scleropogon ambryon Wulp. *Tijdschr. Ent.*, **42**: 54 (1899). Kertesz. *Catalogus Dipterorum*, Budapest, **4**: 75 (1909)[1].

[1]Bengal

2. **aphrices** (Walker)

Dasypogon aphrices Walker, *List diptera British Museum*, London, **2**: 341 (1849); 6 Suppl. **2**: 474 (1854). Wulp, *Catalogue Diptera South Asia*, Hague: 81 (1896).

Habropogon aphrices Walker, *List Diptera British Museum*, London, **2**: 341 (1849).

Dasypogon aphrices Wulp, *Tijdschr. Ent.*, **42**: 54 (1899).

Scleropogon aphrices Kertesz. *Catalogus Dipterorum*, Budapest, **4**: 75 (1909)[1], (*Syn. Dasypogon aphrisus* Wulp). [1]Nepal

3. **damias** (Walker)

Dasypogon damias Walker, *List Diptera British Museum*, London, **2**: 313 (1849). Wulp, *Catalogue Diptera South Asia*, Hague: 81 (1896).

Habropogon damias Walker, *List Diptera British Museum*, London, **2**: 313 (1849).

Scleropogon damias Wulp, *Tijdschr. Ent.*, *42:* 54 (1899). Kertesz. *Catalogus Dipterorum*, Budapest, **4**: 75-76 (1909)[1]. [1]Bengal

4. **echelus** (Walker)

Dasypogon echelus Walker, *List Diptera British Museum*, London, **2**: 314 (1849). Wulp, *Catalogue Dipteras South Asia*, Hague: 81 (1896).

Habropogon echelus Walker, *List Diptera British Museum*, London, **2**: 314 (1849).

Scleropogon echelus Wulp, *Tijdschr. Ent.*, **42**: 54 (1899). Kertesz. *Catalogus Dipterorum*, Budapest, **4**: 76 (1909)[1]. [1]Nepal

5. **imbrex** (Walker)

Dasypogon imbrex Walker, *List Diptera British Museum*, London, **2**: 315 (1849); 6, Suppl. **2**: 473 (1854).

Habropogon imbrex Walker, *List Diptera British Museum*, London, **2**: 315 (1849).

Scleropogon imbrex Wulp, *Catalogue Diptera South Asia*, Hague: 81 (1896); *Tijdschr. Ent.*, **42**: 54 (1899). Kertesz. *Catalogus Dipterorum*, Budapest, **4**: 76 (1909)[1]. [1]Nepal

6. **nigrofasciatus** Brunetti

Stenopogon nigrofasciatus Brunetti, *Ann. Mag. nat. Hist.*, (10), **2**: 76 (1928). Darjeeling (Bengal)

7. piceus (Röder)

Scleropogon piceus Röder, *Ent. Nach.*, **19**: 234 (1893). Wulp, *Catalogue Diptera South Asia*, Hague: 77 (1896). Kertesz. *Catalogus Dipterorum*, Budapest, **4**: 76 (1909)[1]. — [1]Ceylon

8. pulverifer (Walker)

Dasypogon pulverifer Walker, *Insecta Saundersiana Diptera*, London, **1**: 98 (1851); *List Diptera British Museum*, London, 6, Suppl. **2**: 474 (1854). Wulp, *Catalogue Diptera South Asia*, Hague: 81 (1896).

Scleropogon pulverifer Wulp, *Tijdschr. Ent.*, **42**: 54 (1899). Kertesz. *Catalogus Dipterorum*, Budapest, **4**: 76 (1909)[1]. — [1]India

9. raven Bromley

Stenopogon raven, Bromley, *Indian J. agric. Sci.*, **8**: 865 (1938). — Kodaicanal (India)

10. subtus Bromley

Stenopogon subtus Bromley, *Rec. Indian Mus.*, **37**(2): 229 (1935). — Dalhousie (Punjab)

Genus 14. **Stichopogon** Loew

Stichopogon Loew, *Linn. Ent.*, **2**: 499, (1847) (Type for the genus: *Dasypogon elegantulus* Wieder-mann). Wulp, *Catalogue Diptera South Asia*, Hague: 76 (1896). Kertesz. *Catalogus Dipterorum*, Budapest, **4**: 83 (1909)., Bezzi, *Ann. Mus. nat. Hung.*, **8**: 134 (1910). Engel, *Lindner's Die Fliegen*, **24**: 394 (1930).

Philammosius Rondani, *Dipterologiae Italalicae*, Prodromus, **1**: 156, Pl. 3 (1856).

Stilopogon Costa, *Atti R. Acad. Sci. Napoli*, (2), **1**: 62 (1884).

Echinopogon Bezzi, *Ann. Mus. nat. Hung.*, **8**: 131 (1910).

Dichropogon Bezzi, *Ann. Mus. nat. Hung.*, **8**: 133 (1910).

Neopogon Bezzi, *Ann. Mus. nat. Hung.*, **8**: 147-53 (1910).

Lissoteles Bezzi, *Boll. Lab. Zool. Porici*, **4**: 175 (1910).

Cryptopogon White, *Proc. roy. Sec. Tasmania*" 76 (1917). Hardy, *Proc. Linn. Soc. N.S.W.*, **46**: 291 (1921);

52: 396 (1927); **53**: 479 (1928); **54**: 354 (1929); **55**: 260 (1930).

1. **maculipennis** Brunetti

Stichopogon maculipennis Brunetti, *Ann. Mag. nat. Hist.*, (10), **2**: 77 (1928). — Ceylon

2. **nicobarensis** Schiner

Stichopogon nicobaransis Schiner, *Dipt. Navara Reise*: 161 (1868). Wulp, *Catalogue Diptera South Asia*, Hague: 76 (1896). Kertesz. *Catalogus Dipterorum*, Budapest, **4**: 84 (1909)[1]. — [1]Nicobar Is. (India)

SUBFAMILY 4. LAPHRIINE

Laphriinae Loew *Linn. Ent.*, **2**: 384 (1847) Schiner, *Verh. zool.-bot. Ges. Wien.* **16**: 649 (1866); **17**: 355 (1867); *Dipt. Navara Reise* (1868). Wulp, *Catalogue Diptera South Asia*, Hague: 81 (1896). Williston, *Manual of N. American Diptera*, New Haven: 202 (1908). Kertesz, *Catalogus Dipterorum*, Budapest, **4**: 160 (1909). Hermann, *Zool. Jena Abt. Syst.*, **43**: 166 (1920). Bromley *Ann. ent. Soc. Amer.*, **27**: 78 (1934); *Lingnan Sci. J.*, **21** (1-4): 90 (1945). Efflatoun, *Mem. Soc. R. ent. Egypte*, **4** (3): 199-443 (1937).

Laphriini Hardy, *Proc. Linn. Soc. N.S.W.*, **52**: 394 (1927), (in part); **53**: 472 (1928); **54**: 355 (1929); *Ann. Mag. nat. Hist.*, (10), **13**: 516 (1934); *Ent. mon. Mag.*, **84**: 118 (1948).

This subfamily resembles Dasypogoninae in having palpi, whereas it resembles Asilinae in having closed cell Rr. It comprises of 9 genera and 38 secies from the Indian Region. The following in the key to all the Indian genera of this family.

KEY TO GENERA

1. Wings with only two submarginal cells (3rd R and 4th R). — 2

   –Wings with three submarginal cells. — *Pogonosoma* Rondani

2. Metanotal slopes or callosities bearing pile or short, stout bristles. — 3

–Mitanotal callosities without pile or bristles.	4
3. Mesonotum without dense hairs concealing the ground colour; front narrowed above; with curved spines on fore tibae; the discal cell (Ist M_2) and 5th posterior cell (3rd M) parallel and closed distally by cross veins	*Atomosia* Macquart
4. First posterior cell (5th R) closed and petiolate; proboscis short	*Nusa* Walker
–First prosterior cell open or nearly closed at wing margin	5
5. Proboscis dorso-ventrally compres-sed	*Andrenosoma* Rondani
–Proboscis laterally compressed.	6
6. Hind pair of tibae much curved; 4th posterior cell (3rd M) and the inferior cell (Ist A) closed at some distance from the border; head very thickly clothed with long hairs; first joint of antennae with long hairs	*Laxenecera* Walker
–Not with this combination of characters	7
7. Bare metailic species: the proboscis long and a ender; palpi very small; head orbicular and much excised behind; moustache confined or almost so, to oral opening; hind, femora swollen, and all the legs, especially in the males with long pubescence	*Maira* Schiner
–More or less hairy species; the proboscis long but stouter; palpi some what larger and longer but still more or less filiform; moustache thick and not only confined to the oral opening but extending upwards; the face usually with a tubercle; (the character of the head being not or only very slightly excised beind seems rather doubtful one).	8
8. The dis a j cell (Ist M_2) and 4th posterior cell (3rd M) closed distally by cross-veins and running in one line, the cells being not parallel	9
–The disca and 4th posterior cells closed distally by cross-veins usually running in one line, the cells being always parallel.	*Anisosis* Hermann

9. Thorax with fine dense pubescence, coarser and longer on the sides and cocealing the halterers; the broad flattened abdomen is laterally borded with a fringe of long hairs and terminates in a fine tufted process suggesting a sting; all legs of much the same length, remarkably hairy and sturdy, particularly the last pair; wings and thorax of bluish-bronze hue; the wing together with large alula giving the superficial resemblance to coupled fore and hind wings and therefore it mimies *Xylocapa* bees — *Hyperechia* Schiner

–Not with this combination of characters, halters not concealed, abdomen cylindrical and not thickly haired — *Laphria* Meigen.

Genus 1. **Andrenosoma** Rondani

Andrenosona Rondani, *Dipterologiae Italicae,* Prodromus, 1: 160 (1856), (Type for the genus: *Asilus atrum* Linnaeus). Wulp, *Catalogue Diptera South Asia,* Hague: 89 (1896). Ricardo, *Ann. Mag. nat. Hist.,* (9), 20: 206 (1927), [this genus must not take the place of *Nusa* Walker in Kertesz's *Catalogus Dipterorum,* Budapest, 4: 197 (1909); genus *Nusa* Walker bears no resemblance to it and therefore should not be confused with it].

1. **choprai** Bromley

Andrenosoma choprai Bromley, *Rec. Indian Mus.,* 37 (2): 227 (1935). — Darjeeling (Bengal)

Genus 2. **Anisosis** Hermann

Anisosis Hermann, *Ent. Mitt.,* 3: 130 (1914), (Type for genus *Laphria* (Osten-Sacken), Bezzi, *Philipp. J. Sci.* (D), **12**: 125 (1017) (*proewce.*). (To be referred to International Commission on Zoological Nomenclature for assigning a new name to this genus).

1. **phalaris** (Osten-Sacken)

Laphria phalaris Osten-Sacken, *Ent. Z.,* **26**: 109 (1882). Wulp, *Catalogue Diptera South Asia,* Hague: 87 (1896). Kertesz's *Catalogus Dipterorum,* Budapest, **4**: 192 (1909)[1].

[1]Is. of Philippines

Anisosis phalaris Hermann, *Ent. Mitt.*, **3**: 130 (1914)[2]. [2]Ceylon

Genus 3. **Atomosia** Macquart

Atomosia Macquart, Dipteres Exotiques, Paris, **1** (2): 73 (1838), (Type for the genus: *Laphria puella* Wiedermann). Wulp, *Catalogue Diptera South Asia*, Hague: 89 (1896). Kertesz's *Catalogus Dipterorum*, Budapest, **4**: 163 (1909). (*Syn. Cormansis* Walker). Meijere, *Tijdschr. Ent.*, **56**, Suppl.: 54 (1914).

Cormansis Walker, *Insecta Saundersiana Diptera*, London, **1**: 154 (1851).

1. **halictides** (Walker)

Cormanis halictides Walker, *Insecta Sasundersiana Diptera*, London, **1**: 154 (1851); *List Diptera British Museum*, London, 7, Suppl. **3**: 570 (1855).

Atomosia halictides Wulp, *Catalogue Diptera South Asia*, Hague: 89 (1896). Kertesz's *Catalogus Dipterorum*, Budapest, **4**: 164 (1909)[1]. [1]East India

2. **purpurata** Westwood.

Atomosia purpurata Westwood, *Trans. R. ent. Soc. Lond.*, **5**: 233, Pl. XXIII, fig. 5 (1849). Walker, *List Diptera British Museum*, London, 7 Suppl. **3**: 570 (1855). Wulp, *Catalogue Diptera South Asia*, Hague: 90 (1896). Kertesz, *Catalogus Dipterorum*, Budapest, **4**: 165 (1909)[1]. [1]East India

Genus 4. Hyperechia Schiner

Hyperechia Schiner, *Verh. Zool.–bot. Ges. Wien*, **16**: 673 (1866), (Type for the genus: *Laphria xylocopiformis* Walker). Wulp, *Catalogue Diptera South Asia*, Hague: 82 (1896). Kertesz, *Catalogus Dipterorum*, Budapest, **4**: 171 (1909).

1. **xylocopiformis** Walker, *List Diptera British Museum*, London, **2**: 385 (1849). Wulp, *Catalogue Diptera South Asia*, Hague: 82 (1896). Kertesz, *Catalogus Dipterorum*, Budapest, **4**: 172 (1909)[1]. [1]Madras

Genus 5. **Laphria** Meigen

Laphria Meigen, *Illiger's Mag. f. Ins.*, **2**: 270 (1803), (Types for the genus: *Asilus gibbosa* Linnacus). Wulp, *Catalogue Diptera South Asia*, Hague: 82 (1896). Kertesz, *Catalogus Dipterorum*, Budapest, **4**: 176 (1909). (*Syn Choerades* Walker, *Lapria* Meigen and *Rhopalocera* Meigen). Ricardo, *Ann. Mag. nat. Hist*, (8), 11: 153 (1913). Enderlein, *Zool. Anz.*, **44**: 251 (1914). White, *Proc. roy. Soc. Tasmania*: 164 (1916); 84 (1917). Edward, *J. F. M. S. Mus.*, **8** (3): 32 (1919), (subgenera *Pagidolaphria* Harmann and *Choerades* Walker). Malloch, *Proc. Linn. Soc. N.S.W.*, **53**: 298, 609 (1928). Hardy, *Proc. Linn. Soc., N.S.W.*, **54**: 355 (1929); *Ann. Mag. nat. Hist.*, (10), **13**: 517 (1934) (*Syns. Nusa* Ricardo and *Andrenosoma* Malloch).

Lapria Meigen, *Nouvelle classification*, **25**: 40 (1800). *Rhopalocera* Meigen, *Litt. apud. Meigen.*

Nusa Ricardo, Ann. Mag. nat. Hist., (8), **11**: 149 (1913).

Andrenosoma Malloch, *Proc. Linn. Soc. N.S.W.*, **53**: 299 (1928).

Subgenus 1. **Laphria** Meigen *sens. strict.* (Type for the subgenus: *Asilus gibbosa* Linnacus)

1. **abscissa** Walker

Laphria abscissa Walker, *Trans. R. ent. Soc. Lond.*, (n). **5**: 282 (1860). Wulp, *Catalogue Diptera South Asia*, Hague: 84 (1896). Kertesz, *Catalogus Dipterorum*, Budapest, **4**: 176 (1909)[1].

[1]Burma

2. **chrysotelus** Walker

Laphria chrysotelus Walker, *List Diptera British Museum*, London, 7: Suppl. **3**: 552 (1855). Wulp, *Catalogue Diptera South Asia*, Hague: 84 (1896). Kertesz, *Catalogus Dipterorum*, Budapest, **4**: 179 (1909)[2].

[1]East India
[2]India

3. **claripennis** Bigot

Laphria claripennis Bigot, *Ann. Soc. ent. Fr.*, (5), **8**: 224 (1878). Wulp, *Catalogue Diptera South Asia*, Hague: 86 (1896). Kertesz, *Catalogus Dipterorum*, Budapest, **4**: 179 (1909)[1].

[1]Ceylon

4. **dimidiata** Macquart

Laphria dimidiata Macquart, *Dipteres Exotiques,* Paris, Suppl. **1**: 72 (1846); **3**: 24 (1848). Walker, *List Diptera British Museum,* London, 7, Suppl. **3**: 552 (1855). Osten-Sacken, *Ent. Z.,* **26**: 109 (1882). Wulp, *Catalogue Diptera South Asia,* Hague: 83 (1896)[1]. Kertesz, *Catalogus Dipterorum,* Budapest, **4**: 180 (1909)[2], (*Syn. Laphria taphius* Walker).

[1]Is. of Philippines, Java, Celebes and Ceram
[2]Amboina and India

Laphria Taphius Walker, *List Diptera British Museum,* London, **2**: 380 (1849); 7, Suppl. 3: 550 (1855); *Proc. Linn. Soc. Lond.,* **4**: 105 (1860); **5**: 147 (1861) **7**: 232 (1864). Osten-Sacken, *Ann. Mus. Civ. Genova,* **16**: 429 (1882). Wulp, *Catalogue Diptera South Asia,* Hague: 83 (1896).

5. **flava** Linne

Asilusflava Linne, *Fauna Suec., Ed. Alt. 469* (1767); *Systema Naturae,* Ed. 12, **2**: 1007 (1767). Fabricius, *Systema Ent.,* 793 (1775). DeGeer, *Mem. pour serv. a* l' *hist.d'* **6**: 240, Tab, 13, fig. 10-14 (1776). Sulzer, *Abgek. Gesch. d. Insekt.*: 224, Tab. 28, fig. 20 (1776). Fabricius, *Species Insect.,* **2**: 462 (1781); *Mantissa Insect.,* **2**: 359 (1787). Herbst, *Gemeinnutz. Naturgesch.,* **8**: 119, Tab. 346, fig. 3 (1787). Olivier, *Encyclop. Method.,* **4**: 265 (1789). Gmel., *Systema Naturae,* **5**: 2897 (1792). Fabricius, *Ent. System.,* **4**: 380 (1784). Schrank, *Fauna Boica,* **3**: 156 (1803).

Laphria flava Meigen, *Klassif.* **1**: 261, Tab. 13, fig. 20-21 (1804). Fabricius, *Systema Antliat.*: 156 (1805). Latreille, *Gen. Crust. et. Ins.,* **4**: 299 (1809). Fallen, *Dipt. Suec. Asilici,* **4**: 2 (1814). Meigen, *System. Besechreib.,* **2**: 288 (1820). Macquart, *Rec. Soc. Sci. Agric.*: 359 (1826); *Suit. a Buffon,* **1**: 281 (1834). Zetterstedt, *Insecta Lappon Dipt.*: 506 (1838). Macquart, *Dipteres, Exotiques, Paris,* **1**: 2, Tab. 6, fig. 3 (1838). Zeller, *Isis,* **1**: 74 (1840). Zetterstedt, *Dipt. Scand.,* **1**: 161 (1842). Loew, *Linnaea Ent.,* **2**: 545 (1847). Walker, *List Diptera British Museum,* London, **2**: 369 (1849). Zetterstedt. **8**: 2966 (1849). Scholz, *Z. Ent.,* **5**: (17): 45, (1851). Schiner, *Verh. zool.-bot. Ver. Wien,* **4**: 385, 426, 428 (1854). Walker, *List Diptera British Museum,* London, 7, Suppl. **3**:

514 (1855). Schiner, *Verh. zool.-bot. Ver. Wien,* **6**: 419 (1856). Bonsd., *Finlands traving. Ins.,* **1**: 121 (1861). Schiner, *Fauna Austriaca Diptera Wien,* **1**: 138 (1862). Jaennicke, *Ent. Z.,* 11: 87 (1867). Vice, *Scot. Natural.,* **2**: 120 (1873). Portisch., *Trud. russk. Ent.,* **10**: 156 (1877). Becker, *Denkschr Akad.* Wien, **45**: 146, Tab. 3, fig 6 (1882). Hansen, *Fabrica oris Dipt.,* tab. 4, fig. 6-7 (1883). Bigot, *Ann. Soc. ent. Fr.,* (6), **4** Bullet., 115, *var.,* (1884). Neuhaus, *Dipt. Marchica*: 60 (1886). Beck., *Ent. Z.,* **31**: 108 (1887). Strobl, *Mitt. naturw. Ver. Steierm.* (1892) **29**: 31 (1893). L. Coucke, *Ann. Soc. Ent. Belgique,* **38**: 438 (1894). Strobl. *Glasn, zem. Mus. Bosn. Herc.,* **10**: 416 (1898); *Wissensch. Mittheil. Bosnienu. Herzegov.,* **7**: 574 (1900). Yerbury, *Ent. mon Mag.,* (2), **15** (40): 211 (1904). Lampa, *Ent. Tidskr.,* **25**: 213, Tab. 1, fig. 11 (1904). Kertesz, *Catalogus Dipterorum,* Budapest, **4**: 182-83 (1909)[1], (*Syn. Asilus flava* L., *Dioctria Flava* Lw., *Erax conopsoides* Scop., *E. ferox* Scop., *Asilus eplippium* Panz., A. *fulva* Fourer., *A. paedemontanus* Fabr., A. *pedemontana* Meig., and var. *escorialensis* Strobl.)

[1]Europe and Hispania

Dioctria flava Loew, *Progr. Prosen,* **15**: 1 (1840); *Isis,* **7**: 538 (1840).

Erax conopsoides Scopoli, *Ent. Carinol.*: 361 (1763).

Asilus ephippium Panzer, Fauna Germ., **39**: 24 (1797).

Erax feroxi Scopoli, *Ent. Carinol.*: 360 (1763).

Asilus fulva Fourcr., *Ent. Paris,* **2**: 459 (1785). Villers., *Ent. Linn.,* **3**: 601 (1789).

Asilus paedemontanus Fabricius, *Ent. System.,* **4**: 378 (1794).

Asilus pedemontana Meigen, *Klassif.,* **1**: 266 (1804). Fabricius, *Systema Antliat.*: 158 (1805).

Asilus pedemontana var. *escorialensis* Strobl., *Mem. R. Soc. Espan. Hist. Nat.,* **3**: 300 (1906).

6. **furva** Wulp

Laphria furva Wulp, *Tijdschr. Ent.,* **41**: 127 (1898): **42**: 48 (1899). Kertesz. *Catalogus Dipterorum,* Budapest, **4**: 185 (1909)[1].

[1]India

7. **gilvoides** Wulp

Laphria gilvoides Wulp, *Tijdschr. Ent.*, **41**: 124, Pl. IV, fig. 1-2 (1898); **42**: 48 (1899). Kertesz. *Catalogus Dipterorum*, Budapest, **4**: 187 (1909)[1]. — [1]India

8. **nigrohirsuta** Lichtwardt

Laphria nigrohirsuta Lichtwardt, *Ent., Z.*: 123-27 (1909). — Kashmir

9. **semifulva** Bigot

Laphria semifulva Bigot, Ann. Soc. ent. Fr.., (5), **8**: 232 (1878). Wulp, *Catalogue Diptera South Asia,* Hague: 86 (1896). Kertesz. *Catalogus Dipterorum,* Budapest, **4**: 194 (1909)[1]. — [1]India or East India

Subgenus 2. **Choerades** Walker

Choerades Walker, *Insecta, Saundersiana Diptera,* London, **1**: 109 (1851), (Type for the subgenus: *Choerades aurigena* Walker). Enderlein, *Zool. Anz.*, **44**: 247 (1914). Edwards, *J. F. M. S. Mus.*, **8** (3); 32 (1919), (subgenus of *Laphria* Meigen; *syn. Epholkiolaphria* Hermann).

1. **aurigena** Walker

Choerades aurigena Walker, *Insecta, Saundersiana Diptera,* London, **1**: 109, Pl. IV, fig. 8 (1851). — [1]East India

Laphria aurigena Walker, *List Diptera British Museum,* London, 7, Suppl. **3**: 355 (1855). Wulp, *Catalogue Diptera South Asia,* Hague: 84 (1896). Kertesz. *Catalogus Dipterorum,* Budapest, **4**: 178 (1909)[1]. Enderlein, *Zool. Ans.*, **44**: 249 (1914). (variety of *Choerades vulcanus* Wiedermann). — [1]Sumatra ? Java ?

Subgenus 3. **Pagidolaphria** Hermann

Pagidolaphria Hermann, *Ent. Mitt.*, **3**: 85 (1914), (Type for the subgenus: *Dasypogon gigas* Wiedermann). Edwards, *J.F.M.S. Mus.*, **8** (3): 32 (1919), (subgenus of *Laphira* Meigen; *Syn, Saliomima* Enderlein).

Saiomima Enderlein, *Zool. Anz.*, **44**: 243 (1914).

1. *alternans* Wiedermann.

Laphria alternans Wiedermann, Aussereuropaische zweiflugelige Insecten, Hamm **1**: 511 (1828). Walker, *List Diptera British Museum,* London, 7, Suppl. **3**: 555 (1855); *Proc. Linn. Soc. Lond.*, **1**: 10, 113, (1856).

Schiner, *Verh. zoo,-bot. Ges. Wien,* **16**: 691 (1866). Wulp, *Tijdschr. Ent.,* (2, **5**(13): 211 (1870); **7**(15): 189 (1872); *Catalogue Diptera South Asia,* Hague: 82 (1896)[1]. Kertesz, *Catalogus Dipterorum,* Budapest **4**: 177 (1909)[2], (*Syn. Laphria senomera* Macquart). [1, 2]Bengal, Java, Borneo and Malay-Peninsula

Saliomimas alternans Enderlein, *Zool. Anz.,* **44**: 246 (1914).

Sameringolaphira alternans Bezzi, *Philipp. J. Sci.,* **12**: 125 (1917)[3]. [3]Philippines Is. and widely spread in oriental region.

Laphria senomera Macquart, *Dipteres Exotiques,* Paris, **1**(2): 65, Pl. VI, fig. 2 (1838). Walker, *List Diptera British Museum, London* 7, Suppl. **3**: 556 (1855).

2. *gigas* Macquart.

Laphria gigas Macquart, *Dipteres Exotiques,* Paris, **1**(2): 65, (1838). Walker, *List Diptera British Museum, London* 7, Suppl. **3**: 556 (1855). Wulp, *Tijdschr. Ent.,* (2), 7(15): 165 , Pl. IX, fig. 11-13 (1872); *Catalogue Diptera South Asia,* Hague: 83 (1896)[1]; *Tijdschr. Ent.,* **42**: 1236 (1898). Kertesz, *Catalogus Dipterorum,* Budapest **4**: 186 (1909)[2], (*Syn. Laphria horrida* Walker). [1]East India [2]India or, Borneo, Java and ? Sumatra.

Saliomima gigas Enderlein, *Zool. Anz.,* **44**: 246 (1914).

Laphria horrida Walker, *List Diptera British Museum,* London, 7, Suppl. 3: 551 (1855).

3. **reinwardtii Wiedermann.**

Laphria reinwardtii Wiedermann, *Aussereuropaische zweiflugelige Insecten,* Hamm **1**: 503 (1828). Macquart, *Didpteres Exotiques,* Paris, Suppl. **1**: 72 (1846). Walker, *List Diptera British Museum,* London, **2**: 373 (1849); 7, Suppl. **3**: 554 (1855); *Proc. Linn. Soc. Lond.,* **1**: 10, 113, (1856). Schiner, *Verh. zool.bot. Ges. Wien,* **16**: 690 (1866); *Dipt. Novara Reise:* 172 (1865). Wulp, *Tijdschr. Ent.,* (2), **5**(13): 211 (1870); **7**(15): 167 (1872); *Catalogue Diptera South Asia,* Hague: 82 (1896)[1]; *Tijdschr. Ent.,* **41**: 124 (1898). Kertesz, *Catalogus Dipterorum,* Budapest **4**: 193 (1909)[2], (*Syn. Asilus opicata* Walker, *Laphria fervens* Walker and *L.saeva* Walker). Meijere, *Tijdschr. Ent.,* **54**: 305 (1911). [1]Malay Penin., Java and Borneo, [2]Asia mer. and India.

Edwards, *J. F. M. S. Mus.*, **8** (3): 32 (1919)[3], [*Laphria* (*Pagidolaphria*)]. [1, 3]Sumatra

Saliomima reinwardtii Enderlein, *Zool. Anz.*, **44**: 143, (1914).

Asilus apicata Walker, *List Diptera British Museum,* London, **2**: 436 (1849).

Laphria fervens Walker, *List Diptera British Museum,* London 7, Suppl. **3**: 554 (1855).

Laphria Saeva Walker, *List Diptera British Museum,* London 7, Suppl. 3: 554 (1855).

Genus 6. **Laxenecera** Macquart.

Laxenecera Macquart, *Dipteres Exotiques,* Paris, **1**(2): 77 (1838), (Type for the genus: *Laphria bengalensis* Wiedermann). Wulp, *Catalogue Diptera South Asia,* Hague: 89 (1896). Kertesz, *Catalogus Dipterorum,* Budapest, **4**: 169 (1909), (*Syn. Acurana* Walker and *Dyseris* Loew).

Acurana Walker, *Insecta Saundersiana Diptera,* London, **1**: 107 (1851).

Dyseris Loew *Ofvers. kongl. Vet. Akad. Forh.,* **14**: 357 (1858).

1. **albibarbis** Macquart

Laxenecera albibarbis Macquart, *Dipteres Exotiques,* Paris, **1**(2): 78, Pl. VII, fig. 4 (1838), Walker, *List Diptera British Museum,* London, 7 Suppl. 3: 572 (1855). Wulp, *Catalogue Diptera South Asia,* Hague: 89 (1896)[1], (*Syn. ? Acurana sexfasciata* Walker). Kertesz, *Catalogus Dipterorum,* Budapest, **4**: 169-70, (1909)[2]. [1]East India [2]India

Acurana sexfasciata Walker, *Insecta Saundersiana Diptera,* London, **1**: 107 (1851).

2. *bengalensis* Wiedermann.

Laphira bengalensis Wiedermann, *Dipteres Exotiques,* Paris: 237, (1821): *Assuereuropaische*

Laxenecera bengalensis Schiner, *Verh. zool.–bot. Ges. Wien,* **16:** 691 (1866). Wulp, *Catalogue Diptera South Asia,* Hague: 89 (1896). Kertesz, *Catalogus Dipterorum,* Budapest, **4**: 170, (1909)[1]. [1]Bengal

3. **flavibarbis** Macquart

Laxenecera flavibarbis Macquart, *Dipteres Exotiques,* Paris, **1**(2): 77 (1838). Walker, *List Diptera British Museum,* London, 7 Suppl. **3**: 572 (1855). Wulp, *Catalogue Diptera South Asia,* Hague: 89 (1896). Kertesz, *Catalogus Dipterorum,* Budapest, **4**: 170, (1909)[1]. (*Syn. Laphria hirticornis* Guerin). [1]Bengal

?Laphria hirticornis Guerin, *Icongr. d. Regne Anim.,* **7**: 536, Pl. XCIV, fig. 1 (1835).

Genus 7. **Maira** Schiner

Maira Schiner, *Verh. zool.–bot. Ges. Wien,* **16**: 673 (1866) (Type for the genus: *Laphria spectabilis* Guerin). Wulp, *Catalogue Diptera South Asia,* Hague: 87 (1896). Kertesz, *Catalogus Dipterorum,* Budapest, **4**: 201, (1909).

1. **aenea** (Fabricius)

Laphria aenea Fabricius, *Systema Antliat:* 161 (1805). Wiedermann, *Dipteres Exotiques,* Paris: 238 (1821); *Aussereuropaische zweiflugelige Insection,* Hamm. **1**: 515 (1828).

Lampria aenea Macquart, *Dipteres Exotiques,* Paris, **1**(2): 61, Pl. VI, fig. 1 (1838); Suppl. **3**: 21 (1848). Walker, *List Diptera British Museum,* London, 7, Suppl. 3: 511 (1855). Hutton, *Catalogue Diptera* New Zealand: 31 (1881).

Maira aenea Schiner, *Verh. zool.–bot. Ges. Wien,* **16**: 602 (1866). Wulp. *Tidjdschr. Ent.,* (2), 7(15): 202, Pl. X, fig. 10 (1872); *Catalogue Diptera South Asia,* Hague: 87 (1896)[1]; *Tijdschr. Ent.,* **41**: 130 (1898). Meijere, *Nova Guinea V. Zool.:* 77 (1906). Kertesz, *Catalogus Dipterorum,* Budapest, **4**: 201, (1909)[2]. (*Syns. Laphria cyanea* Macquart, *L. colorata* Boisduval, *L. pellucida* Doleschall, *L. comes* Walker, *L. consors* Walker).

[1]Java & Ambonia Is.
[2]Asia mer. India, New Guinea & N. Selandia.

Maria aenea var. nota Bigot, *Ann. Soc. ent. Fr., (5),* **8**: 233 (1878).

?Laphria cyanea Macquart, *Suit a Buffon,* **1**: 286 (1834). Doleschall, *Natuurk. Tijdschr. Ned. Ind.,* (3) **4** (14): 395 (1857).

Laphria colorata Boisduval, *Voyage de l' Astrolabe Ent.*, **2**: 661, Pl. XII, fig. 11 (1835).

?*Laphria pellucida* Doleschall, *Natuurk. Tijdschr. Ned. Ind.*, (3), **4** (14): 396, Pl. I, fig. 3 (1857). Schiner, *Verh. zool.–bot. Ges. Wien*, **17**: 382 (1867).

Laphria comes Walker, *Proc. Linn. Soc. Lond.*, **3**: 85 (1859); **5**: 147, 279, (1861); **6**: 6 (1862). Bigot, *Ann. Soc. ent. Fr.*, (5), **8**: 218 (1878).

Laphria consors Walker, *Proc. Linn. Soc. Lond.*, **3**: 85 (1859); 5: 279 (1861); **7**: 206 (1864).

2. **aurifacies** (Macquart)

Laphria aurifacies Macquart, *Dipteres Exotiques*, Paris, Suppl. **3**: 22. Pl. II, fig. 5 (1848). Walker, *List Diptera British Museum*, London, 7, suppl. **3**: 552 (1855); *Proc. Linn. Soc. Lond.*, **1**: 10, 113 (1856); **3**: 84 (1859).

Maira aurifacies Schiner, *Dipt. Novara Reise*: 174 (1868). Wulp. *Tijdschr. Ent.*, (2), 7(15): 212 (1872); *Catalogue Diptera South Asia*, Hague: 87 (1896)[1]. Kertesz, *Catalogus Dipterorum*, Budapest, **4**: 202 (1909)[2]. Meijere, Tijdschr. Ent., **54**: 306 (1919).

[1]Malay Penin., Borneo, Sumatra and Moluccos
[2]India and Is. of Nicobar.
[2]Aru Is.

3. **compta** (Walker)

Laphria compta Walker, *Proc. Linn. Soc. Lond.*, 5: 146 (1861).

Maira compta Wulp, *Tijdschr. Ent.*, (2), 7(15): 205 (1872); *Catalogue Diptera South Asia*, Hague: 88 (1896)[1]. Kertesz, *Catalogus Dipterorum*, Budapest, **4**: 202 (1909)[2].

[1]Java
[2]India, Ambonia and New Guinea.

4. **elegans** (Walker)

Laphria elegans Walker, *List Diptera British Museum* London, 7, Suppl. **3**: 551 (1855); *Proc. Linn. Soc. Lond.*, **1**: 10 (1856).

Maira elegans Wulp, *Catalogue Diptera South Asia*, Hague: 87 (1896)[1]. Kertesz, *Catalogus Dipterorum*, Budapest, **4**: 202 (1909)[2].

[1]Sumatra
[2]East India an Malay Peninsula

5. **gloriosa** (Walker)

Laphria gloriosa Walker, *Proc. Linn. Soc. Lond.*, **3**: 84 (1859).

Maira gloriosa Osten-Sacken, *Ann. Mus. Cip. Genova*, **16**: 431 (1882). Wulp, *Catalogue Diptera South Asia*, Hague: 88 (1896)[1]. Kertesz, *Catalogus Dipterorum*, Budapest, **4**: 202 (1909)[2]. (Syns. *Laphria consobrina* Walker, *L. socia* Walker, *Maira spectabilis* Schiner and *M. kollari* Wulp).

[1]Timor Laut and Obi.
[1, 2] New Guinea.

Laphria consobrina Walker, *Proc. Linn. Soc. Lond.*, **3**: 84 (1859). Bigot, *Ann. Soc. ent. Fr., (5)*, **8**: 218 (1878). Kirby, *Ann. Mag. nat. Hist., (5)*, **13**: 459 (1884).

[2]Amboina and India

Laphria socia Walker, *Proc. Linn. Soc. Lond.*, **3**: 84 (1859); **5**: 147 (1861); **6**: 18 (1862); **7**: 232 (1864). Bigot, *Ann. Soc. ent. Fr., (5)*, **8**: 218 (1878).

Maira spectabilis Schiner, *Verh, zool.–bot. Ges. Wien*, **17**: 381 (1867) (nec. Guerin).

Maira kollari Wulp, *Tijdschr. Ent.*, (2), **7**: (15): 201 (1872) (nec. Doleschall).

6. **hispidella** Wulp

Maira hispidella Wulp, *Tijdschr. Ent.*, (2), 7 (15): 213 (1872); *Catalogue Diptera South Asia*, Hague: 88 (1896)[1]. Meijere, *Nova Guinea V. Zool*: 78 (1906); *Tijdschr. Ent.*, 56: Suppl.: 50 (1914); 5**8**, suppl.: 21 (1915). Kertesz, *Catalogus Dipterorum*, Budapest, **4**: 203 (1909)[2].

[1]Java

[2]India, New Guinea and Sumatra

7. **kubinyii** Dolleschall

Maira Kubinyii Doleschall, *Natuurk. Tijdschr. Ned. Ind.*, (3), **4** (14): 397, Pl. 1. fig. 1 (1857). Schiner, *Dipt. Novara Reise*: 173 (1868). Kertesz, *Catalogus Dipterorum*, Budapest, **4**: 203 (1909)[1]. (*Syn. Maira kurbinyi* Wulp).

[1]Nicobar Is. (India), Amboina, Molucci and New Guinea

Maira Kubinyi Wulp, *Tijdschr. Ent.*, (2), **7** (15): 204, Pl. X. fig. 11 (1872); *Catalogue Diptera South Asia*, Hague: 88 (1896). Meijere, *Nova Guinea V. Zool:* 78 (1906)[2].

[2]Halmaheira

8. longirostra Bromley

Maira longirostra Bromley, *Rec. Indian Mus.* **37**(2): 227 (1935). Darjeeling (Bengal)

9. nycthemera Wulp

Maira nycthemera Wulp, *Tijdschr. Ent.*, (2), **7** (15): 208, Pl. XI, fig. 1 (1872); *Catalogue Diptera South Asia*, Hague: 88 (1896)[1]. Kertesz, *Catalogus Dipterorum*, Budapest, **4**: 203 (1909)[2].

[1]Java and Borneo
[2]India

10. paria Bigot

Maira paria Bigot, *Ann. Soc. ent. Fr.*, (5), **8**: 230 (1878). Wulp, *Catalogue Diptera South Asia*, Hague: 88 (1896). Kertesz, *Catalogus Dipterorum*, Budapest, **4**: 203 (1909)[1].

[1]India or East India

11. requisita (Walker)

Laphria requisita Wulp, *Proc. Linn. Soc. Lond.*, **4**: 105 (1860); **7**: 206, 232 (1864).

Maira requisita Wulp, *Catalogue Diptera South Asia*, Hague: 88 (1896)[1]. Kertesz, *Catalogus Dipterorum*, Budapest, **4**: 204 (1909)[2].

[1]Celebes, Mysol & Ceram
[2]India

Genus 8. **Nusa** Walker

Nusa Walker, *Insecta Saundersiana Diptera*, London, **1**: 105 (1851) (Type for the genus: *Nusa aequalis* Walker). Kertesz, *Catalogus Dipterorum*, Budapest, **4**: 197 (1909). (Syn. *Elaeotoma* Costa). Ricardo, *Ann. Mag. nat. Hist.*, (9), **20**: 205 (1927) (This genus must not be confused with *Andrenosoma* Rondani, to which it bears no resemblance; Syn. *Dasythrix* Loew).

Dasythrix Loew, *Progr. Realesche Meseritz*: 21 (1861); *Dipt. Sud-Alfrik.*, 1: 126 (1860). Kertesz, *Catalogus Dipterorum*, Budapest, **4**: 206 (1909). (Syn. *Halictosoma* Rondani).

Elaeotoma Costa, *Ati R. Aad. Napoli*, **1**: 49 (1863).

Halictosoma Rondani, *Ann. Mus. Civ.* Genova, **4**: 298 (1873).

1. acqualis Walker

Nusa aequalis Walker, *Insecta Saundersiana Diptera,* London, **1**: 105, Pl. IV, fig. 4 (1851); *List Diptera British Museum,* London, 7, Suppl. **3**: 564 (1855). Kertesz, *Catalogus Dipterorum,* Budapest, **4**: 107 (1909)[1]. Richard *Ann. Mag. nat. Hist.,* (9), **20**: 207 (1927). — [1]India or East India

Andrenasoma aequalis Wulp, *Catalogue Diptera South Asia,* Hague: 89 (1896).

2. Albibasis Ricardo

Nusa albibasis Ricardo, *Ann. Mag. nat. Hist.,* (9), **20**: 209 (1927). — Upper Burma

3. elva (Walker)

Asilus elva Walker, *List Diptera British Museum,* London, **2**: 445 (1849).

Laphria elva Walker, *List Diptera British Museum,* London, 7, Suppl. **3**: 553 (1855). Wulp, *Catalogue Diptera South Asia,* Hague: 83 (1896). Kertesz, *Catalogus Dipterorum,* Budapest, **4**: 181 (1909)[1]. — [1]Bengal

Nusa elva Ricardo, *Ann. Mag. not. Hist.,* (9), **20:** 207 (1927). Bromley, *Rec. India Mus.,* **37** (2): 230 (1935)[2]. — [2]Assam

4. formio Walker

Nusa formio Walker, *Insecta Saundersiana Diptera* London, **1**: 106 (1851); *List Diptera British Museum,* London, 7. Suppl. **3**: 564 (1855). Kertesz, *Catalogus Dipterorum,* Budapest, **4**: 199 (1909)[1]. Ricardo, *Ann. Mag. nat. Hist.,* (9), **20**: 208 (1927). — [1]India or East India

5. yerburyi Ricardo

Nusa yerburyi Ricardo, *Ann. Mag. nat. Hist.,* (9), **20:** 208 (1927). — Ceylon

Genus 9. **Pogonosoma** Rondani

Pogonosoma Rondani, *Dipterologiae Italicae Prodromus,* **1**: 160 (1856) (Type for the genus: *Asilus moroccanum* Fabricius). Wulp, *Catalogue Diptera South Asia,* Hague: 82 (1896). Kertesz, *Catalogus Dipterorum,*

Budapest, **4**: 175 (1909). Ricardo, *Ann. Mag. nat. Hist.*, (9), **20**: 210 (1927)

(Subgenus of *Andrenosoma* Rondani). Bromely, *Lingnan, Sci. J.*, **21** (1-4): 96 (1945).

1. **cedrusa** Ricardo

Pogonosoma cedrusa Ricardo, *Ann. Mag. nat. Hist.*, (9), **20:** 211 (1927) (Bred from Cedrus deodare tree). Simla (Punjab), and Khasi Hills (Assam)

www.ingramcontent.com/pod-product-compliance
Lightning Source LLC
LaVergne TN
LVHW020836091026
849796LV00003B/33
* 9 7 8 9 3 8 6 0 7 1 1 0 1 *